Forum for Interdisciplinary Mathematics

Forum for Interdisciplinary Mathematics is a Scopus-indexed book series. It publishes high-quality textbooks, monographs, contributed volumes and lecture notes in mathematics and interdisciplinary areas where mathematics plays a fundamental role, such as statistics, operations research, computer science, financial mathematics, industrial mathematics, and bio-mathematics. It reflects the increasing demand of researchers working at the interface between mathematics and other scientific disciplines.

More information about this series at http://www.springer.com/series/13386

Girish Chandra · Raman Nautiyal ·
Hukum Chandra
Editors

Statistical Methods and Applications in Forestry and Environmental Sciences

Springer

Editors
Girish Chandra
Indian Council of Forestry Research and Education
Dehradun, Uttarakhand, India

Raman Nautiyal
Indian Council of Forestry Research and Education
Dehradun, Uttarakhand, India

Hukum Chandra
Indian Agricultural Statistics Research Institute
New Delhi, Delhi, India

ISSN 2364-6748 ISSN 2364-6756 (electronic)
Forum for Interdisciplinary Mathematics
ISBN 978-981-15-1475-3 ISBN 978-981-15-1476-0 (eBook)
https://doi.org/10.1007/978-981-15-1476-0

This Springer imprint is published by the registered company Springer Nature Singapore Pte Ltd.
The registered company address is: 152 Beach Road, #21-01/04 Gateway East, Singapore 189721, Singapore

Preface

The application of statistics in forestry and environmental sciences is challenging owing to the great amount of variations present in the nature. With a vast and varied spectrum of disciplines spanning the sector and involvement of other areas of knowledge including sociology, economics and natural sciences, the application of statistical tools in forestry and environment does not follow a thumb rule logic and requires innovative and imaginative thinking. The technological advancements in other areas of science have contributed to the evolution of forestry from merely targeting productivity to cutting-edge areas, like climate change, carbon sequestration and policy formulation. With the emerging interest in forests and environment, the need of the day is to base policy formulation and decision making on strong foundations of science. With the involvement of people in forest management, forestry has become a fusion of social and basic sciences. Thus, the need to integrate a wide range of statistical tools in research has become the necessity of the day.

This present book consists of 17 chapters, providing a broad coverage of statistical methodologies and their applications in forestry and environmental sector. The main aim of the book is to enlarge the scope of the forestry statistics, as only limited books are available in this area. The topics included in this volume are designed to appeal to applied statisticians, as well as students, researchers, and practitioners of subjects like sociology, economics and natural sciences. The inclusion of real examples and case studies is, therefore, essential. We are sure that the book will benefit researchers and students of different disciplines of forestry and environmental sciences for improving research through the practical knowledge of applied statistics.

The Chapter 'Statistics in Indian Forestry: A Historical Perspective' traces the history of quantification of forest resources and the use of statistical methods for the development of the forestry sector in India. The Chapter 'National Forest Inventory in India: Developments Toward a New Design to Meet Emerging Challenges' provides a brief overview of the National Forest Inventory (NFI) in India, vis-à-vis some other developed countries, and highlights the proposed changes in plot design while revising NFI in India. The concept of Internet of Things with possible use in forestry and environmental sciences is presented in the Chapter 'Internet of Things in Forestry and Environmental Sciences'. The Chapter 'Inverse Adaptive Stratified Random Sampling' gives a new sampling design based on adaptive cluster sampling useful for assessing rare plant species. The Chapter 'Improved Nonparametric Estimation Using Partially Ordered Sets' discusses the use of ranked set sampling in improving nonparametric estimation by using partially ordered sets. The robust Bayesian method to analyze forestry data, when samples are selected with probability proportional to length from a finite population of unknown size, is presented in the Chapter 'Bayesian Inference of a Finite Population Mean Under Length-Biased Sampling'. The Chapters 'Calibration Approach-Based Estimators for Finite Population Mean in Multistage Stratified Random Sampling' and 'A Joint Calibration Estimator of Population Total Under Minimum Entropy Distance Function Based on Dual Frame Surveys' deal with the development of calibration estimators of the population mean under two-stage stratified random sampling and joint calibration estimation approach under the entropy distance function for dual frame surveys, respectively. The Chapter 'Fusing Classical Theories and Biomechanics into Forest Modelling' narrates some of the fusing classical theories and biomechanics in forest modeling. The Chapter 'Investigating Selection Criteria of Constrained Cluster Analysis: Applications in Forestry' provides current techniques of partial redundancy analysis and constrained cluster analysis to explore how spatial variables determine structure in a managed regular spaced plantation. The Chapter 'Ridge Regression Model for the Estimation of Total Carbon Sequestered by Forest Species' describes the ridge regression technique in the presence of multicollinearity. Two methods of construction of mixture experiments with process variables have been detailed in the Chapter 'Some Investigations on Designs for Mixture Experiments with Process Variable'. The Chapter 'Development in Copula Applications in Forestry and Environmental Sciences' specifically examines the review on the development of copula models and its applications in the area of forestry and environmental sciences. The Chapter 'Forest Cover-Type Prediction Using Model Averaging' presents the methodology to apply model averaging technique in multinomial logistic regression model by using the forest cover type dataset. Small-area estimation methods to handle zero-inflated, skewed, spatially structured data for the continuous target response are given in the Chapters 'Small Area Estimation for Skewed Semicontinuous Spatially Structured Responses', and 'Small Area Estimation for Total Basal Cover in the State of Maharashtra in India' which describe the small-area estimation technique to produce small-area estimates of the total basal cover for trees, shrubs, and herbs for the state of Maharashtra in India. The applications of three different sampling procedures for assessing the abundance of elephants in the elephant reserves of Kerala, India, are presented in the Chapter 'Estimation of Abundance of Asiatic Elephants in Elephant Reserves of Kerala State, India'. The book ends with a short note on integrated survey scheme with particular relevance to forests in Bangladesh.

Most of the chapters of this book are the outcome of the three-day national workshop on 'Recent Advances in Statistical Methods and Application in Forestry and Environmental Sciences' organized by the Division of Forestry Statistics, Indian Council of Forestry Research and Education (ICFRE), Dehradun, India, from 23–25 May 2018. We would like to thank the Ministry of Statistics and Program Implementation, Government of India and ICFRE for the funding support. The guidance rendered by Dr. Suresh Gairola, Director General, ICFRE, and Shri Arun Singh Rawat, Deputy Director General (Administration), ICFRE, provided impetus and motivation in publishing this book. We extend our thanks and appreciations to the authors for their continuous support during the finalization of the book. We would like to express our sincere thanks to Shamim Ahmad, Senior Editor, Mathematical Sciences, Springer Nature, for his continuous support and cooperation from planning to the finalization of this volume. We would like to thank the anonymous referees for their valuable comments and suggestions for the improvement of this book.

Dehradun, India Girish Chandra
Raman Nautiyal
New Delhi, India Hukum Chandra

Contents

About the Editors

Girish Chandra, PhD, is a scientist at the Division of Forestry Statistics, Indian Council of Forestry Research and Education (ICFRE), Dehradun, India. He previously worked at the Tropical Forest Research Institute, Jabalpur, and Central Agricultural University, Sikkim. He has published over 25 research papers in various respected journals and one book. He was awarded the Cochran–Hansen Prize 2017 by the International Association of Survey Statisticians, the Netherlands. He is also a recipient of the ICFRE Outstanding Research Award 2018 and the Young Scientist Award in Mathematical Sciences from the Government of Uttarakhand. He has presented his research papers at various international conferences and workshops in India and abroad. He has organised two national conferences and is a member of various scientific institutions, including the International Statistical Institute, International Indian Statistical Association, Computational and Methodological Statistics, and the Indian Society for Probability and Statistics. His research interests include sample surveys, probability theory and numerical methods.

Raman Nautiyal is head of the Division of Forestry Statistics, ICFRE. He previously worked as a scientist at the Institute of Forest Genetic and Tree Breeding, Coimbatore. He has handled projects on forestry statistics funded by International Tropical Timber Organization, Japan, Central Statistical Office, NOVOD Board, and the Ministry of Environment, Forest and Climate Change, Government of India.

Hukum Chandra is a National Fellow at the ICAR-Indian Agricultural Statistics Research Institute, New Delhi, India. He holds a PhD from the University of Southampton, United Kingdom, and has completed his postdoctoral research at the University of Wollongong, Australia. His main research areas include sample survey design and analysis, small area estimation, bootstrap methods, statistical modelling and data analysis, and statistical methodology for the improvement of agricultural statistics. He has received numerous awards and recognition for his research contributions, including the National Award in Statistics from the Ministry of Statistics and Programme Implementation, Government of India; Cochran–Hansen Award from the International Association of Survey Statisticians; Young Researcher/ Student Award from the American Statistical Association; Lal Bahadur Shastri Outstanding Young Scientist Award from ICAR; Recognition Award from the National Academy of Agricultural Sciences; Professor P.V. Sukhatme Gold Medal Award; and Dr. D.N. Lal Memorial Award from the Indian Society of Agricultural Statistics. He also received a scholarship from the Commonwealth Scholarship Commission in the United Kingdom.

A council member of the International Association of Survey Statisticians, Dr Chandra has worked as an expert member of various committees at a number of institutions and ministries in India. As an international consultant for the FAO, he has worked in Sri Lanka, Ethiopia and Myanmar. He is an elected member of the International Statistical Institute and Fellow of the National Academy of Agricultural Sciences, India. He has published more than 100 journal papers, three books, and several technical bulletins, project reports, book chapters, working papers, and training and teaching manuals.

Statistics in Indian Forestry: A Historical Perspective

Anoop Singh Chauhan, Girish Chandra and Y. P. Singh

Abstract Indian forestry is one of the oldest profession. Quantification of forest resources was an important component of forest management in the latter half of the nineteenth century in India, replacing comprehensive reports of administrators, mostly written by botanists and medicos. The distinctive role of statistics in Indian forests was recognized with the establishment of Indian Forest Department in the year 1864. The application of statistics over time has been extended from forest administration to forestry research (primarily silviculture, trade in forest products, biodiversity conservation, etc.). This chapter traces the history of quantification of forest resources and the use of statistical methods for the development of forestry sector in India.

Keywords Biometrics · Conservation · Silviculture · Tropical forestry

1 Introduction

Forests, being a vast and complex ecosystem, are challenging and difficult to enumerate. There is a history of investigating forests with different objectives but without any quantitative information. Even before the establishment of Forest Department in 1864, the custodian of these forests, primarily relied on comprehensive reports prepared by the administrators, most of them botanists or of medical background. One such notable report on the deteriorating state of the forests was prepared by Huge Cleghorn in 1850. British Association in Edinborough formed a committee to assess the destruction of tropical forests in India. The committee observed that there is no control over a significant portion of the Indian Empire leading to forest destruction. Again in 1854, Dr. Mc. Clelland prepared a report suggesting certain curtailments to

A. S. Chauhan (✉) · G. Chandra
Division of Forestry Statistics, Indian Council of Forestry Research and Education, Dehradun, Uttarakhand 248006, India
e-mail: ascddn@yahoo.com

Y. P. Singh
Division of Forest Pathology, Forest Research Institute, Dehradun, Uttarakhand 248006, India

G. Chandra et al. (eds.), *Statistical Methods and Applications in Forestry and Environmental Sciences*, Forum for Interdisciplinary Mathematics,
https://doi.org/10.1007/978-981-15-1476-0_1

the exploitation by private parties. In response to these reports, Lord Dalhousie laid down the outline of permanent policy for forest administration on August 3, 1855 (Bruenig 2017).

Statistics were finding a place in the governance of Imperial administration. The first significant development in the pre-independence era was the constitution of a Statistical Committee (1862) for the preparation of performance/framework to collect relevant data on different subject areas. Also, Finance Department established Statistical Branch and Statistical Bureau in 1862 and 1895, respectively. For the purpose of collection of agricultural statistics, the agriculture departments were opened in 1881 in various provinces *inter alia* after the recommendations of the Indian Famine Commission. Until 1913, the Director General, Department of Commercial Intelligence and Statistics was responsible for the compilation and publication of almost all the principal statistical information including crop production. In April 1914, a separate Directorate of Statistics came into being.

During the time of expansion of British colonial rule in the eighteenth century, the Indian subcontinent being a vast storehouse of exotic flora, fauna and minerals was not only a perennial resource of raw material but also a natural depository of fundamental knowledge of nature. This early phase of British expansion in India had begun with many important investigations and surveys. Earlier surveys were carried out by forest personnel, not trained foresters, presenting reports of descriptive nature giving the subjective details of specific geographical areas of interest from different parts of the country. It was difficult to comprehend the vast flora and fauna of the country through these reports. The descriptive reports on resources based on empirical results (numerical observations) made the forest management more precise and uniform. The extensive use of natural resources, basically timber and fuel, increased many folds in the wake of industrial revolution and better economic condition of the people. Another key accomplishment was establishment of Indian railways beginning in the 1850s. The challenge to provide voluminous timber and other forest products continuously laid emphasis on the need for scientific management of forests, experimentation and research. The quantitative methods were then inevitable in all forest operations.

The article focuses on forestry only. Wildlife was integral part of forestry before the formation of Wildlife Institute in 1982. In the initial years of formation of Forest Department, wildlife was an attraction for gaming-*shikar* for forest officers from various departments of the Empire; many of them joined the forest service for the adventure and romance of hunting. The wildlife of India kept declining till the end of the nineteenth century with the improvement of guns and gunpowder. The first step to check this problem was taken in 1887 by an Act. Soon after realizing importance of wildlife for the existence of forests and other natural phenomena, number of studies sprung out where use of statistics become inevitable. The estimation of wildlife and their distribution, the animal-animal and plant-animal interaction—all required statistics. Later, advancement in ecological statistics gained more importance.

The growing knowledge gained in forests through observations, experimentation and encountering other management challenges such as forest fires, grazing, illicit felling and diseases made a place for them in the forestry journal, *The Indian*

Forester, probably first of its kind in the World. The journal became a depository of all information ranging from scientific to others in forestry.

Statistics, beyond the use of descriptive statistics, popularized with the publication of *Biometrika* in 1901 by Francis Galton, Karl Pearson and Raphel Weldon to promote the study of biometrics. The scope of statistics intensified in life sciences and, consequently, in forestry as well. Formally, in pace with development in statistics, a separate Statistical Branch in India was inaugurated on August 1, 1947 under the supervision of Prof. K. R. Nair. After the independence in 1947, statistics gained prominence in administration, planning, research, census and other fields of investigation. The main proponent for growth of statistics in India was Professor P. C. Mahalanobis, the founder of the Indian Statistical Institute at Calcutta in the year 1930. Later, various organizations were created catering to varied data needs, including data on natural resources and their management.

2 Transition from Descriptive to Analytical Approach

Contrary to preparing the detailed reports, Dietrich Brandis, a German botanist and founder of Indian Forest Department, in 1852, introduced observations, data and calculation to describe Burmese forests as Conservator. On appointment as Head of Forest Department in 1864, he initiated practices related to naming, classification, counting, measuring and valuing forests (Brandis 1884a).

The precursor to quantification was developing and defining the forestry terminology related to forest treatment, for example, different classes of forests and trees. On the suggestion of Brandis, A. Smythies worked on terminology (Smythies 1876). These terms, also called categorical or qualitative variables, helped in creating uniformity and removing ambiguity in forest reports. These terms are still in use, for example, *forest treatment*: working plan, block, compartment; *thinning*: light, moderate, heavy, cutting; *forest classes*: reserved, protected and un-classed, coppice; *kind of trees*: conifers, leaf trees, shoots, etc. The quantitative understanding enabled to measure the yield of forest timber, so that, it might be possible to estimate sustainable yield from forests in future, keeping their productivity intact.

Other than India, the quantitative science of forestry was developed, mostly in Europe, in the universities and in state forest research stations. In 1892, the first international forestry organization, the International Union of Forest Research Organisations (IUFRO) was formed. Primary interest was not only to investigate the relationship of forests with water and climate, but also research works on silviculture including selection of best species and provenance, nursery techniques, experimenting with the spacing and thinning of tree stands, etc. (www.iufro.org). Other silvicultural research objective was to find methods of growing and harvesting of trees.

Initially, the standard practice was to classify the landscape into the different regions of vegetation types, and within the region, identification of different species, and presenting them in tabular form. This practice helped foresters to take stock of an entire region without ambiguity, like number of square miles of land, average number of trees per square mile, the volume of wood per square mile and the revenue generated by sale of timber in rupees. With the help of these estimations and relationship among measured units, foresters were more specific in advancing management plan. The precision of numbers had now replaced earlier reports of irregular and ambiguous adjectives and superlatives used to described India's vegetation and trees (Agrawal 2005).

This initiation in measurements, data collection and introduction of statistical techniques in forest management and related issues made the reports more objective and precise to comprehend. The frequent use of statistics in the latter half of the nineteenth century became prominent part regarding the implementation of technologies of the British government. As a scientific investigation in forestry, biometry was in use with the development and practice of silvicultural methods. In the absence of a systematic structure of statistical methods, only descriptive statistics were in use up to 1912. With passing years, biometrics has gained prominence as one of the medium of experimental forestry science. Its usefulness in forestry research extends from molecular level to the whole of the biosphere. Now, quantification became the preferred means to represent forests as well as to other activities in the process of their management.

The distinctive and enduring role of statistics in the production of Indian forests is found in historical documents, reports and journals. India's forests magnitude, economic value, services, losses and gains, importance, and place in the national economy, demands for their conservation and management, and concerns about the effects of human interventions—all are measured and understood with in terms of empirical results. The temporal culmination of data revealed more facts regarding the forest-related issues—as their change in extent as a whole or as per species or geographical regions, economic variables like production and trade and plantation operations-survival and growth. This approach helped forest officials in deciding forest operation and management actions more decisively.

3 Demarcation and Surveys: An Earlier Endeavor

The early phase of British expansion in India has begun with many important surveys; Great Trigonometrical Survey (GTS) was one of them. GTS was founded in 1817 (Sarkar 2012) under the direction of William Lambton and George Everest with the aim to provide all possible geographical knowledge into a single, coherent whole using the method of triangulation. This 'mapping' of the Indian territory comprises topographical maps of the GTS and various other geological, botanical and forestry surveys including the collection of data on finance, trade, health, population, crime, etc. helping in the consolidation of the colonial state. These surveys also helped in collecting data and information on the land settlement and tenure practices which was compiled into district settlement reports for providing the basis of the state's revenue assessments (Bennett 2011).

In 1842, Mr. Connolly, District Collector of Malabar, raised a plantation of teak (*Tectonagrandis*) in Nilambur Valley of Madras (Ribbentrop 1900). It is one of the earliest examples of data collection to observe the growth, production and economic value. The underlying reason for the plantation was shortage of good timber for shipbuilding. To guide forestry practice, Mr. Connolly specified seven types of rules that resembled some of the prescriptions for protection that Brandis was to create in Burma a decade later. These rules addressed issues of planting, inventory, felling, contractual arrangements, monitoring, enforcement and personnel. Before Mr. Chatter Menon, as the sub-conservator of the plantations, the sowing and germination of seeds required much experimentation (Logan 2004). Menon's method for germinating teak continued to be used for the ensuing half-century, and he remained incharge of plantations until his death in 1862 (Jeyade 1947).

Enumerating teak forest in Burma was one of the challenges with Indian forestry. Brandis established a system for managing India's forest wealth after taking charge as Conservator in Pegu. His challenge was to ensure a permanent and sustainable yield of teak, also to control human interactions and produce annual surplus revenue. As a result, he formed a detailed procedure on improvement, conservation and revenue generation and termed it as 'sylviculture practices'. Further, in order to assess the condition of teak forest, Brandis adopted linear valuation survey method by laying predetermined transects (a road, a ridge, a stream, or an imaginary line across the area of interest). In this method, he classified teak trees into four girth classes (6 feet and above, 4 feet 6 in. to 6 feet, 1 foot 6 in. to 4 feet 6 in. and less than 1 foot 6 in. and seedlings) and counted them by making notches on pieces of bamboo representing different size classes; and by this count, calculated the amount of timber in each tree class according to established formulae. It was concluded that the number of trees in the first three girth classes was nearly equal in forests. Therefore, using the principle of extraction, it was recommended that only trees belonging to the first girth class to be felled and only as many trees should be felled as would be replaced during a year by the growing stock of second class trees. This was an attempt to make the guideline for conserving forests, inferred from the numerical data collected (Brandis 1884b).

In the middle of the nineteenth century, one of the upheaval tasks, mostly dependent on Indian forests, was the construction of the railway. All the forests of India were affected wholly or partially due to this massive work. A significant amount of wood was needed continuously for the construction and fuel. The solution to the problem was sought in increasing the longevity of sleeper by treating them and finding other suitable species. Around the 1880s, data was collected on sleepers of primarily used species, namely, sal, teak, deodar, hardwood, ironwood and pine. The data collected on the number of sleepers damaged (species-wise) and percent of sleepers to be replaced after the number of average years. However, the descriptive statistics exhibited contradictory results, for example, in a specific railway zone where deodar has the average of 14 years, in another railway zone some other species showed better average life. It was not possible to reach any conclusion regarding the longevity of sleepers based on descriptive statistics in use (Molesworth 1880). Though the data was collected since the 1870s, later, Smythies (1891), eminent forester, reported '... so many factors enter into the question of the durability of railway sleepers, such as climate, ballast, traffic, seasoning of the sleeper before being laid down and others, that it is almost impossible to draw any useful comparison between the results of these experiments, and it will be as well to await their completion before attempting it. This much, however, is clear that deodar is likely to hold its own with any of its competitors, a gratifying result for those who have the management of deodar forests'. Indeed, this problem of multifactor analysis arose in the 1890s. It was possible to address such practical problems later when R. A. Fisher developed comprehensive statistical designs and their analysis methods in the 1920s (Fisher 1925).

4 The First Forestry Journal: The Indian Forester

The forestry journal was founded in 1875 by the second Inspector General of India, William Schlich to create a platform where foresters could record their observations (Editorial, Indian Forester 1875). Initial articles were descriptive, but the information regarding the trade and extraction of timber was quantitative, presented in tabular form. Different provincial forestry departments started preparing and publishing the working plans regularly that condensed the key features of their operations in the form of statistical tables. The first Annual Report of the Forest Administration in India—after the passage of the Forests Act of 1878—contained quantitative information mainly on the area of land under the control of the forest department and the revenues and expenditures of the provincial forest departments. Within the next decade, all forest departments had begun to follow uniform reporting requirements to describe the state of forests under their control and actions in the forests. Quantification had brought desired uniformity in provincial and divisional operations.

By the end of the nineteenth century, foresters were well versed in using numerical figures in official writings and memoranda while describing forest extent, losses, timber extraction, trade, crime or any attribute that could be presented numerically. Foresters had begun to use statistics as their preferred means to convey meaning after the 1860s. By the time, entire organizational machinery began to collect data actively and continuously. Standardized of data collection techniques increasingly became the basis to train foresters. Quantification became a language to depict and interpret the world of trees and vegetation. With the knowledge of statistics, it was easy for forest administrators to reshape the policies that were more understandable to other stakeholders.

5 Toward Experimentation: Establishment of Forest Research Institute

In the 1890s, many experiments were laid to observe the effects of various methods of propagation, growth, forest products and management of various indigenous and newly introduced species (Fernandez 1883). Attempts were also made to investigate the prospects of forest products such as lac, camphor, coffee, cardamom and vanilla. Important outcomes on germination rates, growth rates and volume conversion factors for various timber species and important recommendations were published in the journal.

After the establishment of Imperial Forest Research Institute in 1906, advances in experimental techniques based upon contemporary statistical methods were introduced in most of the provinces. Many official periodicals such as Forest Bulletins, Forest Pamphlets, Forest Leaflets, Forest Records, Forest Memoirs and Forest Manuals were being published regularly.

The knowledge based on quantitative approach was useful in applying silvicultural and other technologies. This is where the statistics coupled with the science of forestry helping to get higher levels of returns per acre of managed forests. Silvicultural knowledge gathered by successive experiments, field experiences and forestry manuals helped foresters in a big way. With the emphasis on conclusions based empirical results, it was easy to understand and address problems related to soil, geomorphology, and human and non-human interactions.

The initial attempts toward statistical investigation were made by R. S. Troup, the first silviculturist of the institute in 1909, when he initiated the Ledger Files to collect all available data by species and subjects. The summaries of the species were compiled in his classic three-volume book 'Silviculture of Indian Trees (1921)' (Troup 1921). Another attempt was made by Mardson in 1915 when he collected a large amount of tree and crop growth statistics in plots of the forest divisions. It was intended to make useful average values. But for the large discrepancies in experimental plots, these values were not of much use. In 1918, first All-India Silvicultural Conference was convened by Mardson. Successive silviculturists worked on the lines of Troup, compiling data on five-yearly measurements in sample plots and enriching crop yield tables. In 1925, H. G. Champion took charge and extended the work through comprehensive survey of different forest types of India, a knowledge required by silviculturists to work in these forests.

Third Silvicultural Conference (1929) marked the starting point for standardizing methods for various forestry researches in India. Two assistant silviculturists were sent to Indian Statistical Institute, Calcutta in 1939 for undergoing special training under Professor P. C. Mahalanobis. Subsequently, in 1940, the President FRI invited Prof. Mahalanobis to Dehradun for advising the researchers on various statistical problems. Thereafter, Statistical Branch was established on August 1, 1947 under the chairmanship of Prof. K. R. Nair with the mandate of planning of experiments and analysis of data for various branches of the institute using the contemporary statistical techniques.

With the passage of time, understandings of forests and their practices allowed foresters to explore other areas of interest in empirical results for decision making. They also used numbers to compare forests of the country by employing uniform measures such as growth rate, allometric relations between age, girth and length of trees, volume of wood or other biomass and revenue. These comparisons of forests were objective and devoid of human bias. The attributes of forests were comparable by referring to specific numbers representing conservation, improvement, biomass per unit area, growth per year per unit area, and the likely profits from harvesting and selling of the standing plantations, etc. The use of statistics to represent crucial features of forests and timber transformed them into comparable entities.

6 Post-independence Scenario

There was a significant growth of statistics in India with the vision of Professor P. C. Mahalanobis and as a result he was appointed as the first Statistical Adviser to the Cabinet, Government of India in January 1949. He was architect of the statistical system of independent India. Professor P. V. Sukhatme, as Statistical Adviser to the Ministry of Agriculture, was responsible for the development of Agricultural Statistics in which forestry was also one of the important sectors. In 1950, National Sample Survey (NSS) came into being with the aim to collect information through sample surveys on a variety of socioeconomic and other aspects.

In 1957, the Directorate of Economics and Statistics under the Ministry of Agriculture, Govt. of India was established with the mandate to deal with various areas including the forestry statistics. It was also responsible for compiling the data received from state forest departments in the form of reports, namely, Indian Forest Statistics and Forestry in India. However, these publications were discontinued after some time.

The chief researchers of Forest Research Institute during the period of the 1950s and 1970s made the important contribution in Indian forestry. 'A Manual on Sampling Techniques for Forest Surveys' (Chacko 1965) is one of them. His other important contributions in sample survey are found in Chacko (1962, 1963, 1966a, b) and Chacko et al. (1964, 1965). During this period, the contribution by K. R. Nair in design of experiments, statistical quality control and importance of statistical methods in forestry research is also important. The reports from him are Nair (1948a, b, 1950a, b, 1953a, b, 1954) and Nair and Bhargava (1951). Some Statistical reporting in the Indian forestry sector can be seen in Kishwan et al. (2008).

In 1994, the World Bank sponsored project, 'Forest Research, Education and Extension Project' emphasized on growing need for comprehensive statistics in forestry sector. As a follow-up, Directorate of Statistics was formed in Indian Council of Forestry Research and Education (ICFRE) with the mandate of collection, collation and compilation of data generated by the states and union territories, forest corporations and other organizations in the country. The data is being published in the form of Forestry Statistics India since 1995.

Some of the organizations in India that are dealing with specific areas of Forestry Statistics are as follows:

State Forest Departments: The main source of forestry information generated at forest division level is compiled and published in the form of Annual Forest Statistics/Administrative Reports. The working plans prepared by the division forest offices are another good source of forestry statistics.
Survey and Utilization Division, Ministry of Environment, Forest and Climate Change (MoEF&CC): It deals with the matters related to forest development corporations, forest survey of India (except establishment), bamboo and rattan, export and import related to wood and wood products, non-timber forest products. It also published national level report on various forestry statistics based on returns from state forest departments.
Ministry of Statistics and Program Implementation (MoSPI): Collects primary and secondary data on forests to estimate gross domestic product contributed by forest sector. Central Statistical Office (CSO) in MoSPI is the nodal agency for a planned development of the statistical system in the country and for bringing about coordination in statistical activities among statistical agencies in the Government of India and State Directorates of Economics and Statistics. It publishes yearly publication 'Compendium of Environmental Statistics' besides many others.
Division of Forestry Statistics, ICFRE: Forest areas according to ownership type, legal status, economic management or exploitation, protection and use for grazing; forest areas surveyed, the boundaries and their progress including forest settlement; afforestation, volume of standing timber and firewood and the production of timber, firewood and minor forest products; revenue and expenditure of the forest department; employment in forestry and forest industries and foreign trade in forest products.
Forest Survey of India, Dehradun: It generates a considerable amount of forestry data using remote sensing technology and field surveys such as extent of forests and its types covering all 14 physiographic zones of the country. These statistics are published as India's State of Forest Reports biennially.
Indian Institute of Forest Management, Bhopal: It carries out research for the sustainable use of management and allied techniques and methods conducive to the development of forestry in the country.
National Wasteland Development Board: It collects data on afforestation, social and farm forestry.
Directorate of Commercial Intelligence and Statistics: It publishes data on the import and export of forest products as part of its overall statistics on foreign trade.
Institute of Remote Sensing, Dehradun: It deals with nationwide forest cover mapping and biome-level characterization of Indian forests biodiversity at landscape level. It is the source of data on growing stock and biomass assessment, wildlife habitat modeling, sustainable development planning, national-level carbon flux measurement and vegetation carbon pool estimation, ecosystem dynamics and hydrological modeling in north-eastern region, wildlife habitat evaluation in Ranikhet and Ranthambore Tiger Reserve, grassland mapping and carrying capacity estimation (https://www.iirs.gov.in/forestryandecology).

In the last three decades, global forests are considered to have a potential role in sequestering increased CO_2, an important source of global warming. Development of statistical models has become imperative in understanding the effect of elevated CO_2 on tree physiology and the growth dynamics of forest stands. Besides, Indian forests are under tremendous stress due to natural and anthropogenic causes. The present-day objectives of forest management such as resource management, biodiversity conservation, management of hydrological resources and recreation amenities are becoming complex as they have to address the demands of multiple stakeholders. In this context, statistical methods would help unveiling useful information from multilevel data. The number of statistical software developed for data analysis has encouraged applying statistical techniques to solve various practical forestry uses.

7 Conclusion

The occupation of Indian Territory by the Imperial forces posed many problems including demarcation of territory, its ownership and administration, unexplored natural wealth (flora and fauna), scientific management of natural resources, conservation, etc. To understand and administer the vast forests of Indian subcontinents, there was a dire necessity of common scientific language and quantification. This is how the language and measurement/data/statistics came into being in the richness of tropical forests. The linkage of imperial ruler with the outside scientific world also had an influence over the scientific developments in the Indian forestry. Later, statistics has increasingly played a significant role in Indian forestry research and management. The use of statistically designed experiments, both in the laboratory and field had improved our understanding of forest regeneration and responses to management interventions. Such scientific efforts were documented in various forms of publications including Indian Forester. Scientific views were also shared through meetings like silviculture conferences and exchange programs among institutes.

References

Agrawal, A. (2005). Environmentality: Technologies of government and the making of subjects (344p). London: Duke University Press.

Bennett, B. (2011). A network approach to the origin of forestry education in India, 1855–1885. In B. Bennett & J. Hodge (Eds.), *Science and empire: Knowledge and networks of science across the British Empire, 1800–1970* (pp. 68–88). Great Britain: Palgrave Macmillan.

Brandis, D. (1884a). Untitled. *Indian Forester, 10*(8), 313–357.

Brandis, D. (1884b). Untitled. *Indian Forester, 10*(11), 493–500.

Bruenig, E. F. (2017). *Conservation and management of tropical rainforest: An integrated approach to sustainability* (2nd ed., 420p). Oxfordshire, UK: CAB International.

Chacko, V. J. (1962). Sampling in forest inventories. *Indian Forester, 88*(6), 420–427.

Chacko, V. J. (1963). Survey of bamboos, canes and reeds. *Indian Forester, 89*(4), 275–279.

Chacko, V. J. (1965). *A manual on sampling techniques for forest surveys* (172p). Delhi: The Manager of Publications.
Chacko, V. J. (1966a). Sequential sampling in forest insect surveys and diseases. *Indian Forester, 92*(4), 233–239.
Chacko, V. J. (1966b). The second stage of statistics in forestry and forest products work. *Indian Forester, 92*(10), 646–652.
Chacko, V. J., Mukerji, H. K., & Mitra, S. N. (1965). A study of the efficiency of line plot surveys in forest enumerations. *Indian Forester, 91*(1), 28–32.
Chacko, V. J., Rawat, A. S., & Negi, G. S. (1964). A point sampling trial with prisms at new forest. *Indian Forester, 90*(6), 348–359.
Fernandez, E. E. (1883). Deodar in the Dhara Gad Valley. *Indian Forester, 9*(10), 493–502.
Fisher, R. A. (1925). *Statistical methods for research workers* (14th ed., 1973, 362p), New York, Hafner Press.
Jeyade, T. (1947). In Memorium (Nilambur Teak Plantations), 1846–1946. *Indian Forester, 73*(11), 499–500.
Kishwan, J., Sohal, H. S., Nautiyal, R., Kolli, R., & Yadav, J. (2008). Statistical reporting in the Indian forestry sector: Status, gaps and approach. *International Forestry Review, 10*(2), 331–340.
Logan, W. (2004). *Malabar manual* (Vols. 1&2, 772p). New Delhi: Asian Educational Services.
Molesworth, G. L. (1880). Durability of Indian railway sleepers, and the rules for making them. *Indian Forester, 6*(10), 97–99.
Nair, K. R. (1948a). Statistical methods and experimental design. *Indian Forester, 74*(6), 248–250.
Nair, K. R. (1948b). On the application of statistical quality control methods in wood based industries. *Indian Forester, 74*(11), 379–382.
Nair, K. R. (1950a). A brief historical sketch of the introduction of statistical methods in Indian silvicultural experiments. *Indian Forester, 76*(2), 67–68.
Nair, K. R. (1950b). Sampling techniques—Adaptation of modern statistical methods to the estimation of forest areas, timber volumes, growth and drain. *Indian Forester, 76*(1), 31–35.
Nair, K. R. (1953a). Statistical methods in forest products research. *Indian Forester, 79*(2), 87–91.
Nair, K. R. (1953b). Statistical methods in forest research. *Indian Forester, 79*(7), 383–389.
Nair, K. R. (1954). Place of statistics in scientific research. *Indian Forester, 80*(4), 240–241.
Nair, K. R., & Bhargava, R. P. (1951). Statistical sampling in timber surveys in India. In *Indian Forest Leaflets* (153p). Dehradun: Forest Research Institute.
Ribbentrop, B. (1900). Forestry in British India, Republished in 1989 with a commentary by Rawat, A. S. (189p). New Delhi: Indus Publishing Company.
Sarkar, O. (2012). The great trigonometrical survey: Histories of mapping, 1790–1850. *ETraverse, The Indian Journal of Spatial Science, 3*(1), 1–6.
Smythies, A. (1876). Forest terminology with reference only to the more important terms. *Indian Forester, 1*(3), 284–293.
Smythies, A. (1891). Durability of railway sleepers. *Indian Forester, 17*(8), 312–313.
Troup, R. S. (1921). *The silviculture of Indian trees* (Vols. 1–3, 1195p). Oxford: Oxford University Press.

National Forest Inventory in India: Developments Toward a New Design to Meet Emerging Challenges

V. P. Tewari, Rajesh Kumar and K. v. Gadow

Abstract National Forest Inventory and forest assessments are attracting increasing attention owing to their role in providing information related to manifold forest functions. There is a high demand for global information about forests and multiple services that these ecosystems provide. Of particular, current interest are forest assessment systems at national level by countries that wish to engage themselves in the REDD+ initiative. The discipline of *Forest Inventory* has developed a versatile toolbox of techniques and methods useful for national-level forest assessments. This chapter presents a brief overview of the National Forest Inventory (NFI) in India *vis-à-vis* some other developed countries and highlights the proposed changes in plot design while revising NFI in India. Some new initiatives by Forest Survey of India have also been highlighted. With an increase in requirement of information and new technological developments, a constant adaptation of the NFI framework and introduction of the new NFI design for India is essential. The new NFI for India has been designed to cope with changing contexts, new technical developments and the unique socio-ecological conditions in India. The details of the new design with a focus on terrestrial assessments using field plots of different sizes and shapes are presented. It is shown why concentric circular plots, which are widely used in the Northern Hemisphere, are found unsuitable for the species-rich forests of India. The particular structure of the new Indian NFI is also briefly discussed and compared with other large-scale NFI's as well as global *Big Data* initiatives. A unique feature of the new Indian NFI is an integrated system of temporary and permanent field plots. Permanent observational plots are designed to monitor forest change and complement the network of temporary field plots.

Keywords Forest inventory · Permanent observation plot · Sampling design

---

V. P. Tewari (✉)
Himalayan Forest Research Institute, Shimla, India
e-mail: vptewari@yahoo.com

R. Kumar
Forest Survey of India, Dehradun, India

K. v. Gadow
Georg-August University, Goettingen, Germany

G. Chandra et al. (eds.), *Statistical Methods and Applications in Forestry and Environmental Sciences*, Forum for Interdisciplinary Mathematics,
https://doi.org/10.1007/978-981-15-1476-0_2

1 Introduction

The status of the existing situation and estimated future requirements triggers the planning process of renewable and non-renewable natural resources at a particular epoch. This understanding drives the information needs for planning and surveys are planned accordingly to generate desired information with acceptable levels of accuracy and/or precision.

The concept of management of forest resources that started during the nineteenth century necessitated the generation of quantitative information about various parameters of forest resources. In 1863, *Manual of Forest Operation* was prepared in India for a systematic collection of data related to the working of the forests which paved the path of today's *Working Plans* (Tewari and Kleinn 2015). The mapping of forest area was also started by Survey of India, and after 1910, this mapping was made ancillary to topographical surveys (Jamir 2014).

Forest inventory is a systematic collection of data on forestry resources within a given area. It allows assessment of the current status and lays the ground for analysis and planning, constituting the basis for sustainable forest management (http://www.fao.org/sustainable-forest-management/toolbox/modules/forest-inventory/basic-knowledge/en/?type=111). The first National Forest Inventories (NFIs) were established between 1919 and 1923 in Finland, Sweden, Norway and New Zealand (Lorenz et al. 2005). The USA followed in 1930 (McRoberts et al. 2005) and India during the 1960s (Pandey 2012). China implemented a National Forest Inventory during the 1970s (Zeng et al. 2015) followed by Japan in 1999 (Iehara 1999), Canada and Brazil in 2006 (Canada 2018; Brazil 2016). The original aim of these NFIs was to assess forest areas and growing stock volumes within a geographical context. More objectives were added gradually to include changes in biodiversity status and land use, carbon stock and ecosystem services, using a combination of remote sensing technology and field sampling (Tewari 2016).

During 1950s, great amount of work was going on in developed countries, in which more emphasis was laid to develop suitable sampling design for forest inventories. Statisticians were trying different sampling designs ranging from simple random sampling, stratified, systematic, cluster and probability proportional to size. Few of them were also trying to use aerial photographs. Most of the statisticians were of the opinion that the best approach is to use aerial photograph plots and ground plots in combination to estimate areas and volumes (Frayer and Furnival 2000). In this methodology, a double sampling plan in which aerial photographs were used to form strata and to estimate their sizes along with a subsample of the plots which were to be measured as field plots was used.

In Forest Inventory, information is generated on the quantity and quality of the forest resources and land area on which the trees are growing. Thus, the reports contained the information not only on the species and diameter class-wise growing stock, actual utilizable wood and bamboo, etc. but also about accessibility, transportation, infrastructure, description of topography, etc. Preservation of survey results by way of mapping, etc. was done which was felt necessary for the industrial units interested in putting up their manufacturing units.

Precision of estimates, which is influenced by manpower, cost, time and permissible error, is the key factor while designing the forest inventory. Similarly, suitable plot size and shape, optimum sample size has its impact on the requirement of manpower, cost and time for a given level of precision (Tewari 2016). All these parameters are functions of heterogeneity of the forest resources available in the study area. To understand the level of heterogeneity, pilot surveys are conducted, and these parameters were estimated to ascertain suitable sampling designs for a particular study area.

NFIs provide essential data for formulating national forest policies, planning forest industry investments, forecasting wood production and monitoring forest ecosystem dynamics. The traditional role of an NFI has been to provide unbiased information about forest resource covering a whole country and including computation of forest statistics (Tewari 2016). NFIs are continuously adapted to changing information needs and technical innovations, but major changes in design need to be carefully evaluated before implementation.

1.1 Temporary Sampling Plots

The standard procedure in field sampling involves a description of the sampling frame (the complete list of possible plot locations) and the number of plots (n) (or the proportion of locations) that are to be visited. The n plots are then assigned to locations, either randomly or based on a systematic design. A specific set of parameters of interest is assessed in each plot. Statistical reports provide estimates of forest areas, volumes and biomass, summarized in various ways (e.g., by forest type, age class, ownership or protection status) for specific geographic units (the whole country, physiographic zones, administrative subdivisions like districts or states).

The advantage of a system of temporary field plots is a cost-effective assessment. It is not necessary to mark the plot location or to assign numbers to individual trees for re-identification. However, such temporary installations are inefficient in measuring change because the covariance between observations taken at successive sampling events is usually high. Temporary sample plots from forest inventories may be representative of the total population, but provide only limited information about forest change, i.e., the response to management and environmental conditions (Köhl et al. 1995).

1.2 Permanent Observation Plots

Permanent observation plots (POPs) provide information on change, but they are not necessarily representative of the total population (Avery and Burkhart 2002; Burkhart and Amateis 2012). POPs are costly to maintain because plot locations and plot boundaries have to be permanently fixed, trees are numbered for re-identification and individual tree positions are usually mapped (Gadow 2017a, b). A temporary plot with summary information does not provide effective observation, even if it is re-measured.

That may be one of the reasons why several new permanent forest observational networks have been established since the turn of the twentieth century in countries where long-term forest observation has been lacking. Individual scientists identified a real need for permanent monitoring in addition to the NFI assessments. An example of such an initiative is the Forest Observational Network of the Beijing Forestry University (Zhao et al. 2014). The network includes very large field plots with mapped trees in the mixed deciduous forests of Northeastern China, large plots in the northern pine forests and in the central and western *Abies-Picea* mountain forests and in the southeastern tropical forests. The large plots provide rare opportunities to evaluate the effects of scale involving beta diversity (Tan et al. 2017), density dependence (Yao et al. 2016), taxonomic structures (Fan et al. 2017) or species-habitat associations (Zhang et al. 2012).

Another example is a new POP network in Mexico's Sierra Madre Occidental. The Mexican network was initiated in 2007 and includes 429 quarter-hectare observational field plots representing major forest types and physiographic regions. All the trees within each plot are numbered and their positions are mapped and many of the first installations have already been re-measured (Corral Rivas et al. 2016). Such individual initiatives are indicative of the need for permanent observation, but there is a risk that plots maintained by an individual scientist are abandoned when that person retires.

2 Brief History of NFI in India

2.1 Forest Inventory During 1965–2002

After India's independence in 1947, a huge forest area came under the control of the government, and therefore, the National Forest Policy was formulated in 1952 which laid emphasis on forest survey and demarcation along with other aspects of forest management and development. Having been in the era of industrial revolution, forestry sector of Government of India also intended to attract wood-based industries. Keeping this in view, a project named *Pre-Investment Survey of Forest Resources* (PISFR) was undertaken by Government of India in collaboration with United Nations Development Program (UNDP)/Food and Agricultural Organization

(FAO) (Singh 2006) in 1965. Three regions were selected for PISFR, mainly because they contained forest tree species of industrial importance.

From 1965 to 1980, generally systematic cluster sampling was used after conducting a pilot survey in the study area. The study area was divided into suitable grid sizes (5 feet × 5 feet or 2.5 feet × 2.5 feet) depending upon the optimum sample size. Within the selected grid, a cluster of 3–8 subplots was considered for recording the data on different parameters of the inventory. Wherever the information on stratification variable was available (generally, pre-stratification on the basis of aerial photographs), stratified random sampling was used otherwise the collected data was post-stratified to increase the precision of estimates (FSI 2015). As the forest inventories carried out in different parts of the country since 1965 were in a different time frame, it was not possible to generate national-level estimates on growing stock, area statistics and other parameters with reference to one point of time (Pandey 2008).

The late 1970s and early 1980s were important regarding national as well as international forest scenarios and influenced a paradigm shift regarding the role of the national forests in India. The forests, which initially had seemed to be an inexhaustible resource, were rapidly depleting under the pressure of rising human and cattle populations. Therefore, strategies were proposed to remove the focus from production forestry to conservation forestry by developing new programs in social and agro-forestry (Tewari 2016).

To formulate suitable strategies for the new scenario, there was a need to have information at national level. Acknowledging the utility of forest resources survey, the National Commission on Agriculture, in its report in 1976, recommended the creation of a *National Forest Resource Survey Organization*. As a result of this recommendation, PISFR was converted into the Forest Survey of India in 1981 (Tewari and Kleinn 2015). Anticipating this transcription, PISFR had started developing national-level forest inventory design. A high-level committee was constituted in 1980 under the chairmanship of Director, Central Statistical Organization which recommended that systematic sampling should be used for National Forest Inventory (Singh 2006).

During the last decade of the twentieth century, the role of forests was further redefined by including additional parameters like carbon sequestration in plant community and in forest soil, regeneration status, etc. Taking it as opportunity to fulfill its dream of conducting NFI, Forest Survey of India conducted a series of workshops and meetings with experts and came out with an integrated approach during 2001–2002 in which it was decided to conduct forest inventory in a cycle of two years, with the provision of capturing all these parameters. This approach is being followed, and the estimates are being improved cycle after cycle (Pandey 2012).

2.2 NFI Since 2002

At the beginning of the Tenth Five-Year Plan in 2002, considering the resource limitation, a new sampling design was adopted to generate national-level estimates of various parameters including those of growing stock from forests (Pandey 2012).

A two-stage sampling design was adopted for national forest inventory. In the first stage, the country was stratified into homogeneous strata called the "physiographic zones," based on physiography, climate and vegetation. The districts within the physiographic zones were taken as sampling units (FSI 2015). The 14 physiographic zones identified in the country are: Western Himalayas, Eastern Himalayas, North East, Northern Plains, Eastern Plains, Western Plains, Central Highlands, North Deccan, East Deccan South Deccan, Western Ghats, Eastern Ghats, West Coast and East Coast (Fig. 1).

A sample of 10% of districts (about 60 districts in the country) distributed all over the physiographic zones, in proportion to their size, are selected randomly for

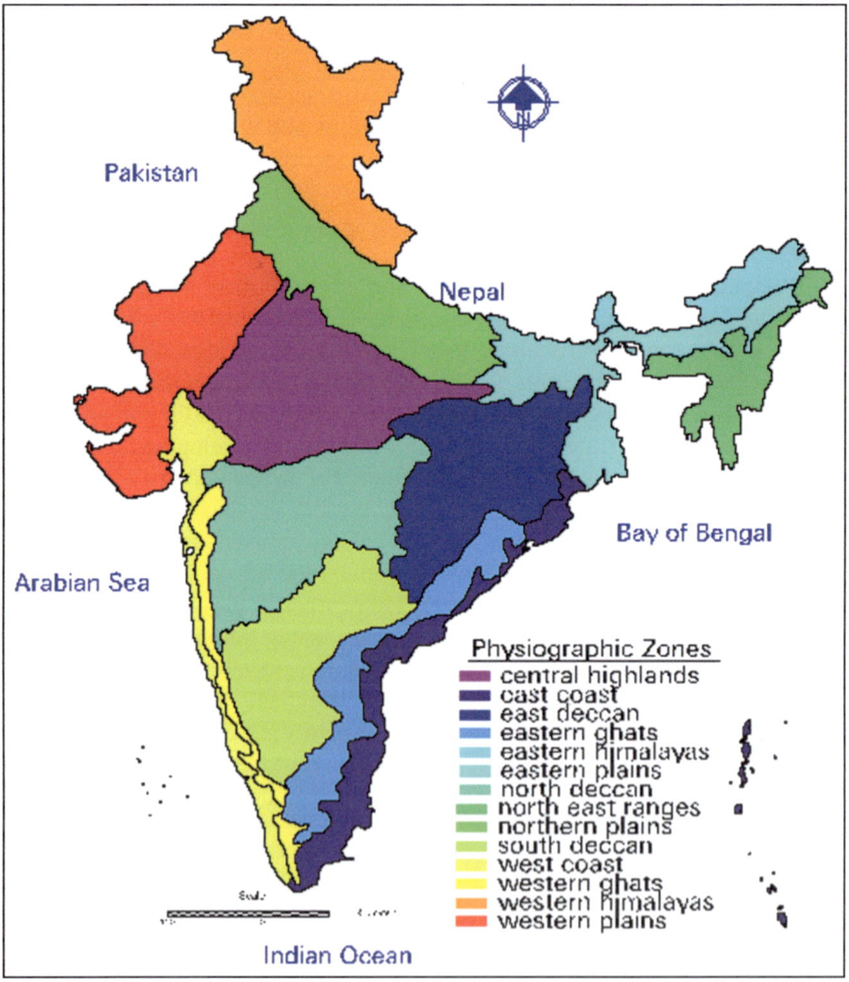

Fig. 1 Physiographic zone-wise map of India

a detailed inventory of forests and trees outside forests (Fig. 2). In the second stage, separate sampling designs are followed for a detailed inventory of forests and trees outside forests (TOF). In the second stage, selected districts are divided into grids of latitude and longitude to form the sampling units, and sample plots are laid out in each grid to conduct field inventory using a systematic sampling design.

Detailed forest inventory is carried out in the 60 districts selected in the first stage. The Survey of India topographic sheets on 1:50,000 scale (size 15 feet × 15 feet, i.e., 15 min latitudes and 15 min longitudes) of the district are divided into 36 grids of 2.5 feet × 2.5 feet which are further divided into subgrids of 1.25 feet × 1.25 feet forming the basic sampling frame. Two of these 1.25 feet × 1.25 feet subgrids are then randomly selected to lay out the sample plots. Other forested subgrids in the districts are selected systematically taking first two subgrids as a random start. If the

Fig. 2 Selected districts of a cycle

center of the selected 1.25 feet × 1.25 feet subgrid does not fall in the forest area, then it is rejected. If it falls in the forest area, it is taken up for inventory. The intersections of diagonals of such subgrids are marked as the center of the plot at which a square sample plot of 0.1 ha area is laid out to record the measurements of diameter at breast height (DBH) and height of selected trees species. In addition, within this 0.1 ha plot, subplots of 1 m × 1 m are laid out at northeast and southwest corner for collecting data on soil, forest floor (humus and litter), etc. The data regarding herbs, shrubs and climbers are collected from four square plots of 1 m × 1 m and 3 m × 3 m size in all four directions along the diagonal 30 m away from center of 0.1 ha plot (FSI 2015).

The data collected in the field are checked to detect any inconsistency or recording error before entering them into computer datasheets. For processing of forest inventory data, the design-based and model-based estimation protocols are employed, and estimates for area and growing stocks are generated.

2.3 *New Initiatives by Forest Survey of India in NFI*

2.3.1 e-Green Watch

Compensatory Afforestation Fund Management and Planning Authority (CAMPA) is the National Advisory Council for monitoring, technical assistance and evaluation of compensatory afforestation and other forestry activities financed by it. Therefore, e-Green Watch is designed and developed as a web-based, role-based workflow applications and integrated information system which shall enable automating of various functions and activities related to monitoring and transparency in the use of CAMPA funds and various works sanctioned in the Annual Plan of Operations (state CAMPA) approved by the state authorities (FSI 2015).

2.3.2 Decision Support System

Decision Support System is a web-GIS-based application which has been developed to provide qualitative and quantitative information with respect to forest area. It enables decision makers to take a well-informed decision based on the information generated by the system. This system helps in a big way for taking decisions with respect to proposals under the Forest Conservation Act (FSI 2015).

It uses different spatial layers for providing information on different issues related to forest and wildlife areas such as "Forest Cover Mapping of Tiger Reserves" and "Real-Time Monitoring of Forest Fires."

3 NFI in Some Other Countries

3.1 Swedish National Forest Inventory

The NFI of Sweden is based on a systematic sample. An inventory is conducted on sample plots, and this forms the basis for both area and volume estimates. The sample plots are circular having radius of 7–10 m. These are, for practical reasons, grouped into clusters which are known as tracts. The tracts may be rectangular or square in shape. Tracts may have different dimensions (300–1800 m) in different parts of the country. The number of sample plots per tract also varies for different parts of the country (Olsson et al. 2007).

The tracts are either permanent or temporary. The permanent tracts are laid out in a systematic grid covering the whole country (with a random start) and are revisited every fifth year (Fig. 3). The temporary tracts, which are visited only once, are randomly distributed but with due consideration of the permanent tracts already laid out in a systematic grid to avoid overlapping.

For estimation protocol, design-based and model-based approaches are used. For small area estimates, satellite imageries are used as well (Olsson et al. 2007).

Fig. 3 Tract distribution of single cycle of 5 years of Swedish NFI

3.2 Finnish NFI

Since the beginning of 1920s, large-area forest resource information has been produced by the NFI of Finland. The NFI design was modified and the inventory cycle is reduced to 5 years from tenth NFI in 2004. Every year measurements are done in the whole country by measuring one-fifth of the plots, and 20% of all plots are measured as permanent (Tomppo 2008).

To respond to current requirements and to optimize the use of the existing resources, sampling design and stand and plot-level measurements have been changed over a period of time. Line-wise survey sampling was adopted in the first NFI taking the line interval as 16 km in most parts of the country, though an interval of 13 km was used in one province and 10 km in another for estimating the error. Line strips of 10 m wide were used for plot measurements keeping plot length as 50 m and the distance between plots as 2 km. Same sampling design with different sampling intensities was used for the next three inventories (Tomppo 2008). The initial four inventories had a cycle of 3 years, but for the next five inventories, it varied from 6 to 9 years.

Presently, systematic cluster sampling is being followed where the sampling units used are clusters, referred to as a tract. The distance between two tracts varies from 6 km × 6 km in the southernmost part of the country to 10 km × 10 km in Lapland. The number of plots per tract varies from 9 to 14, while the distance between adjacent plots within tract is 250–300 m (Tomppo 2008).

For estimation protocol, design-based and model-based approaches are used. For small area estimates, satellite imageries are used as well.

3.3 German NFI

The main aim of NFI in Germany is to provide an overview of forest condition and productivity by using a permanent design which enables re-measurement of the same plots to obtain data on increment and drain. For the first time, a sample-based NFI was carried out between 1986 and 1990, and second NFI took place during 2001–2002 (Polley et al. 2010). The third NFI was completed during 2011–2012.

A periodic survey is being carried out in German NFI in the whole territory at time intervals which is not predefined but has to be determined a new in every repetition. The information provided by the NFI is of immense significance for forest policy, especially in connection with international commitments to report on forest resources. The information generated is also useful for the wood processing industry (Kändler 2008).

A systematic distribution of tracts on regular grids of regionally differing width is used in the inventory design. The primary sampling unit is a quadrangular tract having side length of 150 m, and the tract corners are the centers of permanently marked subplots in which different sampling procedures are adopted for the selection of sample trees and the survey of various characteristics. The General Administrative Regulation prescribes a basic grid of the width of 4 km × 4 km covering entire country with a defined starting point. The sample grid size decreased in some states to a 2.83 km × 2.83 km or 2 km × 2 km. The sample selection on a tract corner is made according to size of plants using different methods like horizontal point sampling and fixed plot sampling. For each inventory tract, almost 150 characteristics are recorded (Kändler 2008).

Different estimators of volume growth are used, and estimates are developed. For estimation protocol, satellite imageries are used as well.

3.4 US NFI

In USA, the Forest Inventory and Analysis (FIA) Program collects, analyzes and reports information on the status, trends and condition of forests. Different kind of information is generated like quantification of existing forest, their location and ownership, change in forest condition, growth of the trees and other vegetation and number of trees died or has been removed (Burkman 2005).

Irrespective of ownership or availability for forest harvesting, a single inventory program includes all forested lands in the country which covers all public and private forest land like reserved areas, wilderness, National Parks, defense installations and National Forests.

As part of annual inventory, measurements are taken on a fixed proportion of the plots in each state every year under FIA. Each portion of the plots is termed as a panel. The legislative mandate requires measurement of 20% of the plots each year in every state which is to be accomplished through a federal-state partnership. Plans have also been formulated for less intensive sampling levels of 15% per year and 10% per year (Burkman 2005).

The plot intensity presumes that sufficient plots are measured to suit precision standards for area and volume estimates that are consistent with historical levels. Individual states have liberty to increase the sample intensity by installing additional plots, but at their own expense, for increasing the precision. The annual inventory system has one advantage that it provides maximum flexibility to states to engage in such intensifications.

To provide a uniform basis for determining the annual set of measurement plots, a nationally uniform cell grid has been super-imposed over old set of sample locations. This system provides a standard frame for integrating FIA and for linking the other data sources of program such as satellite imagery, spatial models and other surveys. The FIA program includes a national set of core measurements collected on a standard field plot. The enhanced FIA program consists of three phases or tiers (Burkman 2005).

In Phase 1, remotely sensed data are collected in the form of aerial photographs/satellite imagery. In this Phase, two tasks (initial plot measurement using remotely sensed data and stratification) are accomplishes. This phase "photograph point" is characterized as non-forest or forest.

In Phase 2, a subset of the photograph points are selected for field data collection which consists of one field sample site for every 6000 acres. Information about forest type, site attributes, tree species, tree size, overall tree condition, etc. is being collected by the field crews.

Phase 3 consists of a subset of Phase 2 sample plots. These plots are measured for a broader set of forest health attributes (like tree crown conditions, lichen community composition, under-story vegetation, down woody debris and soil attributes). An associated sample scheme is also in place to detect and monitor ozone injury on forest vegetation. For every 96,000 acres, there is one Phase 3 plot which means for every 16 Phase 2 plots, there is one Phase 3 plot (Fig. 4).

An FIA plot consists of a cluster of four circular subplots which are spaced in a fixed pattern. The plot is designed to provide a sampling frame for all Phase 2

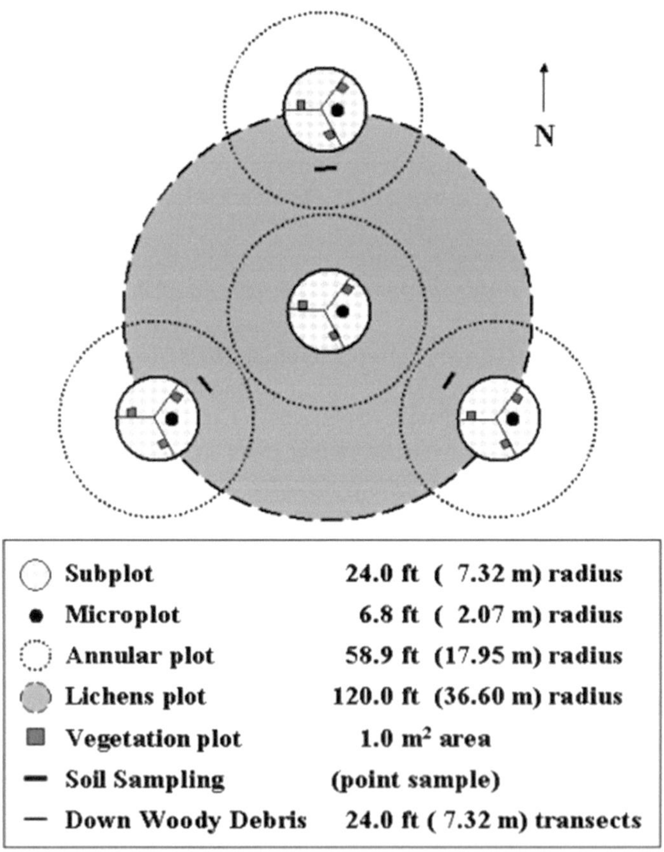

Fig. 4 Phase 2/Phase 3 plot design

and Phase 3 measurements. Most of the measurements are taken within the subplots. Seedlings, saplings and other vegetation are measured in respective defined subplots.

For estimation protocol, design-based and model-based approaches are used. Satellite imageries are also used for stratification, quantification of stratum size and generating estimates of various parameters.

4 National Forest Monitoring and Assessment-FAO's Initiative

FAO initiated an activity to provide support to National Forest Monitoring and Assessment (NFMA) in view of the growing demand for reliable information on forests and tree resources both at country and global levels. The support offered by FAO includes developing a harmonized approach to NFMA, information management, reporting and support to policy impact analysis for national-level decision-making.

NFMA focuses on the establishment of a long-term monitoring system that will be sustainable in time with repeated periodic measurements. This is performed by strengthening national capacity and institutionalizing the inventory process. Field inventory and remote sensing, both are combined in the NFMA methodological approach.

For the NFMA, the systematic sampling design is recommended. The latitude–longitude grids of $1^\circ \times 1^\circ$ (or appropriate size) are prepared to be used as a sampling frame. Sampling units are selected at the intersection of every degree of this grid. Optimum sample size is ascertained depending on country's information needs at required precision. Pre- or post-stratification may be adopted to optimize the precision and costs. The number of sampling units to be surveyed is ascertained by the required statistical reliability of the data, the available resources for the work and with a view of enabling periodic monitoring (Saket et al. 2010).

The major sampling unit is a square tract of 1 km × 1 km, and each sampling unit contains a cluster of four permanent, rectangular, half-hectare sample plots, placed in perpendicular orientations (Fig. 5). Smaller subunits are delineated within each plot, e.g., three sets of subplots, three measurement points and three fallen dead wood transect lines (FAO 2008). The use of satellite imageries is advised for stratification, quantification of size of strata and generating estimates of various parameters.

5 New National Forest Inventory System in India

Generally, the information generated by NFI are used in forest policy making at national and international levels, regional and national forest management planning, planning of forest investments, assessing sustainability of forests, evaluation

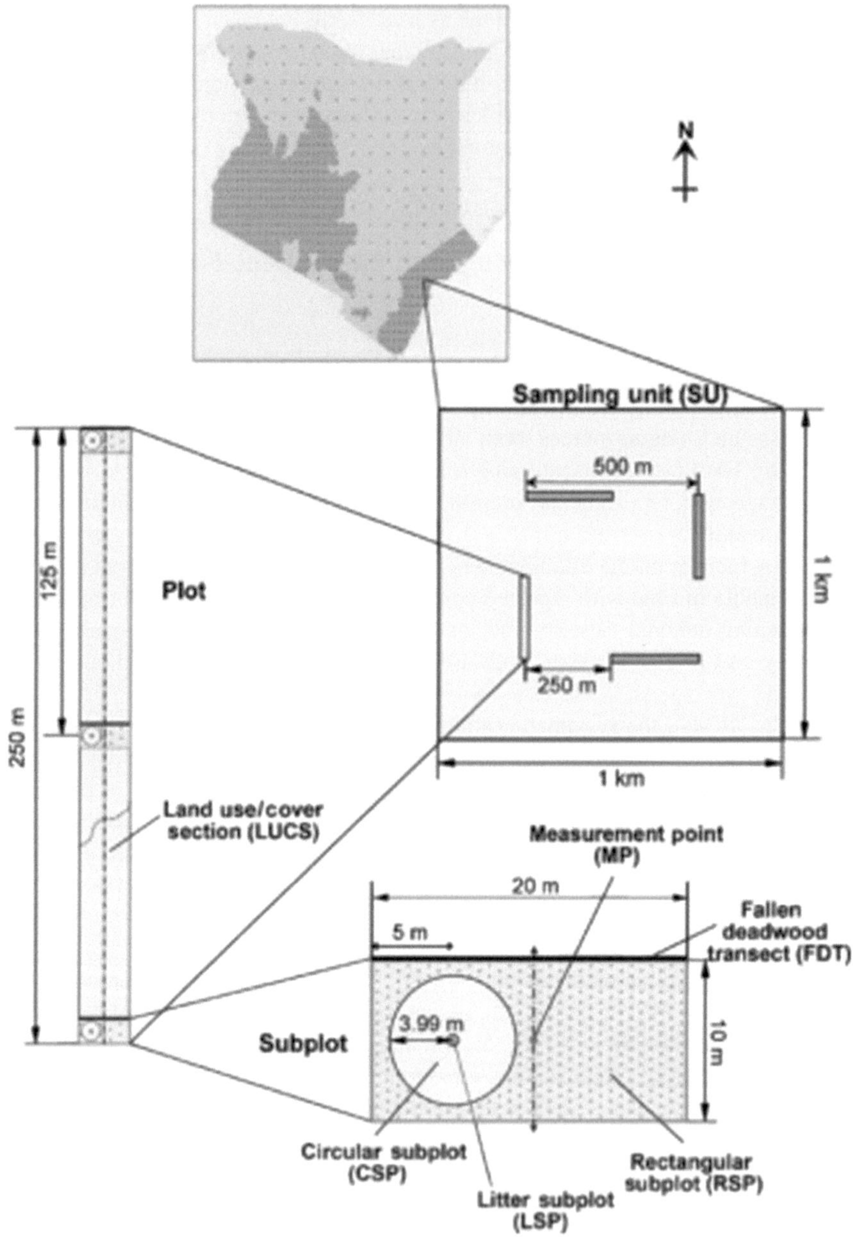

Fig. 5 FAO NFMA design

of greenhouse gas emissions and changes in carbon storage and research, etc. (Tewari 2016).

The considerations related to use of data for planning sustainable forest management require that sampling units are uniformly spread (not a two-stage cluster sampling) and revisited at shorter intervals (five years) to monitor changes and national sample plots, and state plots must have statistical compatibility. It should include new parameters like NTFP, eco-system services, evaluation of green house gas emissions, disturbances in forests, etc. At the same time, it should provide information at state level so that information for Measurement Reporting and Verification (MRV) of REDD+ is generated. Remote sensing data have greater role for state-level applications when integrated with field survey.

Seven main points are crystallized for the NFI which are as: (1) All the countries (and FAO) are following systematic sampling for NFI, (2) nation-wide wall-to-wall grids are considered, (3) cluster of plots are considered, (4) there are permanent and temporary plots, (5) all the nation relevant data are collected, (6) use of geomatics, specially remotely sensed data and (7) there is a provision of repeating the same plot after fixed time period.

The new National Level Continuous Forest Inventory System has been conceived to form a basis for making continuing policy and planning decisions, including the role of forests as an ecosystem (the "conservation view") and its role as resource provider (the "utilization view"), and this holds for all levels of forestry from the local to the global.

The revised NFI is a new generation of NFI system geared for providing comprehensive information to meet above requirements to meet information requirements of sustainable forest management including those under Green India Mission and CAMPA. Parameters for meeting reporting obligations under the conventions on climate change (UN-FCCC), biodiversity (UN-CBD) and combating desertification (UN-CCD) are also included. It may be mentioned that under REDD+, countries may expect considerable payments when they successfully have implemented policies that reduce emissions from deforestation and forest degradation. An important new feature of these payments, however, is that they are strictly performance based, which means that the success of the policies needs to be evidenced by methodologically sound and transparently documented forest monitoring (REDD+ MRV).

Incorporating above suggestions, according to national interest and utility, Forest Survey of India has redesigned the NFI which will provide information about the traditional variables, viz. growing stock, regeneration, land use within Recorded Forest Area (RFA), grazing, size classes, humus classes and origin of stand as well as newer variables, viz. quantification of important NTFP resource species, biodiversity indicators, climate change indicators and carbon storage. It can provide estimates at state level also for the states having more than 10,000 km^2 of forest area.

5.1 *Proposed New Design for NFI*

The NFI is designed to include all forest area of the country which is notified by the government. The digital boundaries of forest area are available for 12 states of the country. For the remaining states, green-wash area (shown as green in Survey of India topographic sheet depicting RFA and other traditional forest areas at the time of preparation of the topographic sheets) will be taken as a proxy of Recorded Forest Area.

The new NFI envisages measurement of about 35,000 sample points across the country in five years. Every year 20% sample points will be covered. In other words, it will measure a fixed proportion of the plots in each state, each year. About 10% of the identified sample plots (may be referred as POPs) will be used for special studies related to biodiversity, climate change, forest soils, etc.

The sample size is optimized so that enough plots are measured to satisfy precision standards for area and volume estimates. There is provision that if states want more precise estimates, then they may choose to increase the sample size by installing additional plots. For NFI, 10% allowable error has been considered at state level for forests and trees outside forest (TOF) separately. To achieve this target, for few states, the number of sampling units has been increased to 2 or 3 per grid.

5.1.1 A Grid-Based Sampling Frame

Remote sensing-based inventories may benefit if the plot-level estimates correspond with the pixel-level information (Ling et al. 2014). The new NFI design involves a change from the current two-stage random sampling approach to a systematic grid-based design using a country-wide uniform grid of 5 $\times$ 5 km pixels. A number (1, 2, 3, 4 or 5) is assigned to each 25 km^2 pixel.

A forest layer will be developed by combining the digital layer of the Recorded Forest Area (RFA) boundaries of the 19 states and additional forest areas (known as "green-wash areas" because of the map coloring) that are not part of the official RFA classification. This forest layer will be overlaid on the grid layer, and the intersection of these two layers will provide the general sampling frame for the NFI. The Forest Survey of India maintains a large database with records of previous NFI sample points which can be utilized in each new NFI.

Each individual grid cell representing a pixel of 25 km^2 will thus be identified regarding one of three possible attributes, namely forested, no forest and TOF. A systematic sampling scheme with random start will be used. A random number will be chosen from 1 to 5 as a random start and thereafter all grids with that number and the attribute "forested" will be identified for data collection. The forest cover map based on satellite-based remote sensing data will be utilized for stratum size calculation.

Within the selected grid pixel, using Geographical Information System (GIS) software, a random point will be selected as the center of a new temporary sample point. On average, every grid cell representing a forest area of 25 km^2 will have one sample point. Field crews will collect data on site attributes, tree species and DBHs on accessible forest land. It is also envisaged that 10% of all plots will be subjected to an additional assessment of indicators of biodiversity, forest health, climate change and soil attributes.

5.1.2 New Plot Configuration

In the existing NFI, Forest Survey of India was laying out single plot of 0.1 ha at the sample point location for tree species. Though, there were multiple subplots for different characteristics, viz. four subplots for regeneration, four subplots for biomass of herbs. Almost all the other leading countries are using cluster of plots for tree measurements. Plots of different sizes are being used even in same country depending upon the variability in different parts of country. Similarly, distance between plots in the cluster varies among countries for obvious reasons.

For new NFI, Forest Survey of India conducted a pilot study, with the following objectives:

- Plot design: Single plot layout or cluster of subplots is better,
- If cluster, then which subplot size is better among 7, 8 and 9 m radius vis-à-vis 0.1 ha plot,
- What should be the distance between subplots: 30 or 40 ms,
- Whether circular subplot is feasible for NFI and
- Time and manpower requirement.

Before the pilot study, it has been decided that the USDA Forest Service FIA plot layout, i.e., cluster of four circular subplots spaced out in a fixed pattern, will be more suitable, if cluster of subplots is to be used.

Pilot study concluded that:

- Among the seven alternatives of plot design, the cluster of subplots with radius 8 m and at 40 m distance has better result compared with existing single plot layout.
- The circular subplot is feasible (the pilot study helped a lot as crew members got convinced while conducting the pilot itself).
- It reduces time and manpower requirement.

Based on outcome of pilot study and discussions thereafter, the following plot design (Fig. 6) is finalized which will address measurement of all the envisaged characteristics of forests.

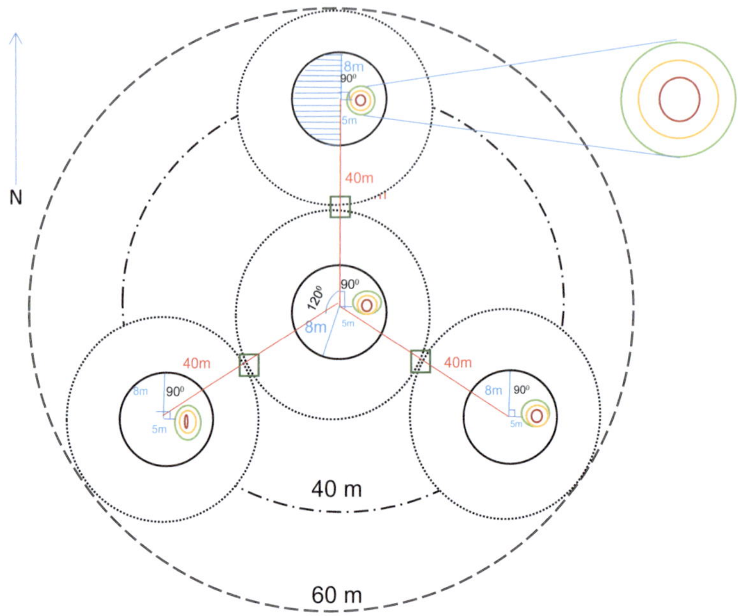

Subplot	8.0 m radius	
Annular plot	20 m radius	
Lichens plot	40 m radius	
Description plot	60 m radius	
Hub vegetation plot	0.6 m radius	
Shrub regeneration litter plot	1.7 m radius	
Deadwood plot	2.8 m radius	
Soil and forest floor sample plot	1m X 1m at mid point between subplots	
Non clump forming bamboo plot		

Fig. 6 Proposed plot design for NFI in India

6 Conclusions

In order to generate information for changing requirements of support to states in sustainable forest management, inventory of NTFP and other new variables, monitoring of change in forest characteristics, inclusion of climate change indicators, significant improvement of precision at the state/regional level, etc., the following changes have been proposed:

- From two-stage sampling design to uniform grid-based (5 km × 5 km) systematic sampling scheme;
- Re-measuring of same plot after 5 years (doubles the workload);
- Single plot design to cluster of subplots;
- Provision of permanent and temporary plots;
- Provision for nationally relevant data;
- Use of GIS and satellite imageries, for enhancing precision of estimates and developing small, is the estimates; and
- Flexibility to increase sample size at state and local levels so that combined effort may yield information at national, regional and local levels.

The value of the network of Mixed Permanent Observation Plots (POPs) will increase with each additional re-measurement because observations collected over long periods of time will increasingly reveal specific responses to climate change and human disturbance. Additional large permanent plots, covering 1 ha or 4 ha, may be established and maintained by the Indian Council of Forest Research and Education to complement the database of Forest Survey of India.

A basic standard for analyzing POP datasets, using the R statistical software, reflects current technology and analytical capabilities. Continuous adaptation is essential, recognizing new developments in hardware and analytical tools, preferably in co-operation with interested research institutes and universities. This requires a national commitment to continuity, a firm decision to ensure that all POPs are re-measured at regular intervals, for example, by assigning a special status to all POP's by an Act or Regulation through the Parliament.

References

Avery, T. E., & Burkhart, H. E. (2002). *Forest measurements* (5th ed., p. 456). New York: McGraw-Hill.

Brazil. (2016). Inventário florestal nacional: principais resultados: Distrito Federal/Serviço Florestal Brasileiro (SFB). Brasília: SFB, Série Relatório Técnico.

Burkhart, H. E., & Amateis, R. L. (2012). Plot installations for modeling growth and yield of loblolly pine plantations. In X. H. Zhao, C. Y. Zhang, K. v. Gadow (Eds.), *Forest observational studies. Proceedings of the International Workshop Held*, September 20–21, 2012 (pp. 17–34). Beijing Forestry University.

Burkman, B. (2005). Forest inventory and analysis—sampling and plot design. FIA Fact Sheet Series. USDA Forest Service. http://www.fia.fs.fed.us/library/fact-sheets/data.../Sampling%20and%20Plot%20Design.pdf. Accessed on February 22, 2018.

Canada. (2018). https://nfi.nfis.org/en/history. Accessed February 22, 2018.

Corral Rivas, J. S., Torres-Rojo, J. M., Lujan-Soto, J. E., Nava-Miranda, M. G., Aguirre-Calderón, O. A., & Gadow, K. v. (2016). Density and production in the natural forests of Durango/Mexico. *Allgemeine Forst-und Jagdzeitung (German Journal of Forest Research), 187*(5–6), 93–103.

Fan, C., Tan L., Zhang, C., Zhao, X., & Gadow, K. v. (2017). Analysing taxonomic structures and local ecological processes in temperate forests in North Eastern China. *BMC Ecology, 17*, 33, 1–11.

FAO. (2008). NFMA-Knowledge reference, dissemination and networking. National forest monitoring and assessment working paper. Food and Agriculture Organization, Rome (p. 12).

Frayer, W. E., & Furnival, G. M. (2000). History of forest survey sampling designs in the United States (pp. 42–49). In M. Hansen & T. Burk (Eds.), *Integrated tools for natural resources inventories in the 21st century: Proceedings of the IUFRO conference*, August 16–20, 1998, Boise, ID. Gen. Tech. Rep. NC–212. St. Paul, MN: U.S. Department of Agriculture, Forest Service, North Central Research Station.

FSI. (2015). *India state of the forest report 2015*. Dehradun, India: Forest Survey of India.

Gadow, K. v. (2017a). The potential of permanent forest observational studies in India. Report prepared for FAO, Rome, July 2017 (p. 38).

Gadow, K. v. (2017b). Permanent forest observational plots-assessment and analysis. Report prepared for FAO, Rome, October 2017 (p. 38).

Iehara, T. (1999). New Japanese forest resource monitoring survey. *Sanrin, 1384,* 54–61. (in Japanese).

Jamir, W. (2014). Mapping of forest cover through remote sensing and geographical information system (GIS), Wokha District, Nagaland IOSR. *Journal of Environmental Science, Toxicology and Food Technology, 8*(4), 97–102.

Kändler, G. (2008). The design of second german national forest inventory (pp. 19–24). In R. E. McRoberts, G. A. Reams, P. C. Van Deusen, & W. H. McWilliams (Eds.), *Proceedings of the Eighth Annual Forest Inventory and Analysis Symposium* (p. 408). Monterey CA, October 19–16, 2006. Gen. Tech. Report WO-79. USDA Forest Service, Washington, DC.

Köhl, M., Scott, C. T., & Zingg, A. (1995). Evaluation of permanent sample surveys for growth and yield studies: A Swiss example. *Forest Ecology and Management, 71,* 187–194.

Ling, D., Tao, Z., Zhenhua, Z., Xiang, Z., Kaicheng, H., & Hao, W. (2014). Mapping forest biomass using remote sensing and national forest inventory in China. *Forests, 5,* 1267–1283.

Lorenz, M., Varjo, J., & Bahamondez, C. (2005). Forest assessment for changing information needs. In G. Mery, R. Alfaro, M. Kanninen, & M. Lobovikov (Eds.), *Forests in the global balance-Changing paradigms* (Vol. 17, pp. 139–150). IUFRO World Series.

McRoberts, R. E., Bechtold, W. A., Patterson, P. L., Scott, C. T., & Reams, G. A. (2005). The enhanced forest inventory and analysis program of the USDA forest service: Historical perspective and announcement of statistical documentation. *Journal of Forestry, 103*(6), 304–308.

Olsson, H., Egberth, M., Engberg, J., Fransson, J. E. S., Pahlén, T. G., Hagner, O., et al. (2007). Current and emerging operational uses of remote sensing in Swedish forestry (pp. 39–46). In R. E. McRoberts, G. A. Reams, P. C. V. Deusen, & W. H. McWilliams (Eds.), *Proceedings of the Seventh Annual Forest Inventory and Analysis Symposium* (p. 319). Portland ME, October 3–6, 2005. Gen. Tech. Report WO-77. USDA Forest Service, Washington, DC.

Pandey, D. (2008). India's forest resource base. *International Forestry Review, 10,* 116–124.

Pandey, D. (2012). National forest monitoring for REDD+ in India. In B. Mora, M. Herold, V. De Sy, A. Wijaya, L. Verchot, & J. Penman (Eds.), *Capacity development in national forest monitoring: Experiences and progress for REDD+* (pp. 19–26). Joint report by CIFOR and GOFC-GOLD. Bogor, Indonesia.

Polley, H., Schmitz, F., Hennig, P., & Kroiher, F. (2010). National forest inventories reports: Germany. In E. Tomppo, T. Gschwantner, M. Lawrence, & R. E. McRoberts (Eds.), *National forest inventories—Pathways for common reporting* (pp. 223–243). Berlin: Springer.

Saket, M., Branthomme, A., & Piazza, M. (2010). FAO NFMA—Support to developing countries on national forest monitoring and assessment. In E. Tomppo, T. Gschwantner, M. Lawrence, & R. E. McRoberts (Eds.), *National forest inventories: Pathways for common reporting* (p. 612). Springer Science, Business Media B.V.

Singh, K. D. (2006). A tribute to Forest Survey of India, Dehradun. Souvenir: Silver Jubilee year 1981–2006. Forest Survey of India, Dehradun, India.

Tan, L., Fan, C., Zhang, C., Gadow, K. v., & Fan, X. (2017). How beta diversity and the underlying causes vary with sampling scales in the Changbai mountain forests. *Ecology and Evolution, 7*(23), 10116–10123.

Tewari, V. P. (2016). Forest inventory, assessment, and monitoring, and long-term forest observational studies, with special reference to India. *Forest Science and Technology, 12*(1), 24–32.

Tewari, V. P., & Kleinn, C. (2015). Considerations on capacity building for national forest assessments in developing countries—With a case study of India. *International Forestry Review, 17*(2), 244–254.

Tomppo, E. (2008). The Finnish national forest inventory. In R. E. McRoberts, G. A. Reams, P. C. V. Deusen, & W. H. McWilliams (Eds.), *Proceedings of the Eighth Annual Forest Inventory and Analysis Symposium* (p. 408). Monterey CA, October 19–16, 2006. Gen. Tech. Report WO-79. USDA Forest Service, Washington, DC (pp. 39–46).

Yao, J., Zhang, X., Zhang, C., Zhao, X., & Gadow, K. v. (2016). Effects of density dependence in a temperate forest in northeastern China. *NATURE Scientific Reports* 6, Article number: 32844.

Zhang, C., Zhao, Y., Zhao, X., & Gadow, K. v. (2012). Species-Habitat associations in a northern temperate forest in China. *Silva Fennica, 46*(4), 501–519.

Zhao, X. H., Corral-Rivas, J. J., Zhang, C. Y., Temesgen, H., & Gadow, K. v. (2014). Forest observational studies-an essential infrastructure for sustainable use of natural resources. *Forest Ecosystem, 1*, 8. https://doi.org/10.1186/2197-5620-1-8.

Zeng, W. -S., Tomppo E., Healey. S. P., & Gadow, K. V. (2015). The national forest inventory in China: History—Results—International context. *Forest Ecosystems, 2*, 23

Internet of Things in Forestry and Environmental Sciences

S. B. Lal, Anu Sharma, K. K. Chaturvedi, M. S. Farooqi and Anil Rai

Abstract Internet of Things (IoT) is a revolutionary technology that aims to interconnect everyday objects equipped with identity, sensors, networking, and processing capabilities and allow them to communicate with one another and with other devices and services over the Internet to accomplish some objective. This is a transition from *interconnected computers* to *interconnected things* that require support for interoperability among heterogeneous devices enabling simplification of new application development for programmers under the infrastructure of IoT. Middleware for IoT is a software layer interposed between the infrastructure and the applications that basically aims to support important requirements for these applications (Yu et al. in Cybern. Inf. Technol. 14(5):51–62, 2014). Generally, the form of communication has been human–human or human–device, but the IoT is a communication as machine–machine. So, it is a network of objects with a self-configured wireless network. These IoT frameworks are used to collect, process, and analyze data streams in real time and facilitate provision of smart solutions. IoT is observed as a natural evolution of environmental sensing systems. This aims to use different sensors to measure key parameters in forest areas in regular basis, with no need of human intervention and to send this information via wireless communication to a central platform. IoT-based environment monitoring technologies and a smart home technology are being accepted by people because they have good prospects for development. IoT products in agriculture include a number of IoT devices and sensors as well as a powerful dashboard with analytical capabilities and in-built reporting features (Yu et al. in Cybern. Inf. Technol. 14(5):51–62, 2014). A networking-based intelligent platform can monitor forest environmental factors in time by applying IoT. This technology has the advantages of low power dissipation, low data rate, and high-capacity transportation.

Keywords Agriculture · Forestry · Environment · Sensor networks · Smart farming · Wireless networks

---

S. B. Lal (✉) · A. Sharma · K. K. Chaturvedi · M. S. Farooqi · A. Rai
Centre for Agricultural Bioinformatics, ICAR-Indian Agricultural Statistics Research Institute, New Delhi, India
e-mail: sblall16@gmail.com

G. Chandra et al. (eds.), *Statistical Methods and Applications in Forestry and Environmental Sciences*, Forum for Interdisciplinary Mathematics,
https://doi.org/10.1007/978-981-15-1476-0_3

1 Introduction

Nowadays, every corner of the world is availing the facility of Internet for carrying out day-to-day activities, business, education, and other fields. The activities using this modern technology include online shopping, trading, ticket booking, chat, discussion, health-related information, and many more uncounted activities. The government is also encouraging the communications made using electronic media. We all are aware that availability of Internet provides connection between one human and another human. Additionally, it is also a connection between a connectible device and any other equipment having a unique identification. These devices have necessary components for being networked. Therefore, it is possible to transfer necessary useful information if these devices are connected together using any networking technology. This will enable the instant decision making for effective and timely use of devices to get a fully automated system (Nandurkar et al. 2014).

When two networkable equipment (things) having a unique identification (ID) are joined to a network (Internet), then managing them by using communication of information between them is called Internet of Things (IoT) (Ashton 2009). Internet is a connection between people; therefore, it is called Internet of People. IoT connects all connectible things, so it is called Internet of Things. IoT is lot more than what an Internet is because it increases the number of connections by any mean of networking technology which includes any connectible electronic component (An et al. 2012). IoT can bring dynamic control of industry and daily life of human being by speeding up the decision-making process for better results and timely actions. Production efficiency and other related activities of industrial sector can experience a phenomenal progress by using IoT. Our daily life can also be equipped with more intelligent activities and well on time actions to gain better living standards. Automatic communication of useful information through connected devices enables every necessary precautionary action very fast. Automatic communication of information among the available connectible devices makes the signal transfer very fast. These devices have unique identification (ID) which is used to connect them together to transfer the useful information. This saves time to have very effective resource utilization ratio. IoT can establish better relationship between human and nature as the integrated sensors or other electronic devices transfer necessary and relevant information very fast resulting in formation of an intellectual entity by integrating human society and physical systems. By utilizing this, an intellectual institution can be formed by integrating human society and physical systems (Khan et al. 2012). It is possible to communicate information among the underlying devices due to the flexible configuration of the software. Mobility can be brought in the field of forest and environmental science too by implementing IoT as it functions as a technology integrator.

In an agriculture environment, sensors measure various pieces of information about the environment and communicate wirelessly through a gateway which receives and transfers the messages and related data to online database. This information is made easily accessible online for analyzing further (Bayne et al. 2017). A large number of wireless sensor networks have been installed in the forest and environmental sectors for monitoring of parameters for growth of plantation such as temperature, humidity, soil moisture, and so on. Estimation of leaf area index is also a crop monitoring process to measure plant growth rates (Qu et al. 2014a, b).

2 Layers of IoT

In the modern era, electronic devices and its usefulness in human life have increased tremendously. All these devices have a very specific ability for their purpose of use (Fig. 1). For example, telephone, mobile phone, laptop, GPS equipment, partial or fully automatic car, electronic watch, and many other devices are designed to fulfill a specific purpose. If these machines are connected by using any suitable medium and the results generated by them are combined together to find an automatic useful result, it can bring a revolutionary change in public life.

There are four layers of an IoT-based system, viz. sensor identification layer, network construction layer, management layer, and integrated application layer. These layers, as shown in Fig. 2, have their specific task to perform to carry out the whole process efficiently. The first layer sense identification layer is responsible for sensing the signal sent by the device (through Wi-Fi or radio frequency) to another device present in the surrounding area and registers its ID. In the second layer, the information sent by these devices is connected to the network for transferring to the next layer. Processing the information obtained is done in the third layer by making it suitable. In the last layer, the work of fulfilling the appropriate response is done by conveying the right information to the right people or devices.

More itemized perspective of four layers of IoT has been represented in Fig. 3. Some important sensor gadgets have been shown in the first layer, for example,

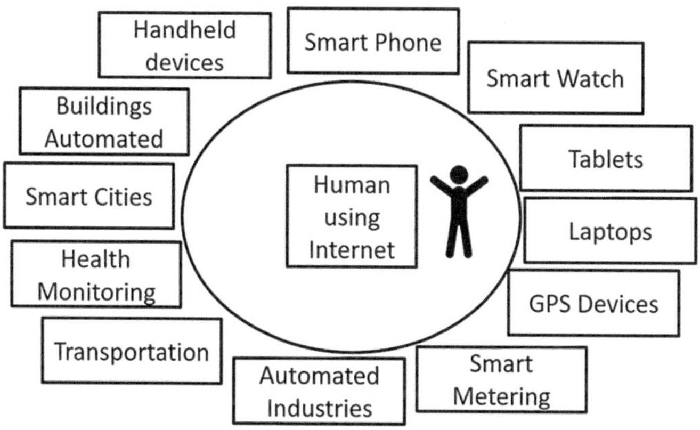

Fig. 1 Use of various instruments in human life

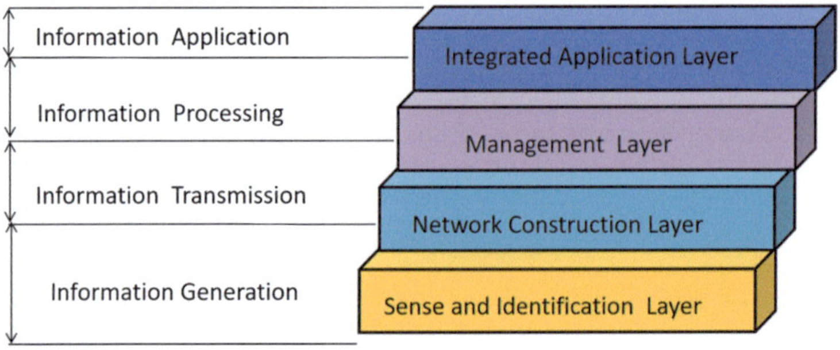

Fig. 2 Four main layers of IoT

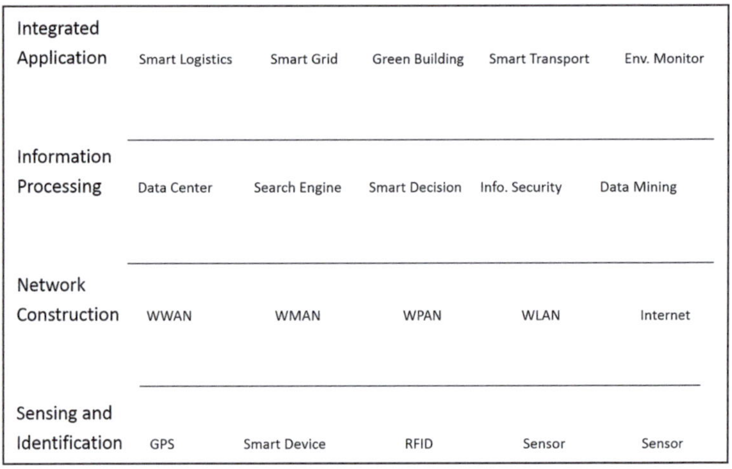

Fig. 3 Detailed drawing of four layers of IoT

GPS device, smart device, radio frequency ID, and other sensor gadgets. All these instruments are utilized to transmit the underlying data to the next layer. The information obtained from the first layer is transported via the network-building layer to the devices present in various processing layers. Some important devices among these are—data center, search engine, smart decision maker, information security, and data mining. An ultimate modern and intelligent system can be developed by taking advantage of the information obtained from these devices and taking appropriate and important decision making in the last layer.

3 Applications of IoT

The use of IoT in different areas has been shown in Fig. 4. Unique IDs for devices connected through a networking technique enable to get the right information on time to avoid any undesirable conditions. Information can be transmitted through networking in the areas as shown in Fig. 4 such as office, home, transport department, equipment present in public places, and sensitized devices available to every person. Having the necessary information in a very short time, it will be easy and highly beneficial to take the next step.

Some of the situations in the era of IoT are found as follows:

(i) If a tree in the forest has many leaves suffering from a color change, a picture is sent to the dashboard with a specific warning color or sound.
(ii) If a person tries to damage a tree in the forest, sensor sends a signal to the dashboard in terms of an alarm or message.
(iii) In agriculture, the amount of water usage can be reduced or optimized if a sensor for measuring soil moisture and weather conditions is installed.
(iv) Measuring climate conditions such as humidity, temperature, light, and so on to effectively utilize the input to enhance the productivity.
(v) Disaster warning: If sensors are able to gather data about the environmental factors, it sends an early warning message about disaster conditions such as earthquakes and tsunamis.
(vi) Environmental quality: Reading information about air quality, radiation, and pathogens at an early stage can save lives of plants.

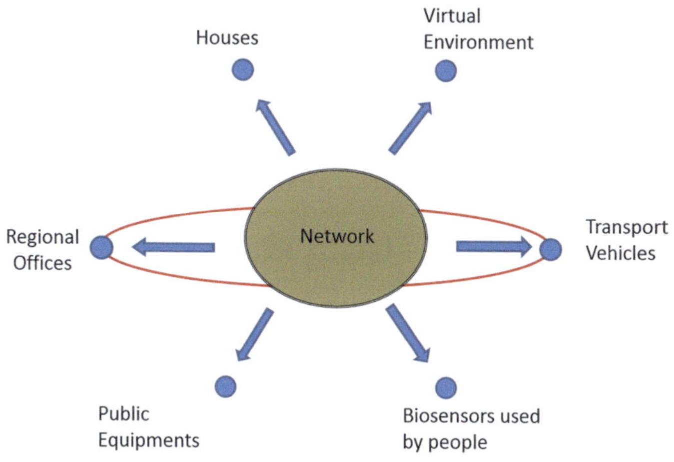

Fig. 4 IoT experiments in various areas

3.1 Benefits of IoT in Agriculture

There can be many benefits of implementation of IoT in agriculture like other areas. For example, if the information of the inputs like soil, water, fertilizers, and pesticides is given to the farmers based on his location and on proper time, then its utility and the efficiency of the farmers can be increased. This will also reduce the cost of production. Accurate input and availability of timely information will increase production. A systematic and stable mechanism can be established through IoT technique, which will enable automated arrangements in the field of agriculture. Its use will eliminate the possibility of failure of mechanisms applied, which will maintain food security in the country. The proper method of applying agriculture and environment will make possible to preserve the environment.

Decrease in level of water tables, availability of water through rivers and other water sources due to urbanization and climate change are posing a challenge to us and forces us to efficiently utilize our water and other resources (Gondchawar and Kawitkar 2016). Wireless sensor networks make possible the monitoring and control of greenhouse parameters for precision agriculture (Kumari and Devi 2013). Algorithms have been developed to control water quantity by use of photovoltaic panels with duplex communication links. Proper irrigation scheduling with optimum water quantity supply was enabled using a web interface (Gutiérrez et al. 2014). A distributed wireless sensor network for remote sensing and controlled irrigation system was developed which aims for customized water application with real-time field sensing for precision agriculture and to maximize production (Kim et al. 2008). With the use of wireless sensor network, it is possible to measure soil parameters such as temperature and humidity. Sensors are installed below the soil and monitored using communication network by monitoring system software installed on a web server (Wang et al. 2010). This system used microcontroller, universal asynchronous receiver transmitter interface, and sensors.

3.2 The IoT for Forest and Environmental Sector

Application of IoT to make any monitoring system smarter and more intelligent includes forest and environment related variables too. Monitoring variables at most appropriate point of time, a ZigBee protocol is worth mentioning as this is a networking based intelligent platform. It has the advantages of low power dissipation, low data rate, high-capacity transportation, low complexity, low rate of compound, and inexpensive and short distance transmission. It is more suitable for the design of the node of forest environmental factors collection platform. ZigBee technology is a kind of wireless connection technology, which can work under radio frequency of 2.4 GHz (Global Pop), 868 MHz (Euro-Pop), and 915 MHz (American Pop) (Yu et al. 2014).

3.2.1 Uses of IoT in Forestry and Environmental Sector

(i) Forestry: The forest ecology is complex, but sensors under IoT can monitor many parameters in real time and can establish new relationships as well as identify new indicators. An ecological research project at a Harvard University forest site has started taking environmental measures and its monitoring by using wireless sensor networks since 2010 (Harris 2015). These sensors capture sound to detect cricket emergence and spring bud burst. These are also linked with atmospheric and climatic measures for enabling better ecosystem modeling and forecasting of likely forest change under climate change scenarios. In commercial forest sector too, sensors can monitor growing conditions, optimization of soil nutrient levels, micro-climate effects, enhancing tree survival, predicting diseases, events of extreme weather or fire, etc. There could be more monitoring using these sensors, namely optimization of milling operations, harvesting, and transport of raw material and equipment usage. Monitoring may also be explored for secondary products such as furniture, milled lumber, their packaging by monitoring production, as well as storage conditions and location throughout the delivery chain (Bayne et al. 2017).

(ii) Sawmilling: Up the value chain, monitoring can be for sawmilling and timber products. To make an appropriate operational decision making, information passing upstream and downstream on proper time would be helpful. Sending right tree or board to the right processing facility by providing proper link between the elements can enhance the efficiency. Using sensors and wireless scanners in sawmills can form a real-time network, and information flow throughout the mill would enable product traceability and product optimization very efficient (Hansen and Leavengood 2016).

(iii) Other processes: In the forest and environmental sector, many more opportunities are present for IoT as under.

   a. Role in inventory monitoring and control.
   b. In situ monitoring of structural members for integrity and moisture content of engineered wood.
   c. Various highly connected components of houses such as smart meters, smart thermostats, and systems where temperature, locks, etc. can be controlled remotely.

4 Data Collection and Monitoring in IoT

Sensors deployed over the area being monitored automate the acquisition of data using the deployed Internet channel simplifying the data collection process. The produced data and logs can be very large in size making its management a complex task.

4.1 *ZigBee Technology*

Smart agriculture using automation and IoT technologies includes smart GPS-based remote-controlled robot to perform tasks such as weeding, spraying, reading moisture, protection from birds and animals, and other monitoring. It also includes efficient irrigation by using the real-time field data. A huge amount of data collected from the installed devices needs effective data management system such as data warehouse. These operations are performed by interfacing sensors, Wi-Fi or ZigBee modules, camera, and actuators with microcontroller and controlled by a smart device connected to the Internet (Gondchawar and Kawitkar 2016). ZigBee devices are highly reliable wireless data transmission network. Transmission distance is within the range from the 75 m (Standard) but can be extended to 100 m, few km, or to any distance. ZigBee is wireless data communication networks, which are made of 65,000 wireless data-transmitting modules. Across the whole ZigBee network, every wireless data-transmitting module can communicate with each other. ZigBee protocol uses IEEE802.15.4, which was originally designed for personal area networks. ZigBee enables three network topologies—a star, a mesh, and cluster tree.

4.2 *Data Collection in ZigBee Technology-Based Infrastructure*

Data collection process in terms of three-layer structures using ZigBee technology has been shown in Fig. 5. These layers are—data collection sensor layer, ZigBee node layer, and ZigBee coordinator layer.

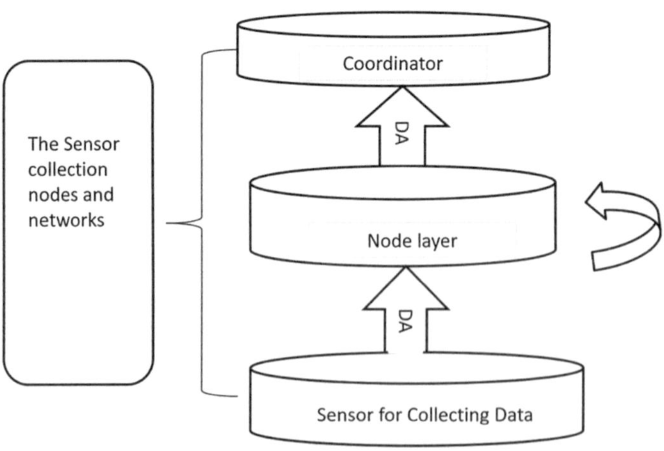

Fig. 5 Three-layer data collection layer for ZigBee technology

4.3 *Data Collection in Other IoT Infrastructure*

For any other IoT-based infrastructure, a large number of sensors (maybe more than 10,000) are deployed. To capture 1000 or more readings every second per sensor, it is expected to be more than 1 GB per second data rate. To stream all of that data, a strategic system is needed to devise. Segmenting the sensor deployment into groups with each group connecting to the Internet via a gateway device having mini data collection, analysis and response machine can help with the overall scaling problem. This would increase the number of gateway devices but solve the data-rate scaling problem. Now with this distributed data, another problem of its aggregation has arisen.

(a) One of the available solutions to this problem is the TICK stack. It is made of four open-source software components designed specifically to make the collection, storage, management, visualization, and manipulation of time series data easy and scalable. These are Telegraf, InfluxDB, Chronograf, and Kapacitor (Simmons 2018).
(b) Riak TS is another high-performance NoSQL database for time series/IoT data. By co-locating data based on time range, it is easier to ingest, transform, store, and analyze sensor and device data. Riak TS scales horizontally with commodity hardware to meet increasing volumes of data.

4.4 *Monitoring Factors*

Timely monitoring of the forest environmental factors is feasible on an IoT-based infrastructure. Nodes of sensor network can interact with their environment by observing and controlling the physical phenomena, such as vibration, humidity, and pressure. Monitoring is done by using a user-friendly interface, to acknowledge the information collected. Data collected by collection nodes includes spatial and a time attribute, the geographic locations of every sensor nodes (longitude and latitude), and the acquisition time of information. Timely monitoring of forest environmental factors according to the geographic locations is done.

4.5 *Challenges Before IoT*

There is a great deal of difficulties being looked in our nation for execution of IoT. In many regions, technical standardization is still disintegrated, because of which there is absence of correspondence and interpersonality. The processing of information obtained from it becomes a daunting task. Managing and promoting rapid innovation is a challenge for governments. Also, confidentiality and security of information are very important points. Sharing information in our country is a cumbersome task, which is due to lack of reliability. Apart from this, the absence of governance (governance) is also a challenge.

5 Conclusions

IoT is an emerging science for forestry that can transform this growing sector. A number of possibilities and opportunities are feasible by installation of sensors in forest environments. A number of technologies are now available for carrying out monitoring using sensors at various points in forests but still more are needed for further developments. In India, the implementation of this technology is gaining pace, but still, the reliability on adaptation of IoT systems needs to be strengthened. IoT systems with more standardized elements with easy to use web-based monitoring system are finding more places globally. The concept of IoT would have the capacity for forest and environmental sector to contain the seamless data stream from field to web. The replacement of sensor nodes does not disturb the radio communication, server management, and software. The potential use of IoT for forest sector would be real-time data monitoring of rainfall, soil moisture and temperature, pest and diseases, and fertilizer uptake. Nutrient and weather monitoring, movements of animals, and people can be recorded using installed wireless sensors. In addition, there are more requirements in terms of technology development to the progress of wireless sensor network in forest sectors such as improved sensor communication technology, data collection techniques, and developing new and improved algorithms for analysis of data gathered through sensors.

References

An, J., Gui, X., & He, X. (2012). Study on the architecture and key technologies for Internet of Things. *Advances in Biomedical Engineering, 11*, 329–335 (IERI-2012).

Ashton, K. (2009). *That 'Internet of Things' thing*. Retrieved May 9, 2017 from https://www.rfidjournal.com/articles/view?4986.

Bayne, K., Damesin, S., & Evans, M. (2017). The internet of things—Wireless sensor networks and their application to forestry. *New Zealand Journal of Forestry, 61*(4), 37–41.

Gondchawar, N., & Kawitkar, R. S. (2016). IOT based smart agriculture. *International Journal of Advanced Research in Computer and Communication Engineering, 5*(6), 838–842.

Gutiérrez, J., Villa-Medina, J. F., Nieto-Garibay, A., & Porta-Gandara, M. A. (2014). Automated irrigation system using a wireless sensor network and GPRS module. *IEEE Transactions on Instrumentation and Measurement, 63*(1), 166–176.

Hansen, E., & Leavengood, S. (2016). Will the internet of trees be the next game changer? *MIT Sloan Management Review*, February 17, 2016. Accessed March 14, 2016 at http://sloanreview.mit.edu/article/will-the-internet-of-trees-be-the-next-game-changer/.

Harris, M. (2015). A web of sensors enfolds an entire forest to uncover clues to climate change. *IEEE Spectrum*, February 26, 2015. Accessed August 12, 2016 at http://spectrum.ieee.org/green-tech/conservation/a-web-ofsensors-enfolds-an-entire-forest-to-uncover-clues-toclimate-change.

Khan, R., Khan, S. U., Zaheer, R., & Khan, S. (2012). Future internet: the internet of things architecture, possible applications and key challenges. In *2012 10th International Conference on Frontiers of Information Technology (FIT): Proceedings* (pp. 257–260). Institute of Electrical and Electronics Engineers Inc. https://doi.org/10.1109/FIT.2012.53.

Kim, Y., Evans, R. G., & Iversen, W. M. (2008). Remote sensing and control of an irrigation system using a distributed wireless sensor network. *IEEE Transactions on Instrumentation and Measurement, 57,* 1379–1387.

Kumari, G. M., & Devi, V. V. (2013). Real- time automation and monitoring system for modernized agriculture. *International Journal of Review and Research in Applied Sciences and Engineering, 3*(1), 7–12.

Nandurkar, S., Thool, V. R., & Thool, R. C. (2014). Design and development of precision agriculture system using wireless sensor network. In *2014 First International Conference on Automation, Control, Energy and Systems (ACES)* (pp. 1–6).

Qu, Y., Fu, L., Han, W., Zhu, Y., & Wang, J. (2014a). MLAOS: A multi-point linear array of optical sensors for coniferous foliage clumping index measurement. *Sensors (Basel), 14*(5), 9271–9289.

Qu, Y., Zhu, Y., Han, W., Wang, J., & Ma, M. (2014b). Crop leaf area index observations with a wireless sensor network and its potential for validating remote sensing products. *IEEE Journal of Selected Topics in Applied Earth Observations and Remote Sensing, 7*(2), 431–444.

Simmons, D. G. (2018). Pushing IoT data gathering, analysis, and response to the Edge. https://dzone.com/articles/pushing-iot-data-gathering-analysis-and-response-to-the-edge. April 19, 18 · IoT Zone · Tutorial.

Wang, Q., Terzis, A., & Szalay, A. S. (2010). A novel soil measuring wireless sensor network. In *2010 IEEE Instrumentation and Measurement Technology Conference Proceedings* (pp. 412–415).

Yu, Z., Xugang, L., & Xue, G. (2014). IoT forest environmental factors collection platform based on ZigBee. *Cybernetics and Information Technology, 14*(5), 51–62.

Inverse Adaptive Stratified Random Sampling

Raosaheb V. Latpate

Abstract In this article, we have proposed a new sampling design which is a combination of stratified random sampling and general inverse adaptive cluster sampling designs. From each stratum, an initial sample of fixed size is drawn. By using the condition of adaptation, we decide the number of successes, and if it includes prefixed number of successes then sampling is stopped and applies the adaptation procedure. Otherwise, sampling will continue till to get prefixed number of successes or reached a fixed upper bound for each stratum. The estimator of the population total is proposed and Monte Carlo study is presented for the sample survey.

Keywords General inverse adaptive sampling · Sequential sampling · Two-stage estimator · Two-stage sampling

1 Introduction

In forestry and environmental sciences, some species of plants and animals are rare and clustered, i.e., abundance of zero's. The traditional sampling methods provide the poor estimates of the population mean/total. In such situations, adaptive sampling Thompson (1990) is useful. In traditional stratified sampling, similar units are grouped a priori into strata, based on prior information about the population. But within a stratum, the population is rare and clumped. Hence, stratified adaptive sampling is useful for estimation of population mean/total (Thompson 1991). In this design, an initial stratified random sample (units) are drawn, and if selected units satisfy the condition of interest then neighboring units are added to the sample. Again, if any of the added units satisfies the condition then their neighbors are added. This process continues till all the neighbors satisfy condition C. The process terminates if all the lastly added units do not satisfy the condition.

R. V. Latpate (✉)
Department of Statistics and Center for Advanced Studies, Savitribai Phule Pune University, Pune, India
e-mail: rvl@unipune.ac.in

G. Chandra et al. (eds.), *Statistical Methods and Applications in Forestry and Environmental Sciences*, Forum for Interdisciplinary Mathematics,
https://doi.org/10.1007/978-981-15-1476-0_4

The adaptive cluster sampling (ACS), proposed by Thompson (1990), is useful for the population such as animal or plant species and mineral or fossil fuel resources. In this design, initial samples (units) are selected by using simple random sampling without replacement (SRSWOR) or simple random sampling with replacement (SRSWR). If selected units satisfy the condition of interest C, then its adjacent units—to left, right, top and bottom are included in the sample. If added units satisfy the condition, then again its neighbors are added and this process continued till all the added units do not satisfy C. We get clusters. Then drop down all the edge units to get the network. Edge units are defined as the units which are added adaptively but do not satisfy C.

In this design, there is a problem of initial sample size. In some instances, we met with all zero occurrences because of the population nature. To overcome this problem, Chirstman and Lan (2001) introduced the inverse adaptive sampling. Let the population consists of set of values $P = \{y_1, y_2, \ldots, y_N\}$ and is divided in two subgroups according to whether the y-values satisfy the condition C as $P_M = \{y : y_i \in C,\ i = 1, 2, \ldots, N\}$ and $P_{N-M} = \{y : y_i \notin C,\ i = 1, 2, \ldots, N\}$, where, M is number of units satisfies the condition and N is population size. In general, $C = \{y : y > c\}$, where c is some constant specified prior to sampling. We do not know which unit belongs to which group until the unit is sampled. In this method, we select units one at a time until we get a predetermined number of rare units. The neighborhood of an initial unit is added to the sample whenever the unit satisfies C. Again, if an added unit satisfies C its neighbors are added and this process continued till all the neighbors do not satisfy C (Salehi and Seber 2004). If the selected unit does not satisfy C, i.e., $y \in P_{N-M}$, there will not be adaptation, and the network includes only an initial sample. In some instances, several units from the same network may be sampled initially. Hence, we consider only distinct networks. The adaptive sampling is most powerful sampling scheme whenever units of P_M are grouped.

Salehi and Samith (2005) proposed two-stage sequential sampling. They studied the population of blue-winged teal and population of two waterfowl species to evaluate the adaptive cluster sampling. They showed that ACS is efficient for these populations. Chirstman and Lan (2001) studied the population of green-winged teal, blue-winged teal and ring-necked ducks. They proposed the inverse adaptive sampling design and its estimators. Gattone et al. (2016) presented adaptive cluster sampling for negatively correlated data. They studied the population of African buffalo in the presence of auxiliary variable and proposed the product type estimator. Moradi et al. (2011) studied regression estimator under inverse sampling. They collected the information about arsenic contamination in soil. Latpate and Kshirsagar (2018a) proposed the two-stage inverse adaptive cluster sampling. They studied the population of thorny plants of Tamini Ghat Maharashtra and obtained the two-stage estimator. Also, they showed the minimum survey cost expression. Latpate and Kshirsagar (2018b) presented the negative adaptive cluster sampling, in which they established the relationship between silica content in the soil and the number of evergreen plants. They are negatively correlated. Also, they proposed the ratio, regression and product type estimator to model the data. Latpate and Kshirsagar (2018c) developed the extension of negative adaptive cluster sampling at second stage. They studied the population of thorny plants and established the relationship between thorny plants and aluminum content in the soil, which is negatively correlated. Latpate et al. (2018) evaluated the expected sample size for ACS.

The organization of the chapter is as follows. Section 2 includes the proposed sampling design. Sample survey is presented in Sect. 3. Results and discussion are included in Sect. 4. Lastly, the concluding remarks are incorporated in Sect. 5.

2 Inverse Adaptive Stratified Random Sampling

Let $U = \{U_{11}, U_{12}, \ldots, U_{1N_1}, U_{21}, U_{22}, \ldots, U_{2N_2}, \ldots, U_{L1}, U_{L2}, \ldots, U_{LN_L}\}$ be the finite population of N units which is first divided into subpopulations of $N_1, N_2, \ldots, N_L$ units. These subpopulations are non-overlapping, and together they comprise the whole populations such that $N_1 + N_2 + \cdots + N_L = N$. These groups are called strata and size of each stratum is known. Initial random samples of size $n_1, n_2, \ldots, n_L$ are drawn independently. Such a procedure is called stratified random sampling. If an initial random sample of size n_h is drawn from the hth strata without replacement. The number of successes $r_h = r\frac{N_h}{N}$, where r is total number of successes (units satisfying the condition C). Each of the selected units is checked with respect to the condition C. If at least $r_h (h = 1, 2, \ldots, L)$, units satisfying the condition C are found in this sample of n_h units then the sampling is stopped. Otherwise, it is continued until either exactly r_h units from P_M are selected or $n_{h_2} (h = 1, 2, \ldots, L)$ (a prefixed number) units in total are selected from the hth stratum, where, $n_h \leq n_{h_2} \leq N_h$. A set of neighboring units is defined for each unit by Thompson (1990). The neighborhood of an initial unit is included to the sample in hth stratum whenever the unit satisfies C. Again if any of the additional units satisfies C, its neighborhood is also included in the sample of hth strata. This procedure continued till the entire neighborhood is included. Terminate the procedure if lastly added units do not satisfy the condition C. If initial sample (unit) does not satisfy the condition, then no further units are added, and the network consists of just the initial unit. Those units adaptively added but do not satisfy the condition is called edge units. For estimation purpose, we drop down the edge units. For hth stratum, each unit i a network A_{hi}. Some instances, several initial units belong to the same network. The distinct networks, labeled by the subscripts $z (z = 1, 2, \ldots, Z)$, from a partition of N_h units. Let m_{hi} denote the size of ith network of hth stratum. The variable of interest related to A_{hi} is $y^*_{hi} = \sum_{j=1}^{m_{hi}} y_{hij}$ and is $\bar{y}^*_{hi} = \frac{\sum_{j=1}^{m_{hi}} y_{hij}}{m_{hi}}$. Repeat the whole procedure to all the strata. The population total can estimated $\tau = \sum_{h=1}^{L} \sum_{i=1}^{N_h} y^*_{hi}$.

P_h is the draw-by-draw selection probability of the hth strata. Such kinds of situations are tackled by using Des Raj estimators (Raj 1956). Murthy proposed the technique to improve the ordered estimator by unordered ones, which is a Rao-Blackwell improvement of Raj's estimator. Salehi and Seber (2001) used it for sequential sampling designs. Let n_{h_1} denote the number of units drawn in total from the hth stratum. This is the final sample size for the hth stratum.

At each stratum, sample selection scheme becomes the sequential sampling. Murthy (1957) showed that corresponding to any ordered estimator of this class, an unordered estimator can be constructed and that estimator is also minimum variance unbiased estimator. Let $t_R(s'_h)$ denote Raj's (1956) estimator of $\tau_h = \sum_{i=1}^{N_h} y^*_{hi}$ which is defined on the basis of an ordered sample s'_h. Using that technique, we can obtain an estimator of τ_h as follows:

$$\hat{\tau}_h = \frac{\sum_{s'_h \in s_h} P(s'_h) t_R(s'_h)}{P(s_h)}$$

where $P(s_h)$ is the probability of obtaining the sample s_h. Murthy has shown that $\hat{\tau}_h$ can be rewritten as:

$$\hat{\tau}_h = \sum_{i=1}^{n_{h_1}} \frac{P(s_h|i)}{P(s_h)} y^*_{hi} = \sum_{i=1}^{N_h} \frac{P(I_{hi} = 1; s_h)}{P(s_h) P_{hi}} y^*_{hi},$$

where $P(s_h|i)$ is the conditional probability of getting sample s_h given that ith unit was selected at the first draw,

$$I_{hi} = \begin{cases} 1 & \text{if unit } j \text{ is selected at the first draw in the sample from } h\text{th strata} \\ 0 & \text{otherwise.} \end{cases}$$

Salehi and Seber (2001) showed that Murthy's estimator given above can be obtained from a trivial estimator by using Rao-Blackwell theorem.

A trivial unbiased estimator of τ_h is given by

$$\hat{t}_h = \sum_{i=1}^{N_h} \frac{y^*_{hi}}{P_{hi}} I_{hi}, \quad \text{provided } P_{hi} > 0 \quad \text{for } i = 1, 2, \ldots, N_h.$$

If s_h denotes the final sample set from the primary unit h then using Rao-Blackwell theorem, we get

$$\hat{\tau}_h = E\left(\hat{t}_h | s_h\right) = \sum_{j=1}^{n_{1u}} \frac{P(s_h|i)}{P(s_h)} y^*_{hi}$$

When $n_{h_1} = n_h$ and $n_{h_1} = n_{h_2}$ then

$$\frac{P(s_h|i)}{P(s_h)} = \frac{N_h}{n_h} \text{and } \frac{N_h}{n_{h_2}} \text{ respectively}$$

When $n_h < n_{h_1} < n_{h_2}$ and $i \in S_{M_h}$ then

$$\frac{P(s_h|i)}{P(s_h)} = \frac{N_h(r_h - 1)}{(n_{h_1} - 1)r_h},$$

When $n_h < n_{h_1} < n_{h_2}$ and $i \in S_{N_h - M_h}$ then

$$\frac{P(s_h|i)}{P(s_h)} = \frac{N_h}{(n_{h_1} - 1)},$$

and $\widehat{M}_h = \frac{N_h(r_h-1)}{(n_{h_1}-1)}$ (Lehmann 1983).

Thus we get,

$$\hat{\tau}_h = \begin{cases} \frac{N_h}{n_h}\sum_{i=1}^{n_h} \bar{y}^*_{hi} & \text{if } n_{h_1} = n_h \\ \widehat{M}_h \bar{y}^*_{M_h} + \left(N_h - \widehat{M}_h\right)\bar{y}^*_{N_h - M_h} & \text{if } n_h < n_{h_1} < n_{h_2} \\ \frac{N_h}{n_{h_2}}\sum_{i=1}^{n_{h_2}} \bar{y}^*_{hi} & \text{if } n_{h_1} = n_{h_2} \end{cases}$$

The unbiased estimator for the population total is given as follows:

$$\hat{\tau} = \sum_{h=1}^{L} \hat{\tau}_h$$

Mean of the observations on the units satisfying C from the hth stratum is given by,

$$\bar{y}^*_{M_h} = \frac{\sum_{i \in S_{M_h}} \bar{y}^*_{hi}}{r_h},$$

Mean of the observations on the units not satisfying C from the hth stratum is given by

$$\bar{y}^*_{N_h - M_h} = \frac{\sum_{i \in S_{N_h - M_h}} \bar{y}^*_{hi}}{(n_{h_1} - r_h)},$$

The variance of $\hat{\tau}$

$$V(\hat{\tau}) = \sum_{h=1}^{L} V(\hat{\tau}_h),$$

We get.

$$\widehat{\mathrm{V}}\mathrm{ar}(\hat{\tau}_h) = \begin{cases} N_h^2\left(1-\frac{n_h}{N_h}\right)\frac{S_{oh}^{*2}}{n_h} & \text{if } n_{h_1}=n_h \\ a_h S_{M_h}^{*2} + \widehat{V}(\widehat{M}_h)\left(\bar{y}_{M_h}^* - \bar{y}_{N_h-M_h}^*\right)^2 + b_h S_{N_h-M_h}^{*2} & \text{if } n_h < n_{h_1} < n_{h_2} \\ N_h^2\left(1-\frac{n_{h_2}}{N_h}\right)\frac{S_{2h}^{*2}}{n_{h_2}} & \text{if } n_{h_1}=n_{h_2} \end{cases}$$

where

$$S_{M_h}^{*2} = \frac{\sum_{i\in S_{M_h}}(\bar{y}_{hi}^* - \bar{y}_{M_h}^*)^2}{(r_h-1)},\ S_{N_h-M_h}^{*2} = \frac{\sum_{i\in S_{N_h-M_h}}(\bar{y}_{hi}^* - \bar{y}_{N_h-M_h}^*)^2}{(n_{h_1}-r_h)},$$

$$\widehat{V}(\widehat{M}_h) = \left(1-\frac{(n_{1h}-1)}{N_h}\right)\left(\frac{\widehat{M}_h\left(N_h-\widehat{M}_h\right)}{(n_{h_1}-2)}\right),$$

$$a_h = \frac{\widehat{M}_h^2}{r_h}\left(\frac{(N_h-n_{h_1}+1)(n_{h_1}r_h-n_{h_1}-r_h)-N_h(n_{h_1}-2)}{N_h(n_{h_1}-2)(r_h-1)}\right),$$

$$b_h = \frac{N_h\left((N_h-n_{h_1}+1)(n_{h_1}-r_h-1)\right)}{(n_{h_1}-1)(n_{h_1}-2)},\ S_{oh}^{*2} = \frac{\sum_{i=1}^{n_h}\left(\bar{y}_{hi}^*-\bar{y}_{h_0}^*\right)^2}{(n_h-1)},$$

$$\bar{y}_{h_0}^* = \frac{\sum_{i=1}^{n_h}\bar{y}_{hi}^*}{n_h},\ S_{2h}^{*2} = \frac{\sum_{i=1}^{n_{h_2}}\left(\bar{y}_{hi}^*-\bar{y}_{h_2}^*\right)^2}{(n_{h_2}-1)},\ \bar{y}_{h_2}^* = \frac{\sum_{i=1}^{n_{h_2}}\bar{y}_{hi}^*}{n_{h_2}}.$$

3 Sample Survey

Area of 400 acres in the Tamhini Ghats was divided into 400 plots each of size 1 acre. A satellite image of the area showed clustering of the evergreen plants. Presence of evergreen plants indicates a high percentage of laterite in the soil. The presence of laterite should be above 20% in the soil. So, it was important to estimate the total number of evergreen plants in the area. Higher value of this estimate indicates the ecological balance. In view of that, a sample survey was conducted. The area of 400 acres was divided into five strata C_1, C_2, C_3, C_4, C_5 containing 130, 65, 70, 105 and 30 plots respectively, each of size 1 acre. From these strata, random sample of size $n_h (h = 1, 2, \ldots, L)$ was selected without replacement. These selected plots are shown in Fig. 1 by putting * in it. Each of these plots selected from the different clusters was checked for the condition $C = \{Y > 0\}$; where Y denotes the number of evergreen plants observed on a plot. The total number of units satisfying the condition C is $r = 24$. The number of successes for each stratum depends on the size of stratum. Networks were identified around the plots which satisfy C, by using Thompson's (1990) procedure. The units from these identified networks were added to the sample along with the corresponding edge units. This process of adding the

	0	0	0	0	0	0	0	0	0	0	0	0	0	0	0	0	0	0	244	85	
	170	135	120	0	0	0	0*	0	0	0	0	0	0	0*	0	0	0	0	111	150	
	227	200*	165	0	0	0	0	0	0	0	0*	0	0	0	0	0	0	0	56*	152	
	157	190	35	0	0	0	0	0	0	0	0	0	0	0	0*	0	145	65*	99	85	
C_1	180	46	95	0	0	0	0*	0	0	0	0	0	0	0	0	0	0	0	0	0	
	0	0	0	0	0	0	0	0	0	0	0	0	0	0*	0	0	0	0*	0	0	
	0	0	0	0	0	0	0	0	0	0	0	0	0	0	0	0	0	0	0	0	
	0	0*	0	0	0*	0	0	0*	265	50*	55	0	0	0	0	0*	0	0	0	0	
	0	0	0	0	0	0	0	0	67	34	33	0	0*	0	0	0	0	0*	0	0	
	0	0	0*	0	0	0	0	0	0	0	0	0	0	0	0	0	0	0	0	0	
	0	0	0	0	0	0	0	0	0	0	0	0	0	0	0	0	0	0	0	0	C_4
	217	140	0	0	0	124	164	0	0	0*	0	0*	0	0	0	0	0	120	150	0	
C_2	64*	54	0	0	0	50*	64	0	0	0	0	0	0	0	0*	0	0	155	188	0	
	0	0	0	0*	0	0	98*	0	0	0	0*	0	0	0	0	0	0	68	37*	0	
	0	0	0	0	0*	0	0	0	0	0*	0	0*	0	0	0	0	0	0	0	0	
	0	0	0	0	0	0	0	0*	0	0	0	0	0	0*	0	0*	0	0	0	0*	
	0	165*	35	0	0	0*	0	0	0	0	88*	0	0	0	0	0	0	0	0*	0	
C_3	0	148	139	0	0	0	0	0	0	0	74	0	0	0	0*	189*	165	0	0	0	
	0	0	0	0	0	0*	0	0	0	0*	56	79*	0	0	0	132	0	0*	66*	212	C_5
	0	0	0	0	0*	0	0	0*	0	0	0	0	0	0	0	0*	0*	0	58	208*	

Fig. 1 Number of evergreen plants (Y) on a plot

neighbors is continued in each stratum so that, either at least $r_1 = 8$ plots satisfying the condition C are found or $n_{12} = 20$ plots in total are selected in the sample sequentially from C_1; either at least $r_2 = 4$ plots satisfying the condition C are found or $n_{22} = 20$ plots in total are selected in the sample sequentially found from C_2; either at least $r_3 = 4$ plots satisfying the condition C are found or $n_{32} = 20$ plots in total are selected in the sample sequentially found from C_3. Either at least $r_4 = 6$ plots satisfying the condition C are found or $n_{42} = 20$ plots in total are selected in the sample sequentially found from C_4 and either at least $r_5 = 2$ plots satisfying the condition C are found or $n_{52} = 20$ plots in total are selected in the sample sequentially found from C_5. The networks formed by the units selected are represented in white color in Fig. 1. Values of Y were recorded for all the units included in the final sample and the total number of evergreen plants in that area was estimated by using the proposed estimator $\hat{\tau}_h$. Further, its variance was also estimated by using the formula for $\widehat{V}(\hat{\tau}_h)$, given in Sect. 4. The functioning of the new design is demonstrated in Fig. 1.

Table 1 Simulation results for the set up $n_0 = 10$, $n_2 = 20$, Number of repetitions = 100,000

Number of successes	Inverse adaptive stratified random sampling	
r	$\hat{\tau}$	$SE(\hat{\tau})$
12	6530.77	3656.11
11	6512.01	3649.48
10	6541.33	3660.93
9	6797.64	3954.36
8	6924.74	4086.44

4 Results and Discussions

To verify the efficiency of this new design, Monte Carlo simulation study was performed for 100,000 repetitions. Table 1 gives the result of this study.

Monte Carlo study for inverse adaptive stratified random sampling showed that as the number of successes (r) decreases the bias reduces, and, finally, it converges with true population total $(= 6923)$, but standard error increases. If we use the unbiased estimator, the cost of survey will reduce substantially with less precision. As the number of successes increases, the precision of estimator increases and bias also increases.

5 Conclusions

In this design, we get the ordered estimator. By using Rao-Blackwellization, the unordered estimator is obtained from the ordered estimator. The estimator of inverse adaptive stratified random sampling converges as the number of successes decreases, whenever the population is clumped and rare. Hence, inverse adaptive stratified random sampling requires smaller sample size as compared to the traditional sampling design. Also, the cost of sample survey is very less as compared to traditional sampling design. Inverse adaptive stratified random sampling is useful for the sample survey in Forestry, Environmental Science, Ecology, Health Science, etc.

References

Christman, M. C., & Lan, F. (2001). Inverse adaptive cluster sampling. *Biometrics, 57,* 1096–1105.

Gattone, S. A., Mohamed, E., Dryver, A. L., & Munich, R. T. (2016). Adaptive cluster sampling for negatively correlated data. *Environmetrics, 27,* E103–E113.

Latpate, R., & Kshirsagar, J. (2018a). Two-stage inverse adaptive cluster sampling with stopping rule depends upon the size of cluster. *Sankhya B*. https://doi.org/10.1007/s13571-018-0177-y.

Latpate, R., & Kshirsagar, J. (2018b). Negative adaptive cluster sampling. *Model Assisted Statistics and Applications., 14,* 65–81. https://doi.org/10.3233/mas-180452. (IOS Press).

Latpate, R., & Kshirsagar, J. (2018c). Two-stage negative adaptive cluster sampling. *Communications in Mathematics and Statistics*. doi https://doi.org/10.1007/s40304-018-0151-z.
Latpate, R., Kshirsagar, J., & Gore, S. (2018). Estimation of sample size for adaptive cluster sampling. *Bulletin of Marathwada Mathematical Society, 19*(1), 32–41.
Lehmann, E. L. (1983). *Theory of point Estimation*. New York: Chapman and Hall.
Moradi, M., Salehi, M., Brown, J. A., & Karimi, N. (2011). Regression estimator under inverse sampling to estimate arsenic contamination. *Environmetrics, 22,* 894–900.
Murthy, M. N. (1957). Ordered and unordered estimators in sampling without replacement. *Sankhya, 18,* 379–390.
Raj, D. (1956). Some estimators in sampling with varying probabilities without replacement. *Journal of the American Statistical Association, 51,* 269–284.
Salehi, M. M., & Seber, G. A. F. (2001). A new proof of Murthy's estimator which applies to sequential sampling. *Australian and New Zealand Journal of Statistics, 43*(3), 281–286.
Salehi, M. M., & Seber, G. A. F. (2004). A general inverse sampling scheme and its applications to adaptive cluster sampling. *Australian and New Zealand Journal of Statistics, 46*(3), 483–494.
Salehi, M., & Smith, D. R. (2005). Two-stage sequential sampling: A neighborhood-free adaptive sampling procedure. *Journal of Agricultural, Biological, and Environmental Statistics., 10*(1), 84–103.
Thompson, S. K. (1990). Adaptive cluster sampling. *Journal of the American Statistical Association, 85,* 1050–1059.
Thompson, S. K. (1991). Stratified adaptive cluster sampling. *Biometrika, 78*(2), 389–397.

Improved Nonparametric Estimation Using Partially Ordered Sets

Ehsan Zamanzade and Xinlei Wang

Abstract Ranked set sampling (RSS) is a cost efficient design that has been widely used in agriculture, forestry, ecological and environmental sciences. Frey (Environmental and Ecological Statistics 19(3):309–326, 2012) proposed a sampling scheme based on to allow for partially ordered sets. This scheme permits a ranker to declare ties and then record the tie structure for potential use in statistical analysis. We first introduce two nonparametric maximum likelihood estimators (MLEs) of the population cumulative distribution function (CDF) that incorporate the information for partially ordered sets. We compare the proposed MLEs with the standard nonparametric MLE of the CDF (without utilizing tie information) via Monte Carlo simulation. Motivated by good performance of the new CDF estimators, we further derive two mean estimators for partially ordered sets. Our numerical results from both simulation and real data show that the proposed estimators outperform their competitors provided that the quality of ranking is not low.

Keywords Imperfect ranking · Nonparametric maximum likelihood estimation · Ranked set sampling · Relative efficiency · Ranking ties

1 Introduction

Ranked set sampling (RSS), proposed by McIntyre (1952), is an appropriate sampling technique for use in situations where ranking sample units in a small set is much easier or cheaper than obtaining their precise values. Ranking can be done by personal

The authors wish it to be known that, in their opinion, they both should be equally regarded as the corresponding authors.

E. Zamanzade
Department of Statistics, University of Isfahan, 81746-73441 Isfahan, Iran

X. Wang (✉)
Department of Statistical Science, Southern Methodist University,
3225 Daniel Avenue, Dallas, TX 75275, USA
e-mail: swang@smu.edu

G. Chandra et al. (eds.), *Statistical Methods and Applications in Forestry and Environmental Sciences*, Forum for Interdisciplinary Mathematics,
https://doi.org/10.1007/978-981-15-1476-0_5

judgment, eye inspection or using available values of a concomitant variable, and need not to be completely accurate (perfect). The ranking information is then used by the researcher to draw a more representative sample from the population of interest, and therefore statistical inference based on a ranked set sample should be more efficient than a simple random sample of the same size. RSS has been found to be useful in various fields including agriculture, forestry, ecological and environmental sciences, biology, medicine and so on.

To draw a balanced ranked set sample of size $N \equiv n \times m$ with set size m, one draws nm simple random samples (sets) of size m. The units in each set of size m are then ranked in an increasing magnitude without referring to their precise values. Then from the first n sets of size m, the sample units with rank 1 are selected for actual quantification, from the second n sets of size m, the sample units with rank 2 are selected for actual quantification and so on. For balanced RSS, n is the number of measured sample units with rank i for all $i = 1, \ldots, m$, and is referred to as the number of cycles. By contrast, in unbalanced RSS, one needs to determine a vector $\boldsymbol{n} = (n_1, \ldots, n_m)$ in a way that n_i is the number of measured sample units with rank i and $N \equiv \sum_{i=1}^{m} n_i$ is the total sample size of the ranked set sample.

Virtually, all standard statistical problems have been well addressed in the RSS literature including estimation of the population mean (Takahasi and Wakimoto 1968; Wang et al. 2006, 2008; Frey 2011), population variance (Stokes 1980; MacEachern et al. 2002; Perron and Sinha 2004), cumulative distribution function (CDF) (Stokes and Sager 1988; Kvam and Samaniego 1994; Huang 1997; Duembgen and Zamanzade 2018), population proportion (Chen et al. 2006, 2007; Zamanzade and Mahdizadeh 2017; Zamanzade and Wang 2017), mean difference (Wang et al. 2016, 2017), distribution-free confidence intervals (Frey 2007), reliability estimation (Mahdizadeh and Zamanzade 2018) and perfect ranking tests (Frey et al. 2007; Zamanzade et al. 2012; Frey 2017).

In RSS, the researcher is required to provide a unique rank for each unit in the set of size m. However, there are situations in which the researcher is not sure about how to rank two or more sample units in the set and he must break the ties at random to implement RSS. Such situations occur frequently in studies in agriculture, forestry, ecological and environmental sciences. To alleviate this difficulty, Frey (2012) proposed a new variation of RSS (denoted hereafter by RSS-t) in which the researcher is allowed to declare ties as he wishes. When implementing RSS-t, the researcher breaks the ties at random, but he also records the tie structure to be used in the estimation process. Let $\left\{X_{[i]j}, i = 1, \ldots, m, j = 1, \ldots, n\right\}$ be a balanced ranked set sample of size $N = mn$, where $X_{[i]j}$ is the jth unit with rank i. RSS-t includes not only the $X_{[i]j}$ values, but also the indicator variables $I_{[i]jk}$, where $I_{[i]jk}$ is one when the jth sample unit with rank i is tied for rank k in its own comparison set, for $i, k \in \{1, \ldots, m\}$, $j \in \{1, \ldots, n\}$. Note that $\sum_{k=1}^{m} I_{[i]jk}$ is always at least one because the sample unit with rank i is always tied for itself. Frey (2012) then developed several mean estimators for RSS-t samples and discussed two models which allow ties in rankings: discrete perceived size (DPS) and tied-if-close (TIC) models. Frey (2012) compared mean estimators with/without utilizing the tie information under the DPS model via Monte Carlo simulation and concluded that using tie information would improve efficiency in estimating the population mean.

We describe three nonparametric maximum likelihood estimators (MLEs) of the population CDF based on RSS-t: one is the standard MLE of the CDF proposed by Kvam and Samaniego (1994) which ignores tie information, and the other two are novel which incorporate tie information into the estimation process. We then compare these three likelihood-based estimators: via Monte Carlo simulation. Motivated by the observation that the two new MLEs of the CDF have better performance than the standard one, we further propose new mean estimators based on these CDF estimators, to make use of the tie information. We compare the proposed mean estimators with those in Frey (2012) using both simulated data and real data of body fat percentage, which consistently shows that the new estimators beat their competitors as long as the quality of ranking is good. Note that to estimate a population CDF $F(t)$, a typical approach is to use the relationship $F(t) = E[I(X \leq t)]$, where $I(\cdot)$ denotes the indicator function and the mean of $I(X \leq t)$ is estimated from the sample. In this paper, we adopt the approach that constructs a nonparametric maximum likelihood estimator of the CDF and then finds the mean estimator using $E(X) = \int_{-\infty}^{+\infty} t dF(t)$. As will be shown in Sect. 3.2, the mean estimators resulting from the NPMLEs of the CDF can perform substantially better than the existing mean estimators when ranking quality is good.

2 CDF Estimation

2.1 Nonparametric Maximum Likelihood Estimators for RSS-t

Maximum likelihood estimation of CDF based on a balanced ranked set sample was developed by Kvam and Samaniego (1994), and its asymptotic behavior was studied by Huang (1997) and Duembgen and Zamanzade (2018). Let $\{X_{(i)j}, i = 1, \ldots, m, j = 1, \ldots, n\}$ be a balanced ranked set sample of size $N = mn$ from a population with CDF $F(t)$, obtained under the assumption of perfect ranking. Then, $X_{(i)1}, \ldots, X_{(i)n}$ are independently and identically distributed, following the distribution of the ith order statistic in a sample of size m from the original population. That is, the CDF of $X_{(i)j}$ is given by $F_i(t) = P(X_{(i)j} \leq t) = B_i(F(t))$ where

$$B_i\left(F(t)\right) = \sum_{r=i}^{m} \binom{m}{r} \left(F(t)\right)^r \left(1 - F(t)\right)^{m-r} = \int_0^{F(t)} m \binom{m-1}{i-1} \left(F(t)\right)^{i-1} \left(1 - F(t)\right)^{m-i}$$

is the CDF of the beta distribution with parameters i and $m+1-i$, evaluated at the point $F(t)$. Therefore, the log-likelihood function of $F(t)$ based on RSS can be written as

$$L(F(t)) = \sum_{i=1}^{m}\sum_{j=1}^{n}\left\{I\left(X_{(i)j} \le t\right)\log\left(B_i\left(F(t)\right)\right) + \left(1 - I\left(X_{(i)j} \le t\right)\right)\log\left(1 - B_i\left(F(t)\right)\right)\right\},$$

where $I(\cdot)$ is the indicator function. Log-concavity of $B_i(F(t))$ in $F(t)$ follows from the fact that a beta distribution with parameters $\alpha, \beta \ge 1$ has an increasing hazard rate (Crowder 2008) and therefore $L(F(t))$ is a concave function in $F(t)$ as well. Thus, the MLE of the population CDF based on RSS is well defined and can be obtained by $\hat{F}_{NM0}(t) = \arg\max_{F(t)\in[0,1]} L(F(t))$. One can use $\hat{F}_{NM0}(t)$ for estimating $F(t)$ based on data from RSS-t by simply ignoring the tie structure. Obviously, $\hat{F}_{NM0}(t)$ is not the true MLE of $F(t)$ unless the rankings are perfect, with no ties allowed.

In what follows, we propose two novel likelihood-based CDF estimators which incorporate tie information from RSS-t into the estimation process. Suppose that $X_{[i]j}$ is tied for two or more units in the set of size m. Then, the CDF of $X_{[i]j}$, say $F_{[i]j,T}(t)$ ("T" in the subscript stands for "tie"), is a mixture of CDFs of beta distributions evaluated at the point $F(t)$, given by

$$F_{[i]j,T}(t) = \frac{\sum_{k=1}^{m} I_{[i]jk} B_k\left(F(t)\right)}{\sum_{k=1}^{m} I_{[i]jk}}.$$

Therefore, the log-likelihood function of $F(t)$ based on data from RSS-t can be written as

$$L_1(F(t)) = \sum_{i=1}^{m}\sum_{j=1}^{n}\left\{I\left(X_{[i]j} \le t\right)\log\left(F_{[i]j,T}(t)\right) + \left(1 - I\left(X_{[i]j} \le t\right)\right)\log\left(1 - F_{[i]j,T}(t)\right)\right\},$$

and the MLE of $F(t)$ based on RSS-t can be obtained as $\hat{F}_{NM1}(t) = \arg\max_{F(t)\in[0,1]} L_1(F(t))$. To guarantee the existence of a unique maximizer of $L_1(F(t))$, we need to show the concavity of $L_1(F(t))$ in $F(t)$. This requires the log-concavity of $F_{[i]j,T}(t)$, which follows from Theorem 2 in Mu (2015).

Another way of incorporating the tie information into the likelihood function is to use a splitting strategy proposed by MacEachern et al. (2004), in which each tied unit is split among the strata corresponding to the ranks for which the unit is tied. Thus, if a sample unit is tied for r different ranks, then we assign it to each of those r strata with equal weight $\frac{1}{r}$. This splitting strategy leads to the following pseudo-likelihood function:

$$L_2(F(t)) = \sum_{i=1}^{m} \left\{ n_i' \hat{F}_{i,sp}(t) \log(F_i(t)) + n_i' \left(1 - \hat{F}_{i,sp}(t)\right) \log(1 - F_i(t)) \right\},$$

where

$$n_i' = \sum_{l=1}^{m} \sum_{j=1}^{n} \frac{I_{[l]ji}}{\sum_{k=1}^{m} I_{[l]jk}} \tag{1}$$

and

$$\hat{F}_{i,sp}(t) = \frac{1}{n_i'} \sum_{l=1}^{m} \sum_{j=1}^{n} \frac{I_{[l]ji}}{\sum_{k=1}^{m} I_{[l]jk}} I\left(X_{[l]j} \le t\right)$$

for $i = 1, \ldots, m$. The corresponding MLE of $F(x)$ based on pseudo-likelihood is then given by $\hat{F}_{NM2}(t) = \arg \max_{F(t) \in [0,1]} L_2(F(t))$.

2.2 Comparison

We now compare the performance of ML-type estimators of the population CDF using RSS-t samples via simulation in which we consider different parent distributions, models for generating ties, quality of ranking and varying design parameters. We assume that ranking is done using a perceptual linear ranking model (Dell and Clutter 1972; Fligner and MacEachern 2006), which assumes the ranking of the variable of interest X in each set of size m is done via a concomitant variable Y, satisfying

$$Y = \rho \left(\frac{X - \mu_x}{\sigma_x} \right) + \sqrt{1 - \rho^2} Z,$$

where μ_x is the mean of X, σ_x is the standard deviation of X, Z is a random variable following the standard normal distribution and the parameter ρ controls the quality of ranking.

We consider two classes of models for generating ranking ties: discrete perceived size (DPS) and tied-if-close (TIC), as proposed by Frey (2012). The DPS model discretizes the values of the concomitant variable Y by rounding Y/c to the largest integer greater than or equal to Y/c. The TIC model declares the ith and jth units to be tied if $|Y_i - Y_j| < c$. Since the transitivity in TIC model is also required, the ith and jth units may be still declared tied even if $|Y_i - Y_j| > c$ as long as there is at least one unit in the set that bridges the gap. In either model, $c > 0$ is a user-chosen model parameter. Frey (2012) pointed out that both DPS and TIC models may show certain undesirable behavior when the parameters of m and c are changed. For DPS models, increasing the value of c does not necessarily lead to more ties in each set, and for TIC models, the number of ties among the units in the set can increase if we add an additional unit to it. Frey (2012) further discussed the differences between two classes of models, but only evaluated the mean estimators under the DPS model and normal distribution.

In the first simulation study, we compare the overall performance of CDF estimators on the real line via the mean integrated square error (MISE), defined as $MISE(\hat{F}) = E\left[\int_{-\infty}^{+\infty}\{\hat{F}(t) - F(t)\}^2 dt\right]$. The relative efficiency (RE) of $\hat{F}_{NMi}$ to $\hat{F}_{NM0}$ is defined as the ratio of their MISEs, i.e., $MISE(\hat{F}_{NM0})/MISE(\hat{F}_{NMi})$ for $i = 1, 2$. We set $N \in \{15, 30\}, m \in \{3, 5\}, \rho \in \{0, 0.5, 0.8, 1\}$, and for each combination of (N, m, ρ), we generate 10,000 RSS-t samples from standard normal ($N(0, 1)$), standard exponential ($Exp(1)$) and standard uniform ($U(0, 1)$) distributions under both DPS and TIC models, respectively. For the DPS model, we set $c \in \{0.5, 1, 2, 4\}$, and for the TIC model, $c \in \{0.25, 0.5, 1, 2\}$. The REs are estimated based on the 10,000 RSS-t samples for each setting. It is worth mentioning that in both DPS and TIC models, whenever ties occur in the ranking process, we assume that the researcher is not aware of actual values of tied units, as is typical in practice, and so he/she selects one of the tied units at random.

Table 1 presents RE values of $\hat{F}_{NM1}$ to $\hat{F}_{NM0}$ under the DPS model. We observe that the efficiency gain using $\hat{F}_{NM1}$ instead of $\hat{F}_{NM0}$ can be as large as 40% for the case of perfect ranking $\rho = 1$, while the efficiency loss is never more than 10% in the case of completely random ranking. When the quality of ranking is fairly good ($\rho \geq 0.8$), the RE never falls below one; when $\rho = 0.5$, $\hat{F}_{NM1}$ is still more efficient than $\hat{F}_{NM0}$ for standard normal and standard exponential distributions, but slightly less efficient than $\hat{F}_{NM0}$ if the population distribution is standard uniform and $N = 15$. This indicates that using tie information, $\hat{F}_{NM1}$ improves the overall performance of CDF estimation as long as the quality of ranking is not bad. Also, note that when $RE > 1$, the RE is generally an increasing function of c.

The general patterns of the estimated REs of $\hat{F}_{NM1}$ versus $\hat{F}_{NM0}$ under the TIC model in Table 2 are similar to those of Table 1, except for two major differences: First, the efficiency gain using $\hat{F}_{NM1}$ instead of $\hat{F}_{NM0}$ can be as large as 66% for the standard exponential distribution and the efficiency loss can be as large as 16%. Second, the REs for the standard uniform distribution are generally lower and fall below one in many cases when $\rho \leq 0.8$.

The estimated REs of $\hat{F}_{NM2}$ versus $\hat{F}_{NM0}$ under the DPS model are presented in Table 3. We observe that although the REs are generally lower than those in Table 1 for $\rho \geq 0.5$, $\hat{F}_{NM2}$ is more robust to ranking errors as compared to $\hat{F}_{NM1}$. The RE values in Table 3 never fall below one when $\rho \geq 0.5$, and even for $\rho = 0$, the maximum efficiency loss over $\hat{F}_{NM0}$ never exceeds 4%.

We also observe from Table 4 that although the REs are generally lower than those in Table 2, $\hat{F}_{NM2}$ is more robust to ranking errors under the TIC model as well and the maximum of efficiency loss using $\hat{F}_{NM2}$ instead of $\hat{F}_{NM0}$ never exceeds 4%.

To examine the point-wise performance of the CDF estimators on the real line, we perform another simulation study in which the CDF estimators are compared via their mean square errors (MSEs) at various points. For a given point t, we define the relative efficiency of $\hat{F}_{NMi}(t)$ to $\hat{F}_{NM0}(t)$ as $RE(t) = MSE(\hat{F}_{NM0}(t))/MSE(\hat{F}_{NMi}(t))$ for

Table 1 Estimating the population CDF under the DPS model: simulated relative efficiencies (defined as ratio of $MISEs$) of $\hat{F}_{NM1}$ versus $\hat{F}_{NM0}$

ρ	m	N	c	$N(0, 1)$				$Exp(1)$				$U(0, 1)$			
				$\frac{1}{2}$	1	2	4	$\frac{1}{2}$	1	2	4	$\frac{1}{2}$	1	2	4
1	3	15		1.03	1.08	1.18	1.27	1.04	1.05	1.10	1.19	1.05	1.11	1.22	1.22
		30		1.03	1.08	1.20	1.32	1.03	1.01	1.10	1.21	1.05	1.11	1.25	1.25
	5	15		1.06	1.11	1.21	1.32	1.05	1.05	1.11	1.23	1.07	1.15	1.24	1.24
		30		1.06	1.12	1.25	1.40	1.03	1.03	1.11	1.27	1.07	1.15	1.25	1.25
0.8	3	15		1.01	1.04	1.09	1.14	1.01	1.01	1.04	1.10	1.02	1.04	1.08	1.10
		30		1.02	1.05	1.13	1.19	1.00	1.01	1.05	1.16	1.02	1.05	1.11	1.11
	5	15		1.02	1.05	1.11	1.16	1.00	1.02	1.04	1.13	1.02	1.04	1.07	1.06
		30		1.03	1.07	1.19	1.24	1.01	1.03	1.08	1.20	1.03	1.06	1.11	1.10
0.5	3	15		1.00	1.01	1.02	1.04	1.00	1.00	1.03	1.07	0.99	0.98	0.97	0.98
		30		1.01	1.03	1.08	1.10	1.01	1.02	1.06	1.12	1.00	1.00	1.01	1.00
	5	15		1.00	1.01	1.04	1.04	1.00	1.01	1.03	1.09	0.99	0.98	0.98	0.96
		30		1.03	1.06	1.12	1.15	1.02	1.04	1.11	1.20	1.01	1.02	1.02	1.02
0	3	15		0.98	0.97	0.95	0.96	0.99	0.97	0.97	0.98	0.96	0.94	0.91	0.90
		30		1.00	1.00	1.02	1.03	1.01	1.03	1.05	1.05	0.98	0.97	0.94	0.95
	5	15		0.99	0.97	0.96	0.97	0.99	0.99	0.98	0.99	0.97	0.93	0.90	0.90
		30		1.01	1.02	1.04	1.06	1.02	1.03	1.07	1.09	0.99	0.99	0.97	0.97

Table 2 Estimating the population CDF under the TIC model: simulated relative efficiencies (defined as ratio of $MISEs$) of $\hat{F}_{NM1}$ versus $\hat{F}_{NM0}$

ρ	m	N	c	$N(0, 1)$				$Exp(1)$				$U(0, 1)$			
				$\frac{1}{4}$	$\frac{1}{2}$	1	2	$\frac{1}{4}$	$\frac{1}{2}$	1	2	$\frac{1}{4}$	$\frac{1}{2}$	1	2
1	3	15		1.03	1.09	1.15	1.12	1.03	1.05	1.06	1.07	1.21	1.09	0.97	0.97
		30		1.04	1.09	1.18	1.21	1.03	1.05	1.06	1.14	1.24	1.14	1.07	1.07
	5	15		1.07	1.17	1.25	1.26	1.06	1.10	1.11	1.21	1.30	1.07	1.02	1.02
		30		1.07	1.18	1.35	1.55	1.05	1.09	1.18	1.42	1.36	1.24	1.23	1.23
0.8	3	15		1.02	1.05	1.11	1.14	1.00	1.02	1.05	1.10	0.97	0.96	0.96	0.97
		30		1.02	1.06	1.15	1.25	1.01	1.02	1.07	1.19	0.99	0.99	1.03	1.06
	5	15		1.03	1.08	1.22	1.28	1.00	1.03	1.13	1.27	0.96	0.93	1.00	1.03
		30		1.03	1.11	1.37	1.58	1.01	1.05	1.23	1.52	1.00	1.02	1.15	1.22
0.5	3	15		1.00	1.01	1.05	1.13	0.99	1.00	1.04	1.14	0.97	0.94	0.90	0.95
		30		1.02	1.04	1.10	1.25	1.01	1.03	1.09	1.26	0.98	0.96	0.96	1.04
	5	15		1.00	1.02	1.13	1.27	1.00	1.00	1.10	1.33	0.96	0.90	0.88	1.01
		30		1.03	1.08	1.28	1.59	1.01	1.07	1.26	1.66	0.99	0.98	1.00	1.20
0	3	15		0.98	0.96	0.96	1.06	0.99	0.97	0.98	1.11	0.96	0.93	0.88	0.91
		30		1.00	1.00	1.04	1.17	1.00	1.02	1.06	1.24	0.98	0.96	0.94	1.00
	5	15		0.98	0.96	0.96	1.21	0.99	0.97	0.99	1.29	0.96	0.90	0.84	0.97
		30		1.01	1.03	1.11	1.50	1.02	1.05	1.17	1.62	0.98	0.96	0.94	1.14

Table 3 Estimating the population CDF under the DPS model: simulated relative efficiencies (defined as ratio of $MISEs$) of $\hat{F}_{NM2}$ versus $\hat{F}_{NM0}$

ρ	m	N	c	$N(0,1)$				$Exp(1)$				$U(0,1)$			
				$\frac{1}{2}$	1	2	4	$\frac{1}{2}$	1	2	4	$\frac{1}{2}$	1	2	4
1	3	15		1.05	1.10	1.15	1.15	1.08	1.15	1.16	1.16	1.06	1.12	1.19	1.19
		30		1.05	1.11	1.15	1.16	1.09	1.14	1.17	1.16	1.07	1.12	1.21	1.21
	5	15		1.07	1.14	1.17	1.16	1.10	1.16	1.19	1.18	1.08	1.16	1.24	1.24
		30		1.08	1.15	1.18	1.16	1.10	1.17	1.18	1.18	1.09	1.17	1.23	1.23
0.8	3	15		1.02	1.05	1.07	1.08	1.03	1.05	1.08	1.07	1.03	1.06	1.11	1.13
		30		1.02	1.05	1.09	1.09	1.02	1.05	1.07	1.09	1.03	1.07	1.12	1.12
	5	15		1.03	1.06	1.09	1.08	1.02	1.06	1.08	1.08	1.03	1.07	1.13	1.13
		30		1.03	1.06	1.10	1.09	1.02	1.05	1.07	1.08	1.04	1.07	1.13	1.13
0.5	3	15		1.00	1.01	1.02	1.02	1.00	1.01	1.03	1.03	1.01	1.02	1.03	1.04
		30		1.00	1.01	1.02	1.02	1.01	1.01	1.02	1.02	1.00	1.01	1.03	1.02
	5	15		1.00	1.00	1.02	1.01	1.00	1.01	1.02	1.02	1.01	1.01	1.04	1.03
		30		1.00	1.00	1.02	1.02	1.00	1.01	1.02	1.02	1.00	1.01	1.02	1.03
0	3	15		0.98	0.97	0.96	0.96	0.98	0.97	0.96	0.96	0.98	0.97	0.96	0.96
		30		0.98	0.97	0.97	0.97	0.99	0.98	0.97	0.97	0.98	0.97	0.95	0.96
	5	15		0.98	0.97	0.96	0.96	0.98	0.97	0.96	0.96	0.98	0.97	0.95	0.96
		30		0.99	0.97	0.97	0.97	0.99	0.98	0.97	0.98	0.98	0.97	0.96	0.96

Table 4 Estimating the population CDF under the TIC model: simulated relative efficiencies (defined as ratio of $MISEs$) of $\hat{F}_{NM2}$ versus $\hat{F}_{NM0}$

ρ	m	N	c	$N(0,1)$				$Exp(1)$				$U(0,1)$			
				$\frac{1}{4}$	$\frac{1}{2}$	1	2	$\frac{1}{4}$	$\frac{1}{2}$	1	2	$\frac{1}{4}$	$\frac{1}{2}$	1	2
1	3	15		1.05	1.12	1.18	1.09	1.08	1.14	1.18	1.11	1.21	1.15	1.04	1.04
		30		1.06	1.12	1.19	1.09	1.08	1.15	1.18	1.10	1.23	1.15	1.03	1.03
	5	15		1.09	1.19	1.21	1.07	1.12	1.20	1.19	1.10	1.28	1.10	1.05	1.05
		30		1.09	1.20	1.21	1.06	1.13	1.21	1.19	1.09	1.29	1.08	1.04	1.04
0.8	3	15		1.03	1.07	1.11	1.08	1.02	1.07	1.11	1.09	1.00	1.02	1.04	1.04
		30		1.03	1.07	1.12	1.07	1.03	1.06	1.12	1.08	1.00	1.01	1.03	1.03
	5	15		1.03	1.09	1.14	1.06	1.03	1.09	1.14	1.09	1.00	1.01	1.06	1.05
		30		1.03	1.09	1.14	1.06	1.03	1.09	1.14	1.08	0.99	1.02	1.04	1.04
0.5	3	15		1.00	1.02	1.04	1.06	1.00	1.02	1.05	1.06	0.99	0.98	0.98	1.03
		30		1.01	1.02	1.04	1.05	1.00	1.02	1.04	1.05	0.98	0.98	0.98	1.02
	5	15		1.00	1.01	1.06	1.05	1.00	1.02	1.06	1.05	0.98	0.97	1.00	1.05
		30		1.00	1.01	1.05	1.05	1.00	1.02	1.06	1.05	0.98	0.98	0.99	1.04
0	3	15		0.98	0.96	0.97	1.01	0.98	0.96	0.97	1.02	0.98	0.97	0.96	1.01
		30		0.98	0.97	0.97	1.01	0.98	0.97	0.97	1.01	0.98	0.96	0.96	1.01
	5	15		0.98	0.97	0.97	1.04	0.98	0.96	0.97	1.03	0.98	0.96	0.97	1.04
		30		0.98	0.97	0.98	1.04	0.99	0.97	0.98	1.03	0.98	0.96	0.96	1.03

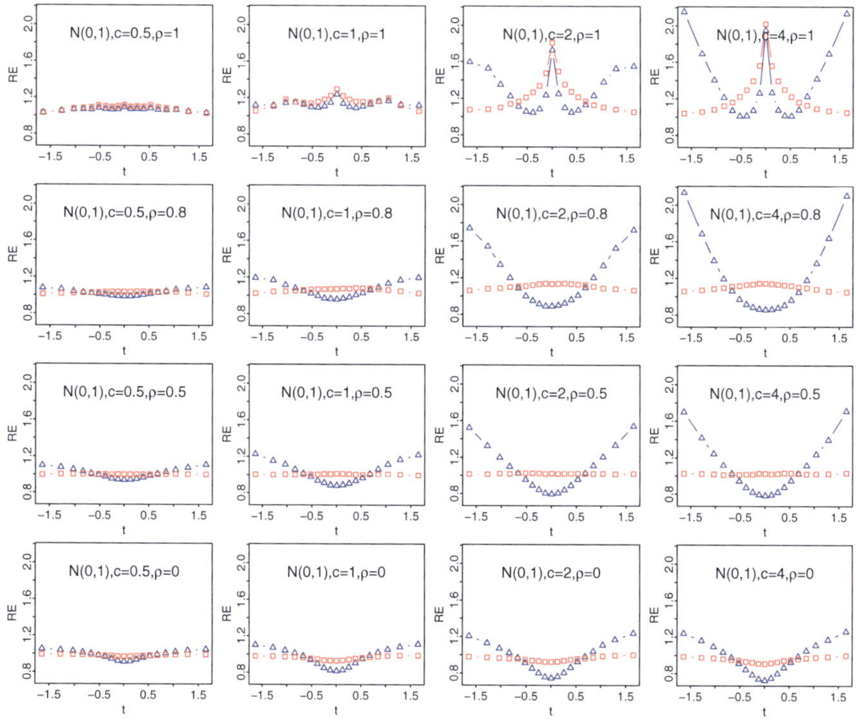

Fig. 1 Estimating the population CDF under the DPS model: simulated relative efficiencies (defined as ratio of MSEs) of $\hat{F}_{NM1}(t)$ versus $\hat{F}_{NM0}(t)$ (represented by $\triangle$ and blue color) and $\hat{F}_{NM2}(t)$ versus $\hat{F}_{NM0}(t)$ (represented by $\square$ and red color) as a function of t when the population distribution is $N(0, 1)$ for $\rho \in \{0, 0.5, 0.8, 1\}$ and $c \in \{0.5, 1, 2, 4\}$

$i = 1, 2$, where $RE(t) > 1$ indicates that $\hat{F}(t)$ is more efficient than $\hat{F}_{NM0}(t)$ at the point t. We set $(N, m) = (30, 5)$, $\rho \in \{0, 0.5, 0.8, 1\}$, $c \in \{0.5, 1, 2, 4\}$ for the DPS model and $c \in \{0.25, 0.5, 1, 2\}$ for the TIC model, $t \in \{Q_{0.05}, Q_{0.1}, \ldots, Q_{0.95}\}$ where Q_p is the pth quantile of the population distribution. For each combination of (ρ, c, t), we estimate $RE(t)$ using 10,000 RSS-t samples randomly generated under each tie-generating model, where the population distribution is set to $N(0, 1)$, and Fig. 1 shows results of the DPS model. We observe that when $c = 0.5$, the performance between $\hat{F}_{NM1}$ and $\hat{F}_{NM2}$ is almost identical but as the value of c increases, the difference between their performances becomes more distinguishable. It is interesting to note that $RE(t)$ of $\hat{F}_{NM1}$ as a function of t has a "W" ("U") shape roughly when the ranking is perfect (imperfect), and its $RE(t)$ falls below one for values of t around zero when the ranking is not perfect. However, the relative efficiency of $\hat{F}_{NM2}$ has an approximate "$\wedge$" shape and more stable patterns than $\hat{F}_{NM1}$ when the quality of ranking varies; it rarely falls below one. This is consistent with what we observe in Table 3. The relative efficiency of $\hat{F}_{NM2}$ is higher (lower) than $\hat{F}_{NM1}$ when the values of t are near the center (tails).

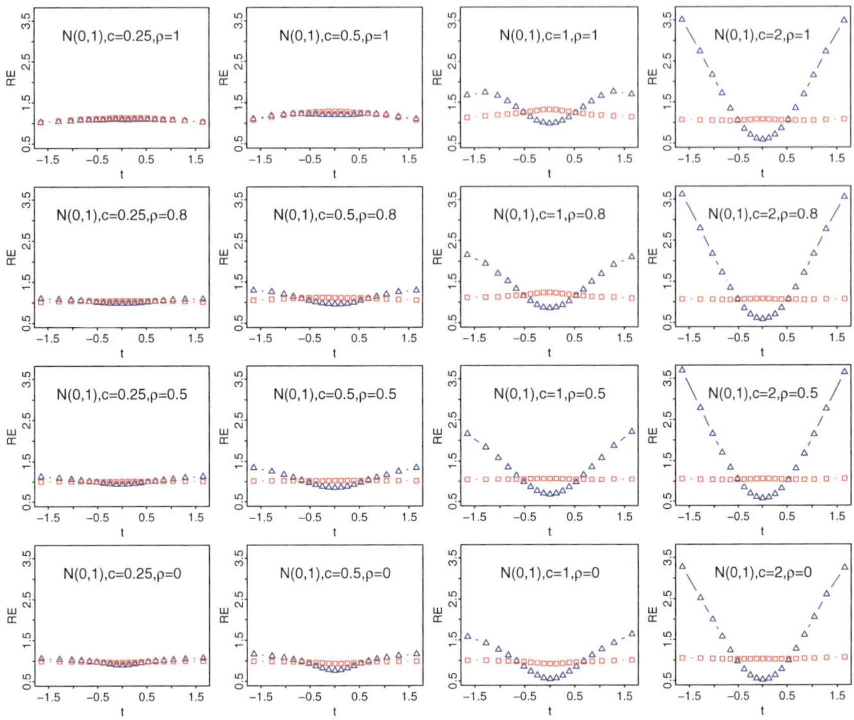

Fig. 2 Estimating the population CDF under the TIC model: simulated relative efficiencies (defined as ratio of MSEs) of $\hat{F}_{NM1}(t)$ versus $\hat{F}_{NM0}(t)$ (represented by $\triangle$ and blue color) and $\hat{F}_{NM2}(t)$ versus $\hat{F}_{NM0}(t)$ (represented by $\square$ and red color) as a function of t when the population distribution is $N(0, 1)$ for $\rho \in \{0, 0.5, 0.8, 1\}$ and $c \in \{0.5, 1, 2, 4\}$

The results under the TIC model can be found in Fig. 2. Here, the RE curves of $\hat{F}_{NM1}$ and $\hat{F}_{NM2}$ are almost identical for $c = 0.25$ and 0.5 but become more distinguishable as the value of c increases. For $c = 1$ and 2, the RE of $\hat{F}_{NM1}$ has a roughly "U" shape and is quite robust to ranking errors for $c = 2$. Again, this is consistent with what we observe in Table 2.

3 Mean Estimation

3.1 *New Nonparametric Estimators Based on MLEs of the CDF*

Let $\{X_{[i]j}, i = 1, \ldots, m, j = 1, \ldots, n\}$ be a balanced ranked set sample of size $N = mn$ from a population with CDF $F(t)$, in which some sample units are tied with some others. We develop several mean estimators based on the ML-type estimators of $F(t)$ described in Sect. 2, using the fact that the population mean can be written as a function of $F(t)$, namely

$$E(X) = \int_{-\infty}^{+\infty} t dF(t). \tag{2}$$

If we replace $F(t)$ in Eq. (2) with any of the MLEs based on data from RSS-t, i.e., $\hat{F}_{NM0}(t)$, $\hat{F}_{NM1}(t)$ and $\hat{F}_{NM2}(t)$, then we can obtain the corresponding ML-based nonparametric estimators of the population mean, denoted by $\hat{\mu}_{NM0}, \hat{\mu}_{NM1}$ and $\hat{\mu}_{NM2}$, respectively.

3.2 Comparison

Below, we compare the ML-based nonparametric estimators of the population mean with the estimators proposed by Frey (2012). For this purpose, we set $N \in \{15, 30\}$, $m \in \{3, 5\}$, $\rho \in \{0, 0.5, 0.8, 1\}$, and then for each combination of (N, m, ρ), we generate 10,000 RSS-t samples under both DPS and TIC models where the population distribution is standard normal (i.e., $N(0, 1)$), standard exponential (i.e., $Exp(1)$) and standard uniform (i.e., $U(0, 1)$), respectively, $c \in \{0.5, 1, \ldots, 3.5, 4\}$ for the DPS model and $c \in \{0.25, 0.5, \ldots, 1.5, 2\}$ for the TIC model. The competing estimators are listed below.

- The standard mean estimator in RSS which ignores tie information and has the form $\hat{\mu}_{st} = \frac{1}{nm}\sum_{i=1}^{m}\sum_{j=1}^{n} X_{[i]j}$.
- The mean estimator based on splitting each tied unit among the strata corresponding to the ranks for which the units were tied. This estimator has the form $\hat{\mu}_{sp} = \frac{1}{m}\sum_{i=1}^{m} \bar{X}'_{[i]}$, where

$$\bar{X}'_{[i]} = \frac{1}{n'_i}\sum_{l=1}^{m}\sum_{j=1}^{n} \frac{I_{[l]ji}}{\sum_{k=1}^{m} I_{[l]jk}} X_{[l]j},$$

  and n'_i is given by Eq. (1).
- The isotonized version of $\hat{\mu}_{sp}$, denoted by $\hat{\mu}_{iso}$. This estimator is obtained using the fact that if the judgment strata are stochastically ordered, then $\mu_{[1]} \leq \cdots \leq \mu_{[m]}$, where $\mu_{[i]}$ is the true mean of the ith stratum. However, this constraint may be violated by their estimates $\bar{X}'_{[1]}, \ldots, \bar{X}'_{[m]}$. Thus, one natural way to improve $\hat{\mu}_{sp}$ is to isotonize the estimates $\bar{X}'_{[1]}, \ldots, \bar{X}'_{[m]}$ using the weighted sample sizes $n'_1, \ldots, n'_m$, and the resulting estimates follow the order constraint. The isotonized version of $\bar{X}'_{[1]}, \ldots, \bar{X}'_{[m]}$ can be given by

$$\bar{X}'_{[i],iso^-} = \min_{1\leq r\leq i} \max_{i\leq s\leq m} \frac{\sum_{l=r}^{s} n'_l \bar{X}'_{[l]}}{\sum_{l=r}^{s} n'_l},$$

or

$$\bar{X}'_{[i],iso^+} = \max_{i \le s \le m} \min_{1 \le r \le i} \frac{\sum_{l=r}^{s} n'_l \bar{X}'_{[l]}}{\sum_{l=r}^{s} n'_l},$$

for $i = 1, \ldots, m$ (see Eqs. (4) and (5) in Wang et al. 2012). As pointed out by Wang et al. (2012), when $n'_i > 0$ for all $i \in \{1, \ldots, m\}$ (i.e., no empty strata exist), $\bar{X}'_{[i],iso^-} = \bar{X}'_{[i],iso^+}$ holds and so we denote the isotonized version of $\bar{X}'_{[1]}, \ldots, \bar{X}'_{[m]}$ by $\bar{X}'_{[1],iso}, \ldots, \bar{X}'_{[m],iso}$. These isotonized estimates can be computed using the pool adjacent violator algorithm (PAVA) (see Chap. 1 of Robertson et al. 1988), and the resulting mean estimator is given by $\hat{\mu}_{iso} = \frac{1}{m} \sum_{i=1}^{m} \bar{X}'_{[i],iso}$.

Remark 1 Frey (2012) described the isotonized version of $\hat{\mu}_{st}$, say $\hat{\mu}_{st,iso}$, and the Rao-Blackwellized (RB) versions of $\hat{\mu}_{st}$ and $\hat{\mu}_{st,iso}$ (note that the RB versions of $\hat{\mu}_{sp}$ and $\hat{\mu}_{iso}$ do not produce new estimators). However, these three lead to new estimators only in unbalanced RSS-t. Since we focus on balanced RSS-t in this paper, we drop them from our comparison set.

In order to compare different mean estimators, we define the relative efficiency of each estimator $\hat{\mu}$ versus $\hat{\mu}_{st}$ by $RE = MSE\left(\hat{\mu}_{st}\right)/MSE\left(\hat{\mu}\right)$. Again, an RE value larger than one indicates the preference of $\hat{\mu}$ over $\hat{\mu}_{st}$, and thus it shows utilizing tie information improves efficiency of mean estimation. Here, we only report the simulated REs from settings with $N = 30$ in Figs. 3 and 4 for the DPS model and in Figs. 5 and 6 for the TIC model. Results from settings with $N = 15$ for both models are not reported for brevity, because the sample size N does not have much impact on performance patterns of these estimators. We further note that although $\hat{\mu}_{iso}$ has higher REs than $\hat{\mu}_{sp}$ in all considered cases, their RE values are very close so that they are hardly distinguishable from each other. Thus, we omit $\hat{\mu}_{sp}$ in our discussion below.

Results Under the DPS Model

From comparing the results in Fig. 3 with those in Fig. 4, we observe that the performance patterns for $m = 3$ are very similar to those for $m = 5$, except that the RE values are generally higher (lower) for $m = 5$ than those of $m = 3$ if they are larger (smaller) than one. It is interesting to see the best estimator depends on the quality of ranking, value of c and the population distribution. If $N(0, 1)$ is the population distribution, then $\hat{\mu}_{iso}$ is the best estimator provided that the quality of ranking is not very good ($\rho \le 0.5$). But for $\rho \ge 0.8$, $\hat{\mu}_{NM1}$ or $\hat{\mu}_{NM2}$ is the best estimator depending on whether c is larger or smaller than 2. This pattern also holds for $Exp(1)$, in which $\hat{\mu}_{iso}$ is the winner for $\rho \le 0.5$, but it is beaten by $\hat{\mu}_{NM1}$ and/or $\hat{\mu}_{NM2}$ for $\rho \ge 0.8$. For $U(0, 1)$, any of the three ML-type mean estimators can be the best, depending on the value of ρ. If the ranking is completely random, then the standard ML-type estimator $\hat{\mu}_{NM0}$ that does not utilize the tie information is the best. If the ranking is imperfect but better than random, then $\hat{\mu}_{NM2}$ is the best and for perfect ranking case ($\rho = 1$), $\hat{\mu}_{NM1}$ is the winner except for $c = 1.5$ in which $\hat{\mu}_{NM2}$ is slightly better.

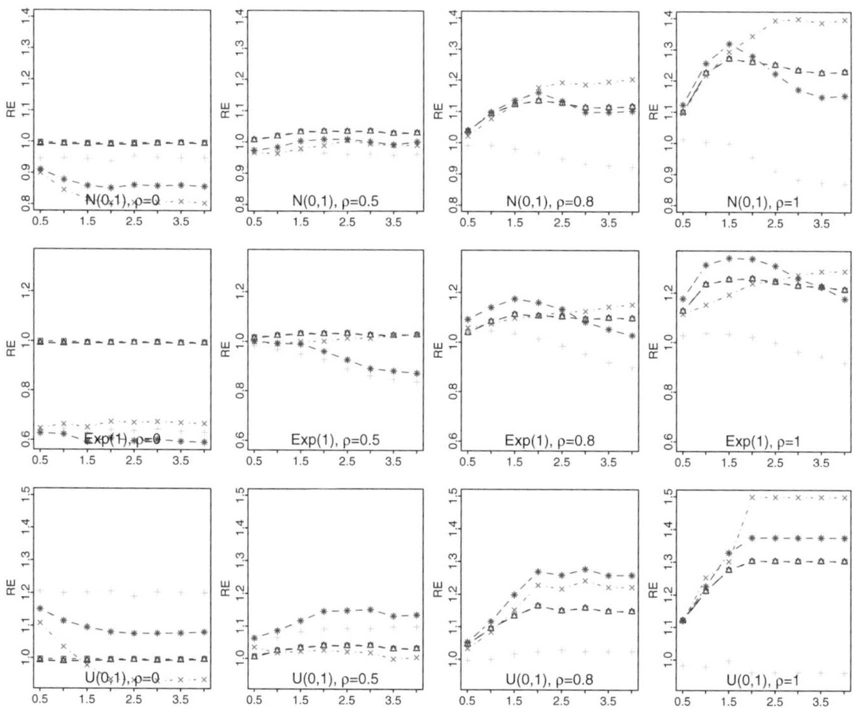

Fig. 3 Relative efficiency of $\hat{\mu}_{sp}$ (represented by $\triangle$ and black color), $\hat{\mu}_{iso}$ (represented by $\square$ and blue color), $\hat{\mu}_{NM0}$ (represented by $+$ and pink color), $\hat{\mu}_{NM1}$ (represented by $\times$ and red color) and $\hat{\mu}_{NM2}$ (represented by $*$ and brown color) to $\hat{\mu}_{st}$ as a function of c under the DPS model for $(N, m) = (30, 3)$, $\rho \in \{0, 0.5, 0.8, 1\}$, when the population distribution is standard normal, standard exponential and standard uniform. Note that the existing mean estimators $\hat{\mu}_{sp}$ and $\hat{\mu}_{iso}$ are represented by symbols with closed shapes $\triangle$ and $\square$. By contrast, all the three ML-type mean estimators are represented by symbols with open shapes, with $+$ for $\hat{\mu}_{NM0}$, $\times$ for $\hat{\mu}_{NM1}$ and $*$ for $\hat{\mu}_{NM2}$ (sort of from simple to more complex shapes); they are also represented by red or like colors (pink, red and brown) from light to dark

Results Under the TIC Model

From Figs. 5 and 6, we find that again the REs are generally increasing or decreasing in m while the other parameters are fixed, depending on whether they are larger or smaller than one and the pattern of REs is almost the same for $m = 3$ and $m = 5$. For standard normal and exponential distributions, $\hat{\mu}_{iso}$ is the best estimator in most cases when $\rho \leq 0.5$, but it is overtaken by $\hat{\mu}_{NM1}$ and/or $\hat{\mu}_{NM2}$ if the quality of ranking is fairly good ($\rho \geq 0.8$). For the standard uniform distribution, the winner always belongs to the set of ML-type estimators. If $\rho = 0$, then $\hat{\mu}_{NM0}$ is the best estimator, followed by $\hat{\mu}_{NM2}$. This pattern also holds for $\rho = 0.5$ except for the cases of $m = 5$ and $c \geq 1.5$ in which $\hat{\mu}_{NM2}$ beats $\hat{\mu}_{NM0}$. For $\rho \geq 0.8$, $\hat{\mu}_{NM2}$ is the best estimator

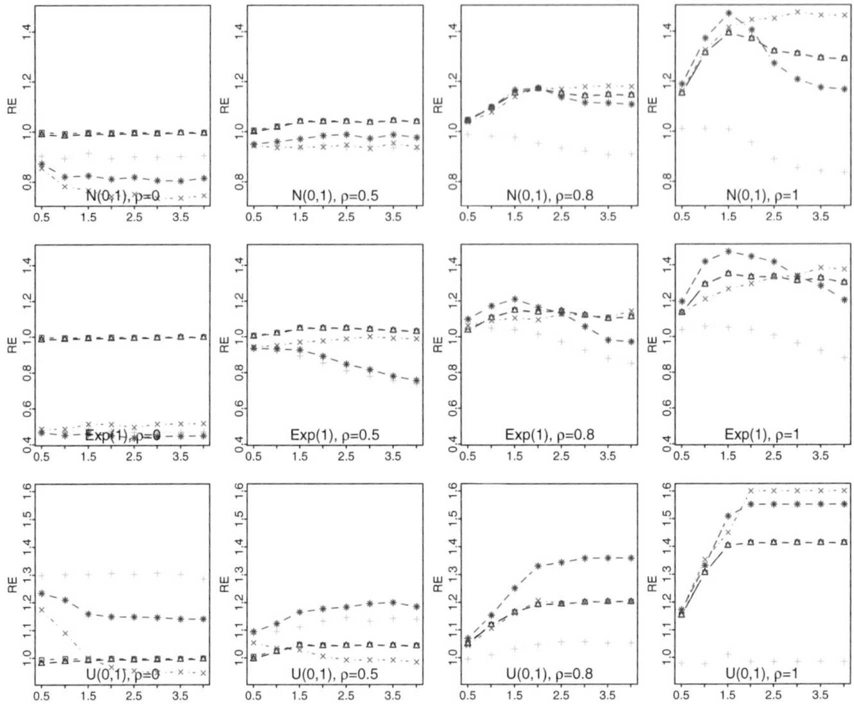

Fig. 4 Relative efficiency of $\hat{\mu}_{sp}$ (represented by $\triangle$ and black color), $\hat{\mu}_{iso}$ (represented by $\square$ and blue color), $\hat{\mu}_{NM0}$ (represented by $+$ and pink color), $\hat{\mu}_{NM1}$ (represented by $\times$ and red color) and $\hat{\mu}_{NM2}$ (represented by $*$ and brown color) to $\hat{\mu}_{st}$ as a function of c under the DPS model for $(N, m) = (30, 5)$, $\rho \in \{0, 0.5, 0.8, 1\}$, when the population distribution is standard normal, standard exponential and standard uniform. Note that the existing mean estimators $\hat{\mu}_{sp}$ and $\hat{\mu}_{iso}$ are represented by symbols with closed shapes $\triangle$ and $\square$. By contrast, all the three ML-type mean estimators are represented by symbols with open shapes, with $+$ for $\hat{\mu}_{NM0}$, $\times$ for $\hat{\mu}_{NM1}$ and $*$ for $\hat{\mu}_{NM2}$ (sort of from simple to more complex shapes); they are also represented by red or like colors (pink, red and brown) from light to dark

except for the case that $c = 0.25$ and $\rho = 1$ in which $\hat{\mu}_{NM1}$ is slightly better than $\hat{\mu}_{NM2}$. In terms of the overall performance of the estimators under the TIC model, we recommend using $\hat{\mu}_{NM2}$ if the quality of ranking is fairly good ($\rho \geq 0.8$).

4 An Empirical Study

In this section, we use a real dataset to compare the performance of different mean estimators for RSS-t samples. It contains measurements of body fat percentage along with several body circumference measurements for 252 men, available at http://lib.stat.cmu.edu/datasets/bodyfat. The histogram of the body fat percentage along with

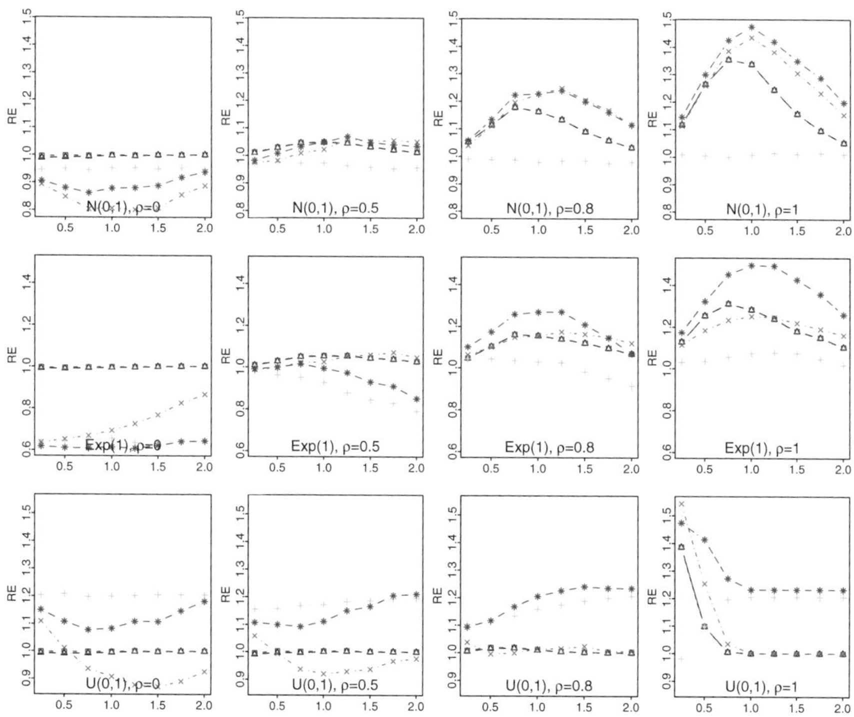

Fig. 5 Relative efficiency of $\hat{\mu}_{sp}$ (represented by △ and black color), $\hat{\mu}_{iso}$ (represented by □ and blue color), $\hat{\mu}_{NM0}$ (represented by $+$ and pink color), $\hat{\mu}_{NM1}$ (represented by $\times$, and red color) and $\hat{\mu}_{NM2}$ (represented by $\ast$ and brown color), to $\hat{\mu}_{st}$ as a function of c under the TIC model for $(N, m) = (30, 3)$, $\rho \in \{0, 0.5, 0.8, 1\}$, when the population distribution is standard normal, standard exponential and standard uniform, respectively. Note that the existing mean estimators $\hat{\mu}_{sp}$ and $\hat{\mu}_{iso}$ are represented by symbols with closed shapes △ and □. By contrast, all the three ML-type mean estimators are represented by symbols with open shapes, with $+$ for $\hat{\mu}_{NM0}$, $\times$ for $\hat{\mu}_{NM1}$ and $\ast$ for $\hat{\mu}_{NM2}$ (sort of from simple to more complex shapes); they are also represented by red or like colors (pink, red and brown) from light to dark

a fitted normal curve is presented in Fig. 7. Although it is slightly right-skewed, the distribution can be roughly approximated by a normal distribution.

We treat the body fat dataset as our hypothetical population and suppose that we are interested in estimating the population mean of the body fat percentage whose true value is $\mu = 19.15$. To draw an RSS-t sample from this population, we assume that ranking is done using standardized values of the concomitant variables, including abdomen circumference, weight and age, under both DPS and TIC models where the parameter c is set as before. The correlation coefficients between the variable of interest and the three concomitant variables are 0.81, 0.61 and 0.29, respectively. So, cases of fairly good ranking ($\rho = 0.81$), moderate ranking ($\rho = 0.61$) and poor ranking ($\rho = 0.29$) are all considered in this study. We also use the standardized values of body fat percentage for ranking, and so the case of perfect ranking ($\rho = 1$)

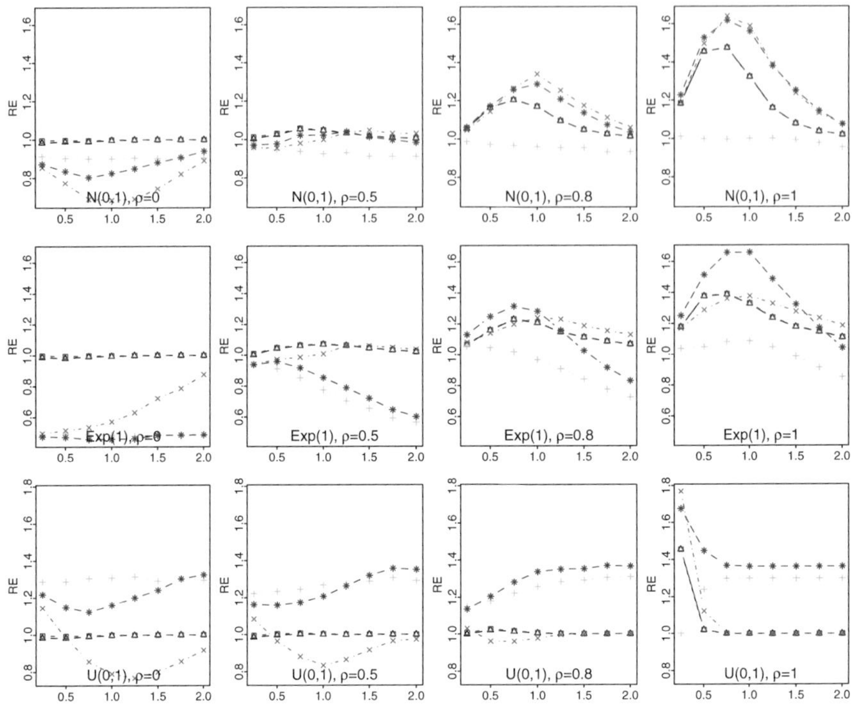

Fig. 6 Relative efficiency of $\hat{\mu}_{sp}$ (represented by $\triangle$ and black color), $\hat{\mu}_{iso}$ (represented by $\square$ and blue color), $\hat{\mu}_{NM0}$ (represented by $+$ and pink color), $\hat{\mu}_{NM1}$ (represented by $\times$ and red color) and $\hat{\mu}_{NM2}$ (represented by $*$ and brown color) to $\hat{\mu}_{st}$ as a function of c under the TIC model for $(N, m) = (30, 5)$, $\rho \in \{0, 0.5, 0.8, 1\}$, when the population distribution is standard normal, standard exponential and standard uniform, respectively. Note that the existing mean estimators $\hat{\mu}_{sp}$ and $\hat{\mu}_{iso}$ are represented by symbols with closed shapes $\triangle$ and $\square$. By contrast, all the three ML-type mean estimators are represented by symbols with open shapes, with "+" for $\hat{\mu}_{NM0}$, $\times$ for $\hat{\mu}_{NM1}$ and $*$ for $\hat{\mu}_{NM2}$ (sort of from simple to more complex shapes); they are also represented by red or like colors (pink, red and brown) from light to dark

is included as well. We set $N = 30$, $m \in \{3, 5\}$; for each combination of (N, m), we draw 10,000 RSS-t samples of size N with replacement from the given population and compute the relative efficiency as defined in Sect. 3. The estimated REs are reported in Tables 5 and 6 for DPS and TIC models, respectively. Clearly, the RE of each estimator decreases as the correlation coefficient ρ decreases in each setting.

For the DPS model, we can see from Table 5 that if $RE > 1$, then it generally increases as the value of m goes from 3 to 5. One of the ML-type estimators is the best estimator which one depends on the quality of ranking and the value of c. When the ranking is perfect ($\rho = 1$), $\hat{\mu}_{NM2}$ is the winner for $c \leq 1$ and it is overtaken by $\hat{\mu}_{NM1}$ for $c \geq 2$. It is interesting to observe that as the quality of ranking decreases, the span of c in which $\hat{\mu}_{NM2}$ beats $\hat{\mu}_{NM1}$ becomes wider and for $\rho \leq 0.61$, $\hat{\mu}_{NM2}$

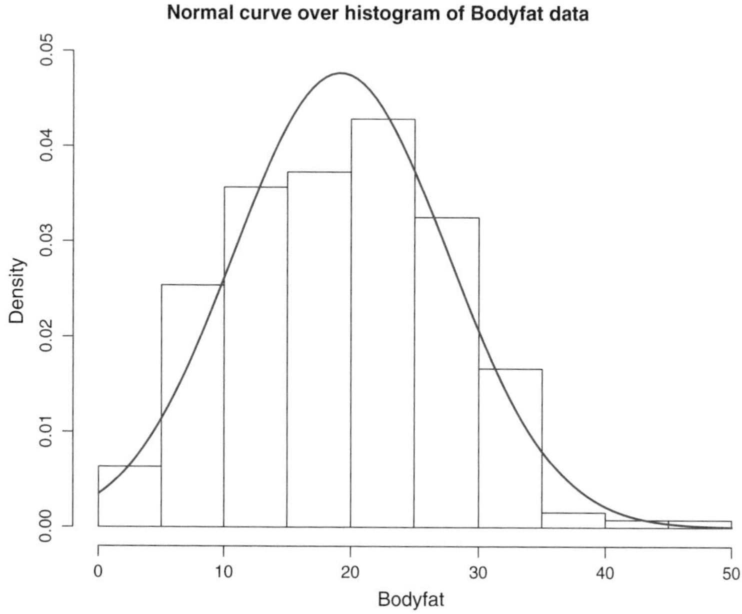

Fig. 7 Histogram of the body fat percentage along with a fitted normal curve

is superior to $\hat{\mu}_{NM1}$ for all considered values of c. This can be justified by the fact that $\hat{\mu}_{NM1}$ is obtained under the assumption of perfect ranking. When the quality of ranking is poor but better than random ($\rho = 0.29$), all estimators except for $\hat{\mu}_{NM1}$ are better than or comparable to $\hat{\mu}_{st}$ and $\hat{\mu}_{NM0}$ is slightly better than the others.

Table 6 shows that for the TIC model, although RE often increases as m increases if $RE > 1$, this is not true for some cases including those with $c = 2$. If the quality of ranking is not low ($\rho \geq 0.61$), then $\hat{\mu}_{NM2}$ is the best mean estimator except for a couple of cases, in which only $\hat{\mu}_{NM1}$ is slightly better than $\hat{\mu}_{NM2}$. When the quality of ranking is poor ($\rho = 0.29$), all estimators except for $\hat{\mu}_{iso}$ and $\hat{\mu}_{NM1}$ have comparable performance; $\hat{\mu}_{iso}$ is slightly better, and $\hat{\mu}_{NM1}$ is a bit worse than the others.

Overall, when the quality of ranking is not inferior (i.e., $\rho \geq 0.61$), either $\hat{\mu}_{NM1}$ or $\hat{\mu}_{NM2}$ has the highest relative efficiency and their gains in RE over the other estimators become more noticeable for large values of c. This is consistent with what we have observed in Sect. 3 for normal data.

Table 5 Estimating the population mean of body fat data under the DPS model: simulated relative efficiencies (defined as ratio of MSEs) of each mean estimator to standard mean estimator for RSS-t samples. The winner of the mean estimators is boldfaced in each setting

	c	$m = 3$					$m = 5$				
		$\hat{\mu}_{sp}$	$\hat{\mu}_{iso}$	$\hat{\mu}_{NM0}$	$\hat{\mu}_{NM1}$	$\hat{\mu}_{NM2}$	$\hat{\mu}_{sp}$	$\hat{\mu}_{iso}$	$\hat{\mu}_{NM0}$	$\hat{\mu}_{NM1}$	$\hat{\mu}_{NM2}$
Body fat ($\rho = 1$)	0.5	1.12	1.12	1.00	1.12	**1.13**	1.16	1.16	1.00	1.17	**1.19**
	1	1.24	1.24	1.00	1.25	**1.27**	1.33	1.33	1.00	1.35	**1.38**
	2	1.27	1.27	0.95	**1.40**	1.31	1.39	1.39	0.96	**1.52**	1.46
	4	1.26	1.26	0.89	**1.45**	1.22	1.33	1.33	0.87	**1.51**	1.26
Abdomen ($\rho = 0.81$)	0.5	1.06	1.06	1.00	1.05	**1.07**	1.07	1.07	1.00	1.07	**1.09**
	1	1.11	1.11	1.00	1.11	**1.13**	1.14	1.14	1.01	1.16	**1.19**
	2	1.15	1.15	1.00	1.21	**1.22**	1.19	1.19	1.00	1.23	**1.27**
	4	1.14	1.14	0.96	**1.24**	1.18	1.16	1.16	0.96	**1.22**	1.19
Weight ($\rho = 0.61$)	0.5	**1.02**	**1.02**	0.99	1.00	**1.02**	**1.02**	**1.02**	0.99	1.00	**1.02**
	1	1.05	1.05	1.01	1.03	**1.08**	1.05	1.06	1.02	1.02	**1.09**
	2	1.05	1.06	1.01	1.04	**1.10**	1.06	1.06	1.02	0.99	**1.09**
	4	1.05	1.05	1.00	1.05	**1.09**	1.06	1.06	1.00	0.99	**1.08**
Age ($\rho = 0.29$)	0.5	**1.00**	**1.00**	**1.00**	0.97	0.99	0.99	**1.00**	**1.00**	0.97	0.98
	1	1.00	1.01	**1.02**	0.98	1.00	1.00	**1.01**	**1.01**	0.98	1.00
	2	1.01	1.01	**1.04**	0.95	1.01	1.00	1.01	**1.05**	0.91	1.00
	4	1.00	1.00	**1.04**	0.93	1.00	1.00	1.01	**1.06**	0.91	1.00

Table 6 Estimating the population mean of body fat data under the TIC model: simulated relative efficiencies (defined as ratio of MSEs) of each mean estimator to standard mean estimator for RSS-t samples. The winner of the mean estimators is boldfaced in each setting

	c	$m = 3$					$m = 5$				
		$\hat{\mu}_{sp}$	$\hat{\mu}_{iso}$	$\hat{\mu}_{NM0}$	$\hat{\mu}_{NM1}$	$\hat{\mu}_{NM2}$	$\hat{\mu}_{sp}$	$\hat{\mu}_{iso}$	$\hat{\mu}_{NM0}$	$\hat{\mu}_{NM1}$	$\hat{\mu}_{NM2}$
Body fat ($\rho = 1$)	0.25	1.11	1.11	1.00	1.11	**1.12**	1.19	1.19	1.00	1.20	**1.22**
	0.5	1.26	1.26	0.99	1.28	**1.29**	1.48	1.48	0.98	**1.53**	1.52
	1	1.35	1.35	1.00	**1.48**	**1.48**	1.32	1.32	1.00	**1.62**	1.58
	2	1.05	1.05	1.06	1.15	**1.26**	1.02	1.02	1.06	1.07	**1.19**
Abdomen ($\rho = 0.81$)	0.25	1.05	1.05	1.01	1.05	**1.07**	1.06	1.07	1.01	1.08	**1.11**
	0.5	1.15	1.15	1.01	1.15	**1.19**	1.18	1.18	1.01	1.20	**1.25**
	1	1.17	1.17	1.01	1.26	**1.31**	1.16	1.16	1.03	1.32	**1.37**
	2	1.03	1.03	1.05	1.11	**1.20**	1.01	1.01	1.03	1.06	**1.14**
Weight ($\rho = 0.61$)	0.25	1.03	1.03	1.00	1.02	**1.04**	1.03	**1.04**	0.99	1.00	**1.04**
	0.5	1.07	1.07	1.00	1.04	**1.09**	1.08	1.08	0.99	1.04	**1.10**
	1	1.08	1.08	1.02	1.11	**1.17**	1.07	1.07	1.00	1.13	**1.17**
	2	1.01	1.01	1.01	1.07	**1.11**	1.01	1.01	0.99	1.03	**1.08**
Age ($\rho = 0.29$)	0.25	1.00	1.00	**1.01**	0.98	0.99	0.99	**1.00**	**1.00**	0.96	0.98
	0.5	**1.01**	**1.01**	1.00	0.96	0.99	1.00	**1.01**	0.99	0.94	0.97
	1	1.01	**1.02**	0.99	0.97	1.00	1.01	**1.02**	0.96	0.92	0.99
	2	1.00	1.00	1.00	1.00	**1.03**	1.00	1.00	0.99	0.99	**1.04**

5 Discussion

We have developed two novel ML-type estimators of the population CDF for RSS samples with tie information available and then used them for constructing new mean estimators. Using Monte Carlo simulation and a real dataset, we have shown that in many situations, the new estimators perform better than their competitors in the literature.

In this paper, we focused on balanced RSS in which tie information is recorded. It would be interesting to investigate the performance of different estimators when tie information is available in unbalanced RSS with empty strata and judgment post-stratification sampling with empty strata which the different versions of isotonized estimators are not identical anymore.

References

Chen, H., Stasny, E. A., & Wolfe, D. A. (2006). Unbalanced ranked set sampling for estimating a population proportion. *Biometrics*, *62*(1), 150–158.

Chen, H., Stasny, E. A., & Wolfe, D. A. (2007). Improved procedures for estimation of disease prevalence using ranked set sampling. *Biometrical Journal*, *49*(4), 530–538.

Crowder, M. (2008). Life distributions: Structure of nonparametric, semiparametric, and parametric families by Albert W. Marshall, Ingram Olkin. *International Statistical Review*, *76*(2), 303–304.

Dell, T. R., & Clutter, J. L. (1972). Ranked set sampling theory with order statistics background. *Biometrics*, *28*(2), 545–555.

Duembgen, L., & Zamanzade, E. (2018). Inference on a distribution function from ranked set samples. *Annals of the Institute of Statistical Mathematics*, 1–29.

Fligner, M. A., & MacEachern, S. N. (2006). Nonparametric two-sample methods for ranked-set sample data. *Journal of the American Statistical Association*, *101*(475), 1107–1118.

Frey, J. (2007). Distribution-free statistical intervals via ranked-set sampling. *Canadian Journal of Statistics*, *35*(4), 585–569.

Frey, J. (2011). A note on ranked-set sampling using a covariate. *Journal of Statistical Planning and Inference*, *141*(2), 809–816.

Frey, J. (2012). Nonparametric mean estimation using partially ordered sets. *Environmental and Ecological Statistics*, *19*(3), 309–326.

Frey, J., Ozturk, O., & Deshpande, J. V. (2007). Nonparametric tests for perfect judgment rankings. *Journal of the American Statistical Association*, *102*(478), 708–717.

Frey, J., & Zhang, Y. (2017). Testing perfect ranking in ranked-set sampling with binary data. *Canadian Journal of Statistics*, *45*(3), 326–339.

Huang, J. (1997). Asymptotic properties of NPMLE of a distribution function based on ranked set samples. *The Annals of Statistics*, *25*(3), 1036–1049.

Kvam, P. H. (2003). Ranked set sampling based on binary water quality data with covariates. *Journal of Agricultural, Biological, and Environmental Statistics*, *8*, 271–279.

Kvam, P. H., & Samaniego, F. J. (1994). Nonparametric maximum likelihood estimation based on ranked set samples. *Journal of the American Statistical Association*, *89*, 526–537.

MacEachern, S. N., Ozturk, O., Wolfe, D. A., & Stark, G. V. (2002). A new ranked set sample estimator of variance. *Journal of the Royal Statistical Society: Series B*, *64*, 177–188.

MacEachern, S. N., Stasny, E. A., & Wolfe, D. A. (2004). Judgement post-stratification with imprecise rankings. *Biometrics*, *60*(1), 207–215.

Mahdizadeh, M., & Zamanzade, E. (2018). A new reliability measure in ranked set sampling. *Statistical Papers*, *59*, 861–891.

McIntyre, G. A. (1952). A method for unbiased selective sampling using ranked set sampling. *Australian Journal of Agricultural Research*, *3*(4), 385–390.

Mu, X. (2015). Log-concavity of a mixture of beta distributions. *Statistics and Probability Letters*, *99*, 125–130.

Perron, F., & Sinha, B. K. (2004). Estimation of variance based on a ranked set sample. *Journal of Statistical Planning and Inference*, *120*, 21–28.

Robertson, T., Wright, F. T., & Dykstra, R. L. (1988). *Order-restricted inferences*. New York: Wiley.

Stokes, S. L. (1980). Estimation of variance using judgment ordered ranked set samples. *Biometrics*, *36*, 35–42.

Stokes, S. L., & Sager, T. W. (1988). Characterization of a ranked-set sample with application to estimating distribution functions. *Journal of the American Statistical Association*, *83*, 374–381.

Takahasi, K., & Wakimoto, K. (1968). On unbiased estimates of the population mean based on the sample stratified by means of ordering. *Annals of the Institute of Statistical Mathematics*, *20*(1), 1–31.

Wang, X., Ahn, S., & Lim, J. (2017). Unbalanced ranked set sampling in cluster randomized studies. *Journal of Statistical Planning and Inference*, *187*, 1–16.

Wang, X., Lim, J., & Stokes, S. L. (2008). A nonparametric mean estimator for judgment post-stratified data. *Biometrics*, *64*(2), 355–363.

Wang, X., Lim, J., & Stokes, S. L. (2016). Using ranked set sampling with cluster randomized designs for improved inference on treatment effects. *Journal of the American Statistical Association*, *111*(516), 1576–1590.

Wang, X., Stokes, S. L., Lim, J., & Chen, M. (2006). Concomitants of multivariate order statistics with application to judgment post-stratification. *Journal of the American Statistical Association*, *101*, 1693–1704.

Wang, X., Wang, K., & Lim, J. (2012). Isotonized CDF estimation from judgment post-stratification data with empty strata. *Biometrics*, *68*(1), 194–202.

Zamanzade, E., Arghami, N. R., & Vock, M. (2012). Permutation-based tests of perfect ranking. *Statistics and Probability Letters*, *82*, 2213–2220.

Zamanzade, E., & Mahdizadeh, M. (2017). A more efficient proportion estimator in ranked set sampling. *Statistics and Probability Letters*, *129*, 28–33.

Zamanzade, E., & Wang, X. (2017). Estimation of population proportion for judgment post-stratification. *Computational Statistics and Data Analysis*, *112*, 257–269.

Bayesian Inference of a Finite Population Mean Under Length-Biased Sampling

Zhiqing Xu, Balgobin Nandram and Binod Manandhar

Abstract We present a robust Bayesian method to analyze forestry data when samples are selected with probability proportional to length from a finite population of unknown size. Specifically, by using Bayesian predictive inference, we estimate the finite population mean of shrub widths in a limestone quarry area with plenty of regrown mountain mahogany. The data on shrub widths are collected using transect sampling, and it is assumed that the probability that a shrub is selected is proportional to its width; this is length-biased sampling. In this type of sampling, the population size is also unknown, and this creates an additional challenge. The quantity of interest is the average finite population shrub width, and the total shrub area of the quarry can be estimated. Our method is assisted by using the three-parameter generalized gamma distribution, thereby robustifying our procedure against a possible model failure. Using conditional predictive ordinates, we show that the model, which accommodates length bias, performs better than the model that does not. In the Bayesian computation, we overcome a technical problem associated with Gibbs sampling by using a random sampler.

Keywords Conditional predictive ordinate, Generalized gamma distribution, Gibbs sampling, Weighted distribution, Random sampling, Transect sampling

1 Introduction

Unequal probability sampling method was first suggested by Hansen and Hurwitz (1943) who demonstrated that the use of unequal selection probabilities frequently allowed more efficient estimators of the population total than did equal probability

Z. Xu · B. Nandram (✉)
Department of Mathematical Sciences, Worcester Polytechnic Institute,
100 Institute Road, Worcester, MA 01609, USA
e-mail: balnan@wpi.edu

B. Manandhar
Department of Mathematics, University of Houston, Houston, USA

G. Chandra et al. (eds.), *Statistical Methods and Applications in Forestry and Environmental Sciences*, Forum for Interdisciplinary Mathematics,
https://doi.org/10.1007/978-981-15-1476-0_6

sampling. The sampling procedure Hansen and Hurwitz (1943) proposed was length-biased sampling. It occurs when the sample selection probabilities are correlated with the values of a study variable, e.g., size variable. This problem falls under the general umbrella of selection bias problems in survey sampling.

Line intercept sampling is a length-biased method used to study certain quantitative characteristics of objects in a region of interest. In general, objects may be of any shape and size and may possess an arbitrary spatial distribution. For example, these objects may be shrubs or patches of vegetation in a field or the projection of logs on the forest floor. The idea of line intercept sampling is to use lines (transects) as sampling units and measuring features of the objects (e.g., widths of shrubs) that crossed by them. A length-biased sampling method produces samples from a weighted distribution. With the underlying distribution of population, one can estimate the attributes of the population by converting the weighted samples to random samples (surrogate samples).

For the estimation of a finite population quantity, the problem is more complex than for a superpopulation parameter because if there is a bias which tends to make the sampled values large, the nonsampled values would tend to be small. Such an adjustment is difficult to carry out. Generally, it has been assumed that the sample size is much smaller than the population size, and this eliminates the finite population estimation problem. Recently, Nandram et al. (2013) proposed a Bayesian non-ignorable selection model to accommodate a selection mechanism for binary data; see also Nandram (2007).

There are several approaches to address the selection bias problem. One approach incorporates the nonsampled selection probabilities in a model. This approach is computer-intensive because the nonsampled part of the population is much larger than the sample; e.g., see Nandram et al. (2006), Nandram and Choi (2010) and Choi et al. (2017). The second approach involves two models, one for the sample, called the survey model, and the other for the population, called the census model. This approach is sometimes called the surrogate sampling approach; e.g., see Nandram (2007) and Nandram et al. (2013). The surrogate sampling approach obtains a surrogate random sample from the census model, and then prediction is done via the census model. The third approach is based on finite population sampling, in which a sample distribution and a sample-complement distribution are both constructed; see Sverchkov and Pfeffermann (2004). It is convenient to use this approach for line intersect sampling. Sverchkov and Pfeffermann (2004) developed design consistent predictors for the finite population total. Essentially, they define the distributions of the sampled values and the nonsampled values as two separate weighted distributions of the census distribution (see Patil and Rao 1978). Yet, another approach is based on quasi-likelihood (Chambers and Skinner 2003) which is difficult to perform in a Bayesian paradigm because the normalization constant is hard to evaluate (typically a complicated function of the model parameters).

Length-biased distributions are a special case of the more general form known as weighted distributions. First introduced by Fisher (1934) to model ascertainment bias, weighted distributions were later formalized in a unifying theory by Rao (1965); see also the celebrated paper of Patil and Rao (1978). Briefly, if the random variable

x has a probability density function (PDF) of $f(x)$ and a nonnegative weight function of $w(x)$, then the corresponding weighted density function is

$$g(x) = \frac{w(x)f(x)}{\int w(x)f(x)dx}.$$

A special case is when the weight function is $w(x) = x$. Such a distribution is known as a length-biased distribution and is given by

$$g(x) = \frac{xf(x)}{\mu},$$

where $\mu = \int xf(x)dx \neq 0$, subject to existence, is the expectation of x. (In Bayesian statistics, we do not use upper and lower cases to differentiate random variables and fixed quantities.)

Various works have been done to characterize relationships between the original distribution and the length-biased distribution. Muttlak and McDonald (1990) suggested using a ranked set sampling procedure to estimate the population size and population mean.

In this chapter, we use a three-parameter generalized gamma distribution as the original distribution to model the widths of shrubs sampled by line intercept method. Line intercept method has been found in widespread applications when estimating particle density, coverage and yields. For example, Lucas and Seber (1977) and Eberhnrdt (1978) derived unbiased estimators for density and percentage cover for any spatial distribution and randomly located transects. McDonald (1980) showed that the Lucas and Seber (1977) estimators for density and percentage cover are unbiased for a simple random sample of unequal length transects. Shrubs can be collected from either randomly located or systematically located transects (Butler and McDonald 1983). It is evident that shrubs with larger widths have higher probabilities of selection.

In Sect. 2, we provide the background of our study. This includes data description and an introduction of the three-parameter generalized gamma distribution, which allows us to robustify our Bayesian method to accommodate the length bias. Section 3 contains the model and the results. The procedure involves the following steps. First, we derive the population size distribution as well as the sample-complement distribution by Bayes' theorem. Next, we propose a random sampling method to generate random parameters from their joint posterior distribution. Then, using each set of parameter values, we obtain a set of samples and the corresponding complement samples (Sverchkov and Pfeffermann 2004). Finally, one sample of the finite population mean can be obtained by taking the average of the pooled samples. The goodness of fit is checked by utilizing conditional predictive ordinates, with results compared for both the models with and without the correction for selection bias. The conclusion is made in Sect. 4.

2 Data and Robustness

2.1 Description of the Data

The data we use were collected using the line intercept sampling method (Muttlak and McDonald 1990). The study was conducted in a limestone quarry area with plenty of regrown mountain mahogany. The study area was defined by the area east of the baseline and within the walls of the quarry, where the baseline was established approximately parallel to the fissures; see Fig. 1 in Appendix 1. By dividing the baseline into three equal parts, three systematically placed transects were established. To ensure uniform coverage over the study area, two independent replications, each

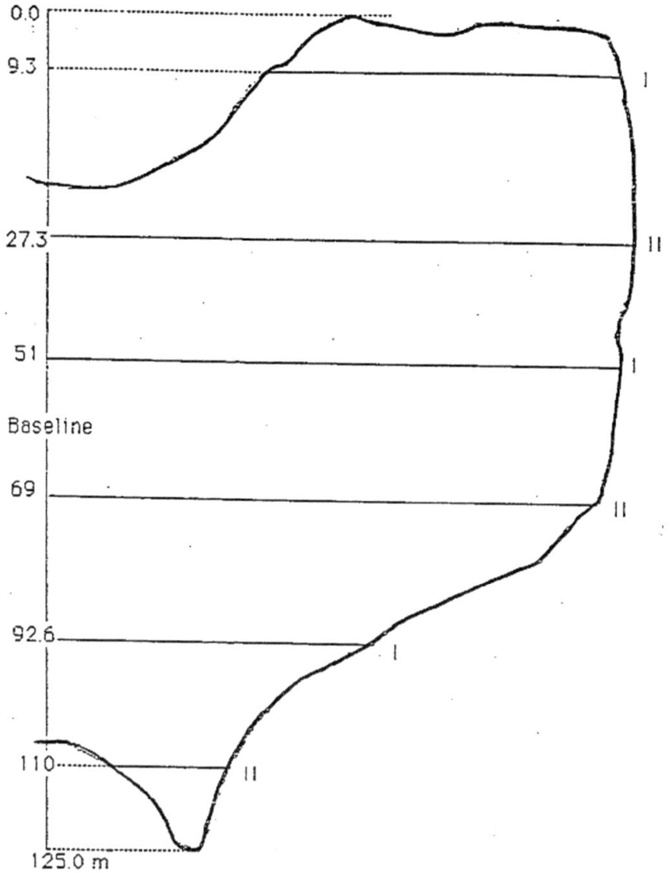

Fig. 1 Sketch of the study area showing the baseline and transects of the two replications (I and II) of systematically located transect lines; random starting points of 9.3 m and 27.3 m were selected with parallel lines separated by 41.66 m; see Muttlak (1988)

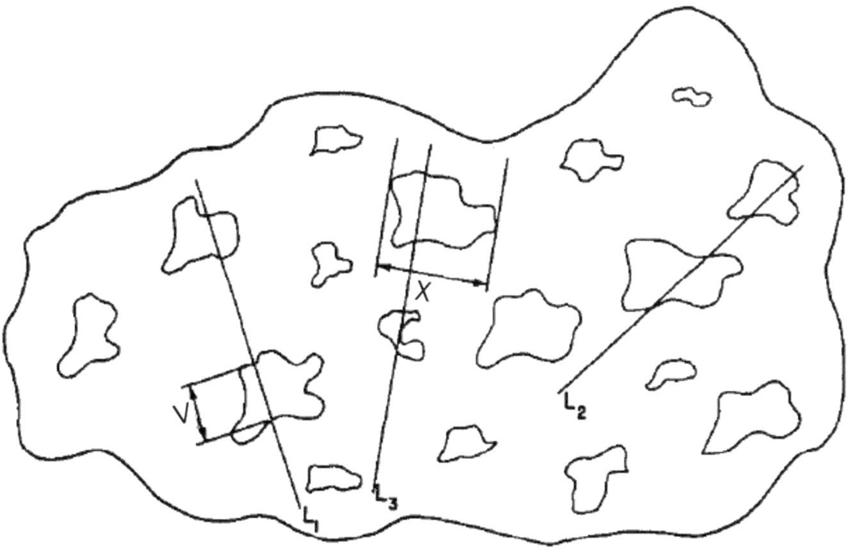

Fig. 2 Sketch of the study area with three transects, L_1, L_2 and L_3: X is the width of the intersected shrub perpendicular to the transect, and V is the length of intersected shrub parallel to the transect; see Muttlak (1988)

with three transects, were selected. One quantity of interest is the mean width of the shrubs in the quarry. So, the variable we study is the width of the projection of the shrub encountered by transects onto the baseline (for illustration, see Fig. 2).

We use the data from both replications; see Tables 1 and 2. The numbers of shrubs counted in three transects are, respectively, 18, 22, 6 and in the two transects of Replication 2 are, respectively, 32, 11 (one transect of Replication 2 has no data). Looking at the box plots of these two replications (Fig. 3), we notice the clear differences in the distributions among the three transects in Replication 1, whereas in Replication 2, the distributions are close. Therefore, when making inferences using

Table 1 Widths (meters) of shrubs in Replication 1

Transect	X_i = width					
I	1.53	0.87	0.79	0.78	1.85	1.45
	0.48	0.52	0.22	0.38	0.59	0.20
	0.42	1.02	0.97	0.56	0.62	0.42
II	1.15	0.87	0.57	0.97	0.57	1.97
	0.58	2.54	1.85	0.35	1.24	1.80
	0.78	0.98	1.30	1.55	1.69	2.12
		1.27	0.75	1.01	1.82	
III	0.71	1.50	1.82	1.86	1.61	1.21

Table 2 Widths (meters) of shrubs in Replication 2

Transect	X_i = width					
I	0.67	0.31	0.83	1.95	1.36	1.45
	0.72	1.15	0.98	1.29	0.88	0.25
	0.63	1.12	0.34	0.21	1.36	0.95
	1.04	0.48	1.05	0.88	0.16	1.08
	0.95	0.25	0.30	1.40	0.58	0.73
			1.30	0.57		
II	0.96	2.08	0.68	1.39	0.50	0.72
	0.19	1.91	0.88	0.48	0.12	

Replication 1, we regard the data from these three transects as from three different strata (they are actually so) and distinguish them in our modeling.

One complication is that we do not know the number of shrubs in the entire quarry. As we intend to use the Bayesian approach, the data from Replication 2 are used to construct a prior distribution of the finite population size and the data from Replication 1 are to provide inference for the population mean.

2.2 *Generalized Gamma Distribution*

The generalized gamma distribution (GG) was first introduced by Stacy (1962). The flexibility of generalized gamma distribution lies in the fact that it has various subfamilies including Weibull distribution, generalized normal distributions and the lognormal as a limit. Khodabin and Ahmadabadi (2010) provided details of the subfamilies of generalized gamma distribution. Because of the flexibility of generalized gamma distribution, we use it as the underlying population distribution in our models.

Some authors have advocated the use of simpler models because of estimation difficulties caused by the complexity of GG parameter structure. For example, Parr and Webster (1965), Hager and Bain (1971) and Lawless (1980) have considered maximum likelihood estimation in the three-parameter generalized gamma distribution. They reported problems with iterative solution of the nonlinear equations implied by the maximum likelihood method. They remarked that maximum likelihood estimators might not exist unless the sample size exceeds 400. (Our sample sizes are much smaller, so we need to be careful.) In our chapter, we perform Bayesian analyses of generalized gamma distribution to overcome this issue.

The probability density of generalized gamma distribution is given by

$$f(x|\alpha, \beta, \gamma) = \frac{\gamma x^{\gamma\alpha-1}}{\beta^{\gamma\alpha}\Gamma(\alpha)} \exp\left[-\left(\frac{x}{\beta}\right)^{\gamma}\right], x > 0, \tag{2.1}$$

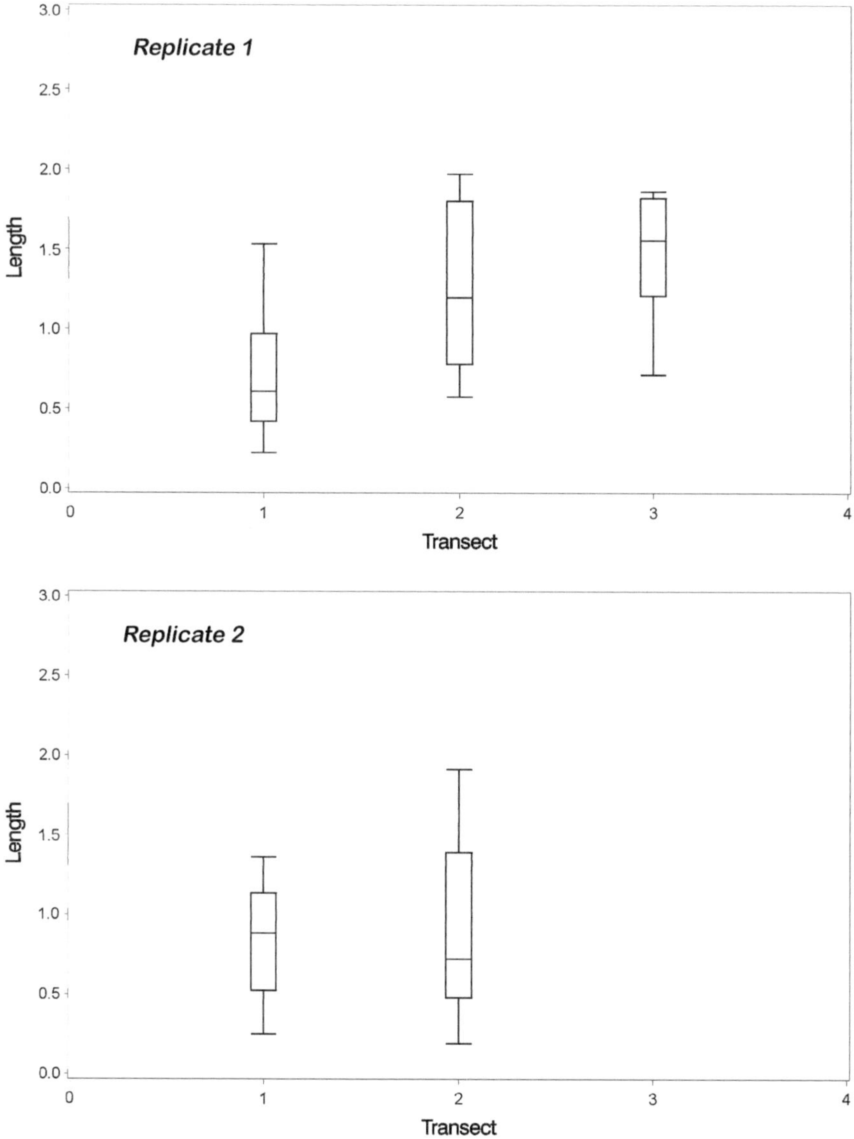

Fig. 3 Box plots of the length-biased data from the two replications

where α β, γ are all positive. It is worth noting that the mean and variance of x are given by

$$E(x) = \beta \frac{\Gamma(\alpha + \frac{1}{\gamma})}{\Gamma(\alpha)}, \quad \text{and} \quad \text{Var}(x) = \beta^2 \left[\frac{\Gamma(\alpha + \frac{2}{\gamma})}{\Gamma(\alpha)} - \left(\frac{\Gamma(\alpha + \frac{1}{\gamma})}{\Gamma(\alpha)} \right)^2 \right],$$

respectively. We write $x \sim \text{GG}(\alpha, \beta, \gamma)$ to denote a random variable with PDF, $f(x|\alpha, \beta, \gamma)$ defined by (2.1), which we call an unweighted generalized gamma distribution.

Note that when $\gamma = 1$, we get the standard gamma distribution, and by making γ differ from 1 many distributions are accommodated, thereby increasing the flexibility of the gamma distribution. It is in this sense we robustify our procedures.

The length-biased distribution of x is $g(x) = \dfrac{xf(x)}{E(x)}$, where $E(x)$ is the expectation of x from the unweighted density function $f(x)$. This can be easily derived as follows. Let I denote the indicator variable; i.e., $I = 1$ if the unit is selected, and $I = 0$ if the unit is not selected. Under length-biased sampling, the probability that the unit has been selected given the value x is $f(I = 1|x) = Cx$, where C is a constant. By Bayes' theorem, the sample PDF $g(x)$ is

$$\begin{aligned} g(x|I = 1) &= \frac{f(I = 1|x)f(x)}{\int f(I = 1|x)f(x)dx} \\ &= \frac{Cxf(x)}{\int Cxf(x)dx} = \frac{xf(x)}{E(X)}, \end{aligned}$$

for convenience we will write $g(x)$ for $g(x|I = 1)$.

Therefore, by using $\text{GG}(\alpha, \beta, \gamma)$ as the population distribution, the sample distribution is

$$\begin{aligned} g(x|\alpha, \beta, \gamma) &= \frac{\frac{\gamma x^{\gamma\alpha}}{\beta^{\gamma\alpha}\Gamma(\alpha)} \exp\left[-\left(\frac{x}{\beta}\right)^{\gamma} \right]}{\beta \frac{\Gamma(\alpha + \frac{1}{\gamma})}{\Gamma(\alpha)}} \\ &= \frac{\gamma x^{\gamma\alpha}}{\beta^{\gamma\alpha+1}\Gamma(\alpha + \frac{1}{\gamma})} \exp\left[-\left(\frac{x}{\beta}\right)^{\gamma} \right], x > 0. \end{aligned} \tag{2.2}$$

It is worth noting that $g(x)$ is also a generalized gamma distribution with parameters $\alpha_g = \alpha + \frac{1}{\gamma}$, $\beta_g = \beta$ and $\gamma_g = \gamma$, denoted by $x \sim \text{GG}(\alpha + \frac{1}{\gamma}, \beta, \gamma)$, with mean and variance adjusted to

$$E(x) = \frac{\beta\Gamma(\alpha + \frac{2}{\gamma})}{\Gamma(\alpha + \frac{1}{\gamma})}, \quad \text{and} \quad Var(x) = \frac{\beta^2\Gamma(\alpha + \frac{3}{\gamma})}{\Gamma(\alpha + \frac{1}{\gamma})} - \left(\frac{\beta\Gamma(\alpha + \frac{2}{\gamma})}{\Gamma(\alpha + \frac{1}{\gamma})} \right)^2.$$

We call it the weighted generalized gamma distribution.

3 Bayesian Methodology

In this section, we derive the population size distribution, the sample-complement distribution, as well as the posterior distribution for the parameters.

Denote ℓ as the number of transects and N_i as the total number of shrubs in the ith transect, $N = \sum_{i=1}^{\ell} N_i$. Note that all the N_i and N are unknown. Prior information about N is needed to carry out a full Bayesian analysis. Denote n_i as the number of shrubs from the ith transect, and $n = \sum_{i=1}^{\ell} n_i$ is the number of samples. Let $x_1, \ldots, x_n$ be the widths of the sampled shrubs and $x_{n+1}, \ldots, x_N$ be the widths for the nonsampled ones, which are to be predicted. The quantity of interest is

$$\bar{X} = \frac{1}{N} \sum_{i=1}^{N} x_i = f\bar{x}_s + (1-f)\bar{X}_{ns},$$

where $f = \frac{n}{N}$ is the sample fraction and $\bar{x}_s = \frac{1}{n}\sum_{i=1}^{n} x_i$ and $\bar{X}_{ns} = \frac{1}{N-n}\sum_{i=n+1}^{N} x_i$ are, respectively, the sample and non-sample means. A posteriori inference is required for $\bar{X}_{ns}$. It is worth noting that $\bar{X}_{ns}$ is not a sufficient statistic and cannot be derived directly from the sample. Therefore, one needs to draw $x_{n+1}, \ldots, x_N$ to predict $\bar{X}_{ns}$.

In many studies, the population size N is unknown and it must be estimated before inference can be made about $\bar{X}$. Our application is no exception; however, this is easy to address as we have two sets of replicated samples. The second replication (samples from the two transects are similarly distributed) can be used to construct a prior for N. The first replication (three transects need to be treated as three strata) is used to estimate the population mean shrub width. In this way, "using the data twice" is avoided.

We assume that the population distributions for different strata are GG, that is

$$x_{ij} \mid \alpha, \beta_i, \gamma \overset{ind}{\sim} \mathrm{GG}(\alpha, \beta_i, \gamma), j = 1, \ldots, N_i, i = 1, \ldots, \ell,$$

accommodated the length bias, the sample distribution is

$$x_{ij} \mid \alpha, \beta_i, \gamma \overset{ind}{\sim} \mathrm{GG}(\alpha + \frac{1}{\gamma}, \beta_i, \gamma), j = 1, \ldots, n_i, i = 1, \ldots, \ell. \quad (3.1)$$

The remaining problem is to find the distribution of the nonsampled values, x_{ij}, $i = 1, \ldots, \ell, j = n_i + 1, \ldots, N_i$, the so-called sample-complement distribution (Sverchkov and Pfeffermann 2004).

In Sect. 3.1, we show how to obtain the prior distribution for N_i. In Sect. 3.2, we describe the sample-complement distribution. In Sect. 3.3, we combine the results of 3.1 and 3.2 to derive the full Bayesian model. In Sect. 3.4, we study the posterior distributions in detail.

3.1 Prior Distribution of the Finite Population Size

We first find the estimate of N based on the sample size. Then, estimates of N_i can be obtained assuming proportional allocation.

The Horvitz–Thompson unbiased estimator of N is

$$\hat{N} = \sum_{i=1}^{n} \frac{1}{\pi_i}$$

where π_i is the probability that the ith unit is selected; see Cochran (1977). Since the line intercept sampling gives the length-biased data, we are actually sampling with probability proportion to width x. Thus, we have

$$\pi_i = Cx_i, \quad i = 1, \ldots, n,$$

where C is a constant and $C = \frac{1}{W}$, where $W = 125$ (meters) is the length of the base line. Then, the estimated value of N under selection bias is

$$\hat{N} = 125 \times \sum_{i=1}^{n} \frac{1}{x_i},$$

Using the data from the Replication 2, we have $\hat{N} = 10,061$. (Note that Replication 1 has $\ell = 2$ strata and Replication 2 has $\ell = 3$ strata.) Then, using proportional allocation in Replication 1, as $n_1 = 18, n_2 = 22, n_3 = 6$, we have $\hat{N}_1 = 3937, \hat{N}_2 = 4812, \hat{N}_3 = 1312$.

Next, we assume

$$n_i \mid N_i, \mu_o \overset{ind}{\sim} \text{Binomial}(N_i, \mu_o), n_i = 0, \ldots, N_i, i = 1, \ldots, \ell,$$

μ_o does not depend on transects because of the nature of proportional allocation method.

Using independent noninformative priors for N_i,

$$\pi(N_i) \propto \frac{1}{N_i}, N_i \geq n_i.$$

We derived the posterior distributions of N_i,

$$\pi(N_i \mid n_i, \mu_o) = \frac{(N_i - 1)!}{(n_i - 1)!(N_i - n_i)!} \mu_0^{n_i} (1 - \mu_0)^{N_i - n_i}, \quad N_i \geq n_i, i = 1, \ldots, \ell, \tag{3.2}$$

which is a negative binomial distribution with $E(N_i \mid n_i, \mu_o) = \dfrac{n_i}{\mu_o}$. By equating $\hat{N}_i$ to $E(N_i \mid n_i, \mu_o)$, we solve for the estimated value of μ_o, which is $\mu_o = \dfrac{n_i}{\hat{N}_i} = 0.0046$.

Therefore, based on Replication 2 our data-based prior distributions of the N_i are independently negative binomial distributions with parameters n_i and $\mu_o = 0.0046$, $i = 1, \dots, \ell$.

3.2 Sample-Complement Distribution

Next, we need to make inference about the nonsampled values. That is, we obtain the sample-complement distribution (Sverchkov and Pfeffermann 2004) and draw samples from it. We consider a single transect first and drop the transect indicator i.

Let $I_j = 1$ if $j \in s$ and $I_j = 0$ if $j \notin s$, where s denotes the sample set. Then,

$$I_j | x_j \sim Ber\left(\frac{x_j}{W}\right) \text{ and } x_j \sim f(x_j)$$
$$\Rightarrow \pi(I_j, x_j) \propto \left[\frac{x_j}{W} f(x_j)\right]^{I_j} \left[\left(1 - \frac{x_j}{W}\right) f(x_j)\right]^{1-I_j}$$
$$\Rightarrow \pi(x_j | I_j = 0) = \frac{\left(1 - \frac{x_j}{W}\right) f(x_j)}{\int \left(1 - \frac{x_j}{W}\right) f(x_j) dx_j}.$$

Thus, the posterior sample-complement distribution given all the parameters is

$$\pi(x_{n+1}, \dots, x_N | \alpha, \underset{\sim}{\beta}, \gamma, x_1, \dots, x_n, N)$$
$$= \prod_{j=n+1}^{N} \frac{\left(1 - \frac{x_j}{W}\right) f(x_j)}{\int \left(1 - \frac{x_j}{W}\right) f(x_j) dx_j} = \prod_{j=n+1}^{N} \left[\frac{1 - \frac{x_j}{W}}{1 - \frac{\mu}{W}}\right] f(x_j), \tag{3.3}$$

where $f(x)$ is GG and μ is the expectation of x, which is $\mu = \frac{\beta \Gamma(\alpha + \frac{1}{\gamma})}{\Gamma(\alpha)}$. We use the sampling importance re-sampling (SIR) algorithm to perform the sampling. The SIR algorithm is ideal because $\prod_{j=n+1}^{N} f(x_j)$ is a good proposal density and samples are easy to draw.

3.3 Full Bayesian Model

For the sample data, our model is

$$x_{ij} \mid \alpha, \beta_i, \gamma \overset{ind}{\sim} GG(\alpha + 1/\gamma, \beta_i, \gamma), j = 1, \dots, n_i$$

and the prior for $\alpha, \beta_i, i = 1, \ldots, \ell$ and γ is

$$\pi(\beta_i) \propto \frac{1}{\beta_i}, i = 1, \ldots, \ell, \pi(\alpha) \propto \frac{1}{(1+\alpha)^2}, \pi(\gamma) \propto \frac{1}{(1+\gamma)^2}.$$

Note that the priors on the β_i are improper and the priors on α and γ are the $f(2, 2)$ distributions ($f(2, 2)$ denotes the f distribution with degrees of freedom being (2, 2)), which are nearly noninformative (no moments exist) but proper.

The posterior sample-complement distribution when incorporating all strata is

$$\pi(x_{ij}, i = 1, \ldots, \ell, j = n_i + 1, \ldots \mid N_i, \alpha, \underset{\sim}{\beta}, \gamma)$$
$$= \prod_{i=1}^{\ell} \prod_{j=n_i+1}^{N_i} [\frac{1 - \frac{x_{ij}}{W}}{1 - \frac{\mu}{W}}] f(x_{ij} \mid \alpha, \underset{\sim}{\beta}, \gamma),$$

where f and μ are defined in the same way as (3.3).

The joint posterior density of $\alpha, \underset{\sim}{\beta}, \gamma$ given $\underset{\sim}{x}_s = \{x_{ij}, j = 1, \ldots, n_i, i = 1, \ldots, \ell\}$ is

$$\pi(\alpha, \underset{\sim}{\beta}, \gamma \mid \underset{\sim}{x}_s) \propto \frac{\gamma^n \left(\prod_{i=1}^{\ell} \prod_{j=1}^{n_i} x_{ij}\right)^{\gamma\alpha}}{\left(\prod_{i=1}^{\ell} \beta_i^{n_i}\right)^{\gamma\alpha+1} \left[\Gamma(\alpha + \frac{1}{\gamma})\right]^n}$$
$$\exp\left[-\sum_{i=1}^{\ell} \sum_{j=1}^{n_i} \left(\frac{x_{ij}}{\beta_i}\right)^{\gamma}\right] \frac{1}{(1+\alpha)^2} \frac{1}{(1+\gamma)^2} \prod_{i=1}^{\ell} \frac{1}{\beta_i}.$$

The posterior density can be simplified by transforming β_i to $\phi_i = 1/\beta_i^{\gamma}, i = 1, \ldots, \ell$. (Note that the Jacobian of the transformation must be included.) Then, the joint posterior density of $\alpha, \underset{\sim}{\phi}, \gamma$ given $\underset{\sim}{x}_s$ is

$$\pi(\alpha, \underset{\sim}{\phi}, \gamma \mid \underset{\sim}{x}_s) \propto \frac{1}{(1+\alpha)^2} \frac{1}{(1+\gamma)^2}$$
$$\times \frac{\gamma^{n-\ell} \left(\prod_{i=1}^{\ell} \prod_{j=1}^{n_i} x_{ij}^{\gamma\alpha}\right) \prod_{i=1}^{\ell} \phi_i^{n_i\left(\alpha+\frac{1}{\gamma}\right)-1}}{\left[\Gamma(\alpha+\frac{1}{\gamma})\right]^n} \exp\left[-\sum_{i=1}^{\ell} \phi_i \sum_{j=1}^{n_i} x_{ij}^{\gamma}\right]. \tag{3.4}$$

This is not a standard posterior density; however, one can fit this model using Markov chain Monte Carlo methods (i.e., to obtain sample of $\alpha, \underset{\sim}{\phi}, \gamma$).

3.4 Further Study of the Posterior Density

One important problem we need to worry about is the posterior propriety of $\pi(\alpha, \underset{\sim}{\phi}, \gamma \mid \underset{\sim}{x}_s)$. First, it is easy to see that

$$\phi_i \mid \alpha, \gamma, \underset{\sim}{x}_s \overset{ind}{\sim} \text{Gamma}\{n_i(\alpha + \frac{1}{\gamma}), \sum_{j=1}^{n_i} x_{ij}^{\gamma}\}, i = 1, \ldots, \ell. \tag{3.5}$$

Then, integrating out the ϕ_i, we get

$$\pi(\alpha, \gamma \mid \underset{\sim}{x}_s) \propto \prod_{i=1}^{\ell} \left\{ \gamma^{n_i - 1} \frac{\left(\prod_{j=1}^{n_i} x_{ij}^{\gamma\alpha}\right)}{\left(\sum_{j=1}^{n_i} x_{ij}^{\gamma}\right)^{n_i(\alpha+\frac{1}{\gamma})}} \frac{\Gamma\{n_i\left(\alpha + \frac{1}{\gamma}\right)\}}{\left(\Gamma(\alpha + \frac{1}{\gamma})\right)^{n_i}} \right\} \frac{1}{(1+\alpha)^2} \frac{1}{(1+\gamma)^2}. \tag{3.6}$$

It is convenient to let $a_i = \sum_{j=1}^{n_i} x_{ij}^{\gamma}/n_i$ and $g_i = \left(\prod_{j=1}^{n_i} x_{ij}^{\gamma}\right)^{1/n_i}$ denote the arithmetic and geometric means of the $x_{ij}^{\gamma}, j = 1, \ldots, n_i, i = 1, \ldots, \ell$. Then, we have

$$\pi(\alpha, \gamma \mid \underset{\sim}{x}_s) \propto \prod_{i=1}^{\ell} \left\{ \left(\frac{g_i}{a_i}\right)^{n_i\alpha} \left(\frac{\gamma^{n_i-1}}{a_i^{n_i/\gamma}}\right) \frac{\Gamma\{n_i\left(\alpha + \frac{1}{\gamma}\right)\}}{n_i^{n_i(\alpha+\frac{1}{\gamma})}\left(\Gamma(\alpha + \frac{1}{\gamma})\right)^{n_i}} \right\} \frac{1}{(1+\alpha)^2} \frac{1}{(1+\gamma)^2}. \tag{3.7}$$

Thus, we essentially have a two-parameter posterior density. Although an overkill, we attempted to fit this model using a Gibbs sampler. There are difficulties in performing the Gibbs sampler (perhaps associated with the difficulties encountered in finding MLEs in generalized gamma distribution) because high correlations are present among the parameters and thinning is not helpful. The problem is essentially high correlations between α and γ. Thus, we consider an alternative algorithm which simply uses the multiplication rule of probability.

However, since $\gamma = 1$ makes the generalized gamma density a standard gamma density, it is sensible to bound γ in an interval centered at 1. That is, we take $a_o^{-1} \leq \gamma \leq a_o$; a sensible choice is $a_0 = 10$ or so. Thus, we replace the prior on γ by $\gamma \sim \text{Uniform}(a_o^{-1},\ a_o)$; the original prior is inconvenient and not helpful.

We prove the theorem below which adds credence to our Bayesian methodology.

Theorem *Assuming that $a_o^{-1} \leq \gamma \leq a_o$, the joint posterior density of $\pi(\alpha, \gamma \mid \underset{\sim}{x}_s)$ is proper.*

Remark Using the multiplication rule of probability,

$$\pi(\underset{\sim}{\phi}, \alpha, \gamma \mid \underset{\sim}{x}_s) = \pi(\underset{\sim}{\phi} \mid \alpha, \gamma, \underset{\sim}{x}_s)\pi(\alpha, \gamma \mid \underset{\sim}{x}_s),$$

the theorem implies that $\pi(\underset{\sim}{\phi}, \alpha, \gamma \mid \underset{\sim}{x}_s)$ is also proper. Clearly, $\pi(\underset{\sim}{\beta}, \alpha, \gamma \mid \underset{\sim}{x}_s)$ is also proper.

Proof We make two observations. First, using the arithmetic–geometric inequality, we have $\left(\frac{g_i}{a_i}\right)^{n_i\alpha} \leq 1$. Second, using $a_o^{-1} \leq \gamma \leq a_o$, the function $\frac{\gamma^{n_i-1}}{a_i^{n_i/\gamma}}$, is bounded uniformly in γ. Therefore, we only need to show that

$$I = \int_{a_o^{-1}}^{a_o} \int_0^{\infty} \left\{ \prod_{i=1}^{\ell} \frac{\Gamma\{n_i \left(\alpha + \frac{1}{\gamma}\right)\}}{n_i^{n_i(\alpha+\frac{1}{\gamma})} \{\Gamma(\alpha + \frac{1}{\gamma})\}^{n_i}} \right\} \frac{1}{(1+\alpha)^2} \frac{1}{a_o - a_o^{-1}} d\alpha d\gamma < \infty.$$

Next, we transform α to $\theta = \alpha + \frac{1}{\gamma}$, keeping γ untransformed. Then, the integral becomes

$$I = \int_{a_o^{-1}}^{a_o} \int_{\frac{1}{\gamma}}^{\infty} g^*(\theta) \frac{1}{(1+\theta-\frac{1}{\gamma})^2} \frac{1}{a_o - a_o^{-1}} d\theta d\gamma < \infty,$$

where

$$g^*(\theta) = \prod_{i=1}^{\ell} \frac{\Gamma(n_i\theta)}{n_i^{n_i\theta}\{\Gamma(\theta)\}^{n_i}}.$$

We only need to show that

$$g_i(\theta) = \frac{\Gamma(n_i\theta)}{n_i^{n_i\theta}\{\Gamma(\theta)\}^{n_i}}$$

is bounded uniformly in θ for any $i = 1, \ldots, \ell$. For convenience, we will drop the subscript, i, momentarily, so we simply need to show that $\Delta(\theta) = \tilde{\Gamma}(n\theta) - n\tilde{\Gamma}(\theta) - n\theta \ln(n)$, where $\tilde{\Gamma}(\cdot)$ is the logarithm of the gamma function, uniformly bounded in θ; see the Appendix 1.

Finally, because $g^*(\theta) \leq A < \infty$, we are left with

$$I \leq A \int_{a_o^{-1}}^{a_o} \frac{1}{(a_o - a_o^{-1})} \left\{ \int_{\frac{1}{\gamma}}^{\infty} \frac{1}{(1+\theta-\frac{1}{\gamma})^2} d\theta \right\} d\gamma$$
$$= A \int_{a_o^{-1}}^{a_o} \frac{1}{(a_o - a_o^{-1})} d\gamma \int_0^{\infty} \frac{1}{(1+\alpha)^2} d\alpha = A.$$

Therefore, our claim on propriety holds.

Table 3 Summary of the parameters and population mean by Gibbs sampler

Name	Min.	First qu.	Median	Mean	Third qu.	Max.
α	0.25	0.28	1.62	2.86	5.36	6.78
β_1	0.25	0.25	0.40	0.55	0.84	1.25
β_2	0.25	0.25	0.62	0.82	1.33	1.91
β_3	0.25	0.25	0.53	0.92	1.33	4.83
γ	0.48	0.58	0.83	1.06	1.35	2.50
$\bar{X}$	0.066	0.088	3.44	13.92	26.01	58.86

4 Bayesian Computations and Data Analyses

In this section, we perform Bayesian analysis of the posterior distributions of population parameters by a numerical method, called random sampler, which performs better than the Gibbs sampler. We then obtain the nonsampled values using the sampling importance re-sampling (SIR) algorithm. Recall that the data we use here are Replication 1, which has three transects, i.e., $\ell = 3$.

4.1 Random Sampler

Since the Gibbs sampler is a Markovian updating scheme, we have shown, in our work not presented in this chapter, that most of the estimated values of population mean are larger than what we expected (see results in Table 3). One of the reasons is that high correlations among these parameters make the Gibbs sampler inefficient in the sense it may take a very large number of iterations to converge in distribution. In this section, we propose a non-Markovian algorithm, called random sampler, in order to avoid the particular issue mentioned above.

Therefore, α and γ cannot be sampled directly from their unbounded parameter space. We use the transformation $\alpha' = \frac{\alpha}{1+\alpha}$ and $\gamma' = \frac{\gamma}{1+\gamma}$. Then,

$$\Pi(\alpha', \gamma' | x_{11}, \ldots, x_{3n_3}) = \int_{\phi_1}\int_{\phi_2}\int_{\phi_3} \Pi(\alpha, \phi_1, \phi_2, \phi_3, \gamma | x_{11} \ldots, x_{3n_3}) d\phi_1 d\phi_2 d\phi_3$$

$$= \left\{ \frac{\gamma^n \left(\prod_{i=1}^{3} \prod_{j=1}^{n_i} x_{ij} \right)^{\gamma\alpha}}{\left[\Gamma(\alpha + \frac{1}{\gamma})\right]^n} \frac{\Gamma\left(n_1(\alpha + \frac{1}{\gamma})\right) \Gamma\left(n_2(\alpha + \frac{1}{\gamma})\right) \Gamma\left(n_3(\alpha + \frac{1}{\gamma})\right)}{\left(\sum x_{1j}^{\gamma}\right)^{n_1(\alpha+\frac{1}{\gamma})} \left(\sum x_{2j}^{\gamma}\right)^{n_2(\alpha+\frac{1}{\gamma})} \left(\sum x_{3j}^{\gamma}\right)^{n_3(\alpha+\frac{1}{\gamma})} \gamma^3} \right\}_{\alpha = \frac{\alpha'}{1-\alpha'}, \gamma = \frac{\gamma'}{1-\gamma'}},$$

$\alpha' \in (0, 1)$, $\gamma' \in (0, 1)$.

Two-dimensional grid method can be applied to draw α' and γ' from their joint distribution. But grid method is computationally intensive in more than one dimension.

We used the multiplication rule to draw samples of α' and γ'.

$$\Pi(\alpha', \gamma'|x_{11} \ldots, x_{3n_3}) = \Pi(\alpha'|\gamma', x_{11} \ldots, x_{3n_3})\Pi(\gamma'|x_{11} \ldots, x_{3n_3}). \tag{4.1}$$

To apply this rule, we fist generated a sample of $\gamma'^{(1)}$ from $\Pi(\gamma'|x_{11} \ldots, x_{3n_3})$ and then generated a sample of $\alpha'^{(1)}$ from $\Pi(\alpha'|\gamma'^{(1)}, x_{11} \ldots, x_{3n_3})$. Repeat this procedure M times to obtain M sets of $\alpha'(\alpha)$ and $\gamma'(\gamma)$. The corresponding $\phi(\beta)$ can also be obtained by sampling from $\Pi(\phi_i|\alpha, \phi_k, \gamma, x_{11}, \ldots, x_{3n_3})$.

The term $\Pi(\alpha'|\gamma', x_{11} \ldots, x_{3n_3})$ in (4.1) is easy to derive.

$$\Pi(\alpha'|\gamma', x_{11} \ldots, x_{3n_3})$$
$$\propto \left\{ \frac{\gamma^n \left(\prod_{i=1}^{3}\prod_{j=1}^{n_i} x_{ij}\right)^{\gamma\alpha}}{\left[\Gamma(\alpha + \frac{1}{\gamma})\right]^n} \frac{\Gamma\left(n_1(\alpha + \frac{1}{\gamma})\right)\Gamma\left(n_2(\alpha + \frac{1}{\gamma})\right)\Gamma\left(n_3(\alpha + \frac{1}{\gamma})\right)}{\left(\sum x_{1j}^{\gamma}\right)^{n_1(\alpha+\frac{1}{\gamma})}\left(\sum x_{2j}^{\gamma}\right)^{n_2(\alpha+\frac{1}{\gamma})}\left(\sum x_{3j}^{\gamma}\right)^{n_3(\alpha+\frac{1}{\gamma})}} \right\}_{\alpha=\frac{\alpha'}{1-\alpha'}},$$

$\alpha' \in (0, 1)$.

The term $\Pi(\gamma'|x_{11} \ldots, x_{3n_3})$ in (4.1) can be derived by integrating $\Pi(\alpha', \gamma'|x_{11} \ldots, x_{3n_3})$ with respect to α'. Unfortunately, it is not possible to integrate $\Pi(\alpha', \gamma'|x_{11} \ldots, x_{3n_3})$ by analytical techniques. For this reason, numerical methods have to be used. We use the 20-point Gaussian quadrature to approximate $\Pi(\gamma'|x_{11} \ldots, x_{3n_3})$.

$$\begin{aligned}\Pi(\gamma'|x_{11} \ldots, x_{3n_3}) &= \int_0^1 \Pi(\alpha', \gamma'|x_{11} \ldots, x_{3n_3})d\alpha' \\ &= \frac{1}{2}\int_{-1}^{1} \Pi\left(\frac{1}{2} + \frac{1}{2}\alpha', \gamma'\right) d\alpha' \\ &\approx \frac{1}{2}\sum_{i=1}^{20} \omega_i \Pi\left(\frac{1}{2} + \frac{1}{2}x_i, \gamma'\right),\end{aligned}$$

where $x_i, i = 1, \ldots, 20$ are the roots of orthogonal polynomials $P_{20}(x)$ for $[-1, 1]$ and $\omega_i, i = 1, \ldots, 20$ are the corresponding Gauss–Legendre weights.

The summary of samples drawn by random sampler for each parameter and the population mean are shown in Tables 3 and 4. It is shown that the population mean has the IQR of (0.67, 0.81), with the median of 0.75. In Fig. 4, it is shown that the posterior distribution of α is bimodal (pointing to the difficulty in estimating α), while those of β_1, β_2, β_3 and γ are skewed to the right. In the next section, we will perform the model checking by conditional predictive ordinate (CPO).

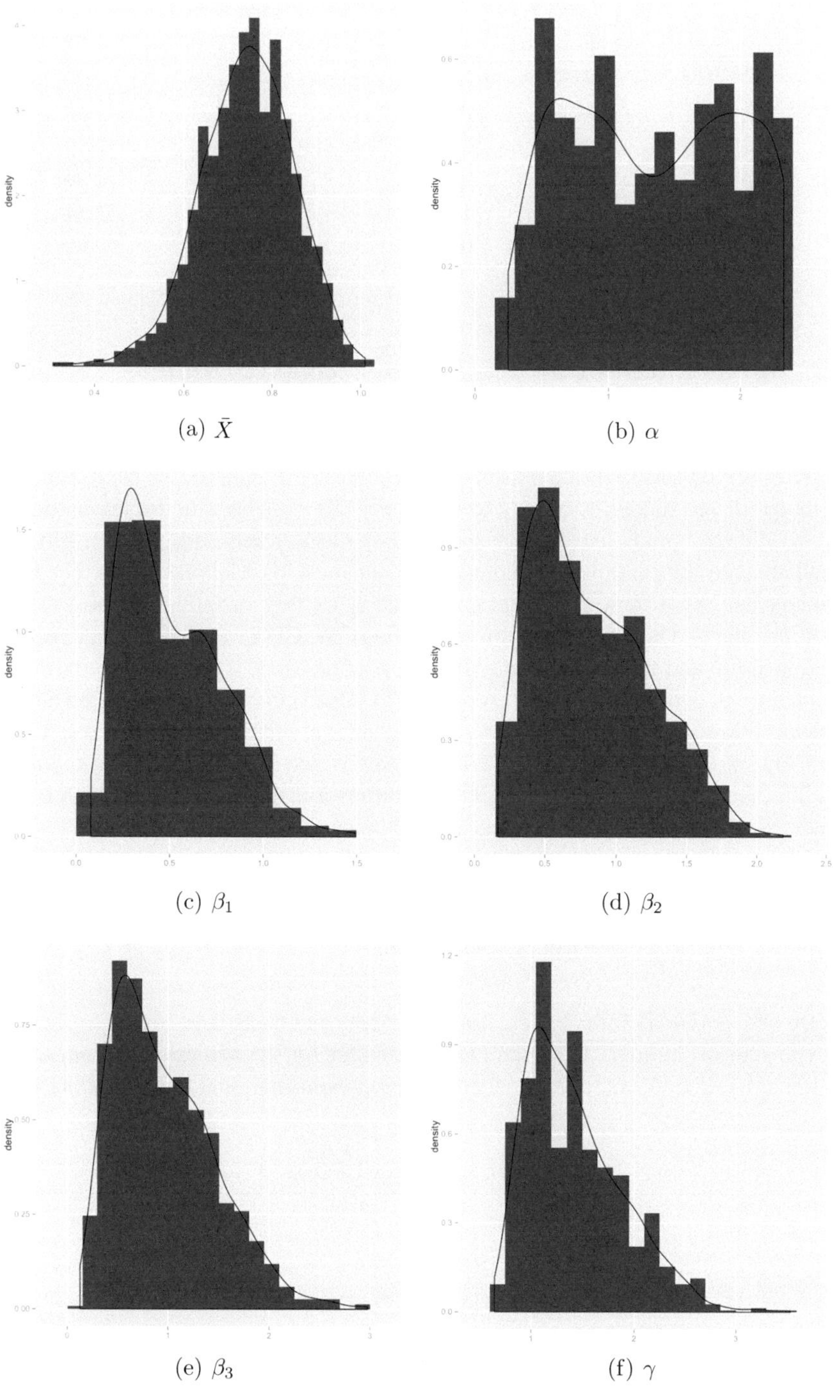

Fig. 4 Posterior distributions of population mean; α; β_1; β_2; β_3; and γ

Table 4 Summary of the parameters and population mean by random sampler

Name	Min.	First qu.	Median	Mean	Third qu.	Max.
α	0.25	0.77	1.37	1.34	1.93	2.33
β_1	0.07	0.29	0.46	0.52	0.71	1.50
β_2	0.17	0.49	0.75	0.83	1.12	2.19
β_3	0.13	0.56	0.87	0.96	1.29	2.98
γ	0.64	1.05	1.36	1.43	1.68	3.54
$\bar{X}$	0.31	0.67	0.75	0.74	0.81	1.01

4.2 *Model Checking by Conditional Predictive Ordinate*

Posterior predictive check is known as comparing the predicted distribution estimated by the observed data to the data itself. Such double use of data causes overestimation of the predicted model. One way to overcome this drawback is the leave-one-out cross-validation predictive density proposed by Geisser and Eddy (1979), which is also known as the conditional predictive ordinate or CPO (Gelfand 1996). The CPO can be easily used for various purposes, such as outliers or influential observation identification and non-nested hypothesis testing. The CPO value of x_i is obtained by leaving x_i out and fitting the model to the rest of the data. It indicates the posterior probability of getting the value x_i. A high CPO value suggests better fit of the model, whereas a low value implies outlier and influential observation. Another way to get the CPO value is through Monte Carlo estimate which is the harmonic mean of the likelihood of x_i. Specifically, the CPO_i is the inverse of the posterior mean of the inverse likelihood of x_i. The Monte Carlo estimate of CPO_i is

$$\widehat{\text{CPO}}_i = \left[\frac{1}{M} \sum_{h=1}^{M} \frac{1}{f(x_i|\tilde{\theta}^{(h)})} \right]^{-1}, i = 1, 2, \ldots, n,$$

where $\tilde{\theta}^{(h)} \overset{iid}{\sim} \Pi(\tilde{\theta}|\tilde{x})$ for $h = 1, \ldots, M$; see Molina et al. (2014).

The sum of the $\log(\text{CPO}_i)$ can be an estimator for the natural logarithm of the marginal likelihood, sometimes called the log pseudo-marginal likelihood (LPML)

$$\text{LPML} = \sum_{i=1}^{n} \log(\widehat{\text{CPO}}_i).$$

Models with larger LPMLs are better. To compare the predictive distributions (both model with correction for selection bias and model without correction for selection bias) using our length-biased sample, we calculated the LPML for both models. The likelihood of x_i under correction for selection bias model is given by

$$f(x_i|\alpha, \beta, \gamma) = \frac{\gamma x_i^{\gamma\alpha}}{\beta^{\gamma\alpha+1}\Gamma(\alpha + \frac{1}{\gamma})} \exp\left[-\left(\frac{x_i}{\beta}\right)^{\gamma}\right],$$

where β is the corresponding parameter for the stratum that x_i is from. The likelihood of x_i under model without correction for selection bias is given by

$$f(x_i|\alpha, \beta, \gamma) = \frac{\gamma x_i^{\gamma\alpha-1}}{\beta^{\gamma\alpha}\Gamma(\alpha)} \exp\left[-\left(\frac{x_i}{\beta}\right)^{\gamma}\right],$$

where β is the corresponding parameter for the stratum that x_i is from. We have seen that the LPML for model with correction for selection bias is larger (LPML $=$ -36.10) than the one for the model without correction for selection bias (LPML $=$ -47.54), therefore showing that the model with correction for selection bias fits length-biased sample better.

4.3 Nonsampled Widths

We define the importance function as

$$\Pi_a(x_{n+1}, \ldots, x_N|N) = \frac{\prod_{i=n+1}^{N} f(x_i)}{\int \prod_{i=n+1}^{N} f(x_i) dx_{n+1} \ldots dx_N}. \tag{4.2}$$

Then, the importance ratios are

$$\frac{\Pi(x_{n+1}, \ldots, x_N|N)}{\Pi_a(x_{n+1}, \ldots, x_N|N)} \propto \prod_{i=n+1}^{N} \frac{1 - \frac{x_i}{W}}{1 - \frac{\mu}{W}}. \tag{4.3}$$

A random sample can now be obtained by re-sampling with probability proportional to the ratios.

The algorithm to obtain the nonsampled values is as follows.

- Step 1. Obtain M sets of $(\alpha, \beta_1, \beta_2, \beta_3, \gamma,)$ using the sampling methods described in Sect. 4.1.
- Step 2. Obtain a sample of N from formula (3.2).
- Step 3. For each set of parameters, generate the vector $\tilde{x_j}$ where $x_{ij}, i = n_j + 1, \ldots, (N_j - n_j),\ j = 1, 2, 3$ from the corresponding generalized gamma distribution.
- Step 4. Compute the population mean and the importance ratio w.
- Step 5. Repeat the step 2 to step 4 $M - 1$ times.
- Step 6. Draw ξM values of the population means with the probabilities proportional to the ratios in (4.3). We choose $\xi = 0.1$.

5 Summary

In this chapter, we have presented a model for estimating population mean under length-biased sampling. To robustify our procedure that accommodates the length-biased sampling, we have used a weighted distribution of the three-parameter generalized gamma distribution. Interest is on the finite population mean of shrub width in the entire quarry. In order to avoid certain technical issues associated with classical inference when using the generalized gamma distribution, we proposed a non-Markovian Bayesian numerical method, called random sampler, which performs better than Gibbs sampler when the population parameters are highly correlated. Posterior population distributions are easily estimated using this method. Conditional predictive ordinate shows that the model with correction for selection bias performs better than the model without correction for selection bias.

While accommodating transect sampling is a challenge, another important challenge in our procedure is to estimate the unknown population size. To ensure a full Bayesian procedure, we have used the data in Replication 2 (two strata) to estimate the finite population size that is usually unknown in this type of problem. The data from Replication 1 were used to estimate the average shrub width of the finite population. This estimate can, in turn, be used to give an estimate of the total shrub area assuming a standard geometry (e.g., a circle with the width being the diameter or a square with the width being length of a side).

An interesting topic for future research would be including covariates to study potential predictors. In Muttlak and McDonald (1990), in addition to the measurement of shrub widths, two more attributes of mountain mahogany, maximum height and number of stems, were measured. Both attributes are important predictors of the average shrub width of an area's vegetation. Semi-parametric linear regression (Chen 2010) or generalized linear regression can be considered to measure this association. We can incorporate the covariates through a gamma type regression model. Let the covariates be $\underset{\sim}{z}_{ij}$, $i = 1, 2, 3, j = n_1, n_2, n_3$ and $\underset{\sim}{\phi}_i$, $i = 1, 2, 3$. Because the mean of each stratum is linearly related to $\beta_1, \beta_2, \beta_3$, respectively, we take $\beta_i = e^{\underset{\sim}{z}_i{}' \underset{\sim}{\phi}_i}$, $i = 1, 2, 3$. For the shrub data, our model is

$$P(\underset{\sim}{x}|\underset{\sim}{z}, \underset{\sim}{\phi}, \alpha, \gamma) = \prod_{i=1}^{3} \prod_{j=1}^{n_i} \frac{\gamma x_{ij}^{\gamma\alpha-1} \left[e^{-z_{ij}\phi_i}\right]^{\gamma\alpha}}{\Gamma(\alpha)} \exp\left[-\left(x_{ij} e^{-z_{ij}\phi_i}\right)^{\gamma}\right].$$

A similar form can be easily written down for the length-biased sampling. Our future plan is to fit a model to accommodate the covariates.

Acknowledgements The authors thank the two referees for their comments. Balgobin Nandram's work was supported by a grant from the Simons Foundation (#353953, Balgobin Nandram).

Appendix 1: Uniform Boundedness of $\Delta(\theta)$

We need to show that

$$\Delta(\theta) = \tilde{\Gamma}(n\theta) - n\tilde{\Gamma}(\theta) - n\theta \ln(n)$$

is uniformly bounded in θ. We will show that $\Delta(\theta)$ is asymptotically flat.

First, differentiating $\Delta(\theta)$, we have

$$\Delta'(\theta) = n\{\psi(n\theta) - \psi(\theta) - \ln(n)\},$$

where $\psi(\cdot)$ is the digamma function. Now, using the duplication property (Abramowitz and Stegun 1965, Chap. 6) of the digamma function, one can show that $\Delta'(\theta) \geq 0$. That is, $\Delta(\theta)$ is monotonically increasing in θ; see also Fig. 5.

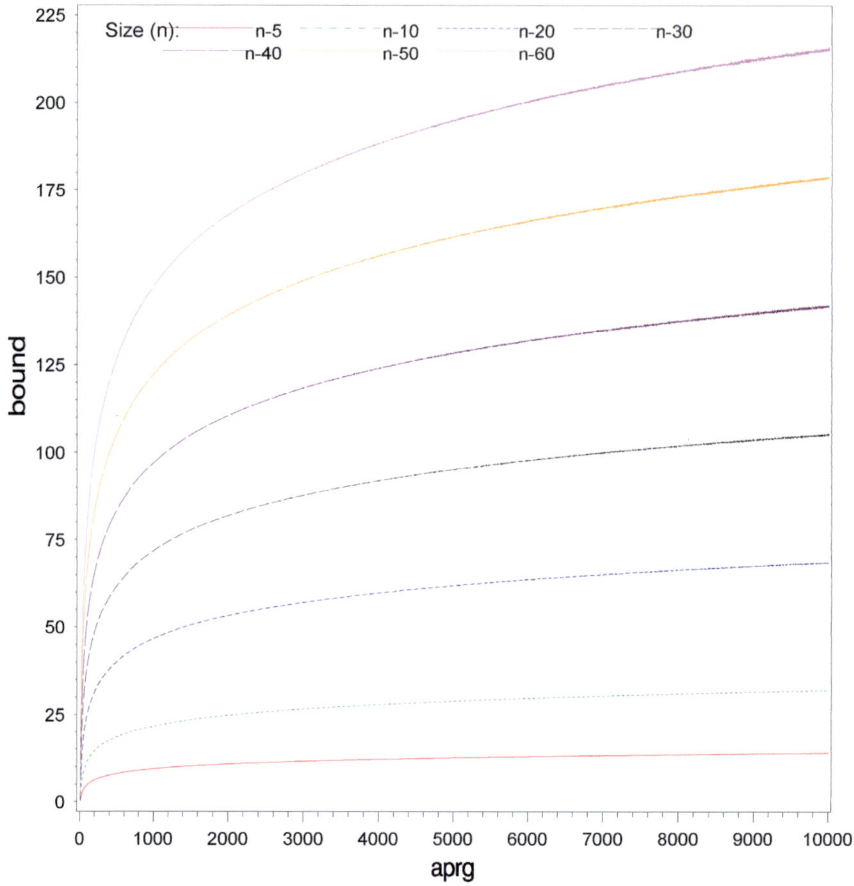

Fig. 5 Line plots of $\Delta(\theta)$ for selected sample sizes (n)

Second, differentiating $\Delta'(\theta)$, we have

$$\Delta''(\theta) = \frac{n}{\theta}\{n\theta\psi'(n\theta) - \theta\psi'(\theta)\}.$$

Using a theorem (Ronning 1986) which states that $x\psi'(x)$ decreases monotonically in x, we have $\Delta''(\theta) \leq 0$. That is, $\Delta(\theta)$ is concave and the rate of increase of $\Delta(\theta)$ decreases; see Fig. 6.

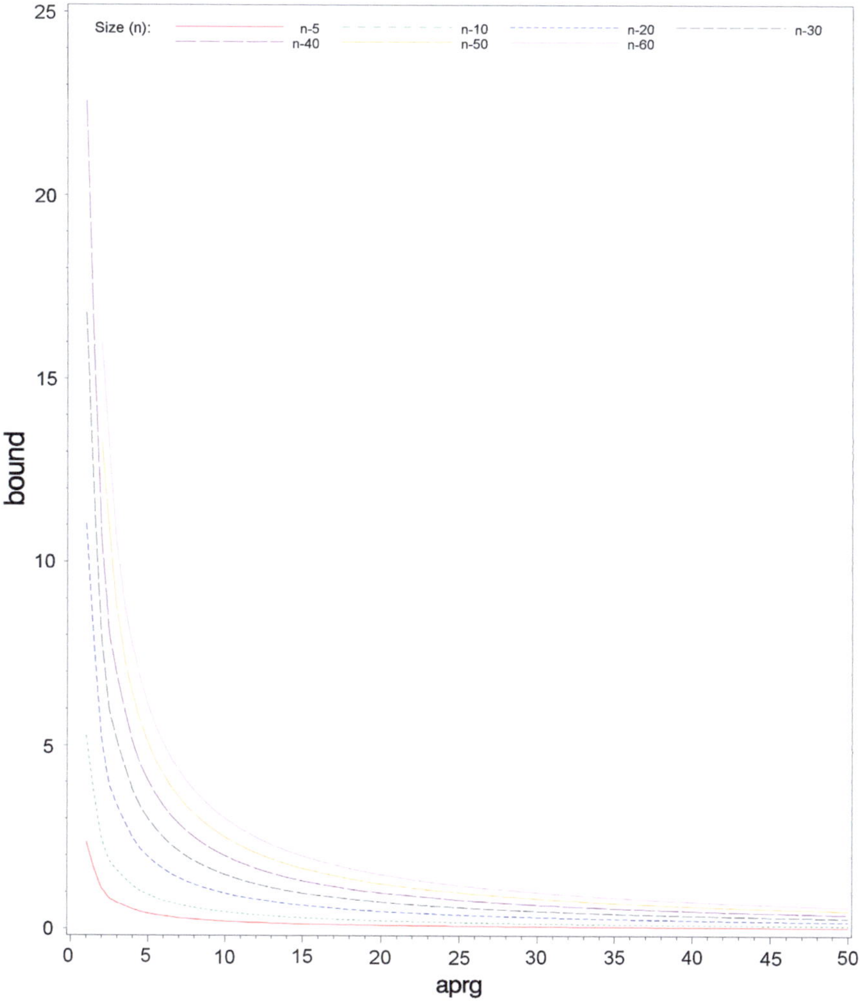

Fig. 6 Line plots of $\Delta'(\theta)$ for selected sample sizes (n)

Therefore, $\Delta(\theta)$ asymptotes out horizontally and $\Delta(\theta)$ must be bounded, so is its exponent.

References

Abramowitz, M., & Stegun, I. A. (1965). *Handbook of mathematics functions*. Mineola, NY: Dover Publications.

Butler, S. A., & McDonald, L. L. (1983). Unbiased systematic sampling plans for the line intercept method. *Journal of Range Management*, *36*, 463–468.

Chambers, R. L., & Skinner, C. J. (2003). *Analysis of survey data*. Hoboken, NJ: Wiley.

Chen, Y. Q. (2010). Semiparametric regression in size-biased sampling. *Biometrics*, *66*, 149–158.

Choi, S., Nandram, B., & Kim, D. (2017). A hierarchical Bayesian model for binary data incorporating selection bias. *Communications in Statistics: Simulation and Computation*, *46*(6), 4767–4782.

Cochran, W. G. (1977). *Sampling Techniques*. Hoboken, NJ: Wiley.

Eberhnrdt, L. L. (1978). Transect methods for population studies. *The Journal of Wildlife Management*, *42*, 1–31.

Fisher, R. A. (1934). The effects of methods of ascertainment upon the estimation of frequencies. *Annals of Eugenics*, *6*, 13–25.

Geisser, S., & Eddy, W. F. (1979). A predictive approach to model selection. *Journal of the American Statistical Association*, *74*, 153–160.

Gelfand, A. E. (1996). Model determination using sampling-based methods. *Markov Chain Monte Carlo in Practice*. London: Chapman and Hall.

Hager, H. W., & Bain, L. J. (1971). Reliability estimation for the generalized gamma distribution and robustness of the Weibull model. *Technometrics*, *13*, 547–557.

Hansen, M. M., & Hurwitz, W. N. (1943). On the theory of sampling from finite populations. *Annals of Mathematical Statistics*, *14*, 333–362.

Khodabin, M., & Ahmadabadi, A. (2010). Some properties of generalized gamma distribution. *Mathematical Sciences*, *4*, 9–28.

Lawless, J. F. (1980). Inference in the generalized gamma and log gamma distributions. *Technometrics*, *22*, 409–419.

Lucas, H. A., & Seber, G. F. (1977). Estimating coverage and particle density using the line intercept method. *Biometrika*, *64*, 618–622.

McDonald, L. L. (1980). Line-intercept sampling for attributes other than coverage and density. *Journal of Wildlife Management*, *44*, 530–533.

Molina, I., Nandram, B., & Rao, J. N. K. (2014). Small area estimation of general parameters with application to poverty indicators: A hierarchical Bayesian approah. *Annals of Applied Statistics*, *8*(2), 852–885.

Muttlak, H. A. (1988). *Some aspects of ranked set sampling with size biased probability of selection* (Ph.D. Dissertation, Department of Statistics, University of Wyoming, pp. 1-96).

Muttlak, H. A., & McDonald, L. L. (1990). Ranked set sampling with size-biased probability of selection. *Biometrics*, *46*, 435–445.

Nandram, B. (2007). Bayesian predictive inference under informative sampling via surrogate samples. In S. K. Upadhyay, U. Singh, & D. K. Dey (Eds.), *Bayesian statistics and its applications* (pp. 356-374). Anamaya, New Delhi (Chapter 25)

Nandram, B., Bhatta, D., Bhadra, D., & Shen, G. (2013). Bayesian predictive inference of a finite population proportion under selection bias. *Statistical Methodology*, *11*, 1–21.

Nandram, B., & Choi, J. W. (2010). A Bayesian analysis of body mass index data from small domains under nonignorable nonresponse and selection. *Journal of the American Statistical Association*, *105*, 120–135.

Nandram, B., Choi, J. W., Shen, G., & Burgos, C. (2006). Bayesian predictive inference under informative sampling and transformation. *Applied Stochastic Models in Business and Industry*, *22*, 559–572.

Parr, V. B., & Webster, J. T. (1965). A method for discriminating between failure density functions used in reliability predictions. *Technometrics*, *7*, 1–10.

Patil, G. P., & Rao, C. R. (1978). Weighted distributions and size-biased sampling with applications to wildlife populations and human families. *Biometrics*, *34*, 179–189.
Rao, C. R. (1965). On discrete distributions populations arising out of method of ascertainment (pp. 320–332). Pergamon Press and Statistical Publishing Society.
Ronning, G. (1986). On the curvature of the trigamma function. *Journal of Computational and Applied Mathematics*, *15*, 397–399.
Stacy, E. W. (1962). A generalization of the gamma distribution. *The Annals of Mathematical Statistics*, *33*, 1187–1192.
Sverchkov, M., & Pfeffermann, D. (2004). Prediction of finite population totals based on the sample distribution. *Survey Methodology*, *30*, 79–92.

Calibration Approach-Based Estimators for Finite Population Mean in Multistage Stratified Random Sampling

B. V. S. Sisodia and Dhirendra Singh

Abstract An effort has been made to develop calibration estimators of the population mean under two-stage stratified random sampling when auxiliary information is available at primary stage unit level. The properties of the developed estimators are derived in terms of design-based approximate variance and approximate consistent design-based estimator of the variance. Some simulation studies have been conducted to investigate the relative performance of calibration estimators over the usual estimator of the population mean without using auxiliary information in two-stage stratified random sampling. It has been found that the two-step calibration estimator has outperformed than the other calibration estimators and the usual estimator without using auxiliary information.

Keywords Auxiliary information · Calibration estimator · Two-step calibration · Two-stage stratified random sampling

1 Introduction

The use of the auxiliary information has been utilized in various ways in the past to obtain improved estimates of population parameters in finite population survey sampling. Some classical known methods are ratio- and regression-type methods of estimation (Sukhatme et al. 1984). Model-assisted approach of Cassel et al. (1976) led to generalized regression (GREG) estimator. Deville and Särndal (1992) used known population mean/total of set of auxiliary variables related to the study variate to develop calibration estimators, class of estimators appealing a common base of auxiliary information, by calibrating sampling design weights using certain calibration equations. It uses calibrated weights, which are supposed to be close to original sampling design weight $d_k = \pi_k^{-1}$, where π_k is inclusion probability of kth unit in the sample s of size n. It is given according to the given distance measure

B. V. S. Sisodia (✉) · D. Singh
Department of Agricultural Statistics, N. D. University of Agriculture and Technology, Kumarganj, Faizabad, Uttar Pradesh 224229, India
e-mail: bvssisodia@gmail.com

G. Chandra et al. (eds.), *Statistical Methods and Applications in Forestry and Environmental Sciences*, Forum for Interdisciplinary Mathematics,
https://doi.org/10.1007/978-981-15-1476-0_7

while respecting some set of constraints, known as calibration equations. The usual Horvitz–Thompson (1952) estimator of population total $t_y = \sum_{k=1}^{N} y_k$ of the study variate y is given by

$$\hat{t}_{y\pi} = \sum_{k=1}^{n} d_k y_k$$

They defined the calibration estimator of t_y as

$$\hat{t}_{yc} = \sum_{k=1}^{n} w_k y_k \tag{1.1}$$

where w_k is calibrated weight as close as possible to d_k in an average sense for a given distance measure while respecting a calibration equation

$$\sum_{k=1}^{n} w_k x_k = t_x \tag{1.2}$$

where t_x is the known population total of an auxiliary variable x related to y. They used the chi-square distance measure $\sum_{k=1}^{n} \left((w_k - d_k)^2\right)/d_k q_k$, where $\frac{1}{q_k}$ is some known positive weight unrelated to d_k. This distance measure is minimized with respect to w_k subject to the calibration Eq. (1.2), which gives w_k as

$$w_k = d_k \left[1 + \frac{x_k q_k}{\sum_{k=1}^{n} d_k x_k q_k} \left(t_x - \sum_{k=1}^{n} d_k x_k \right) \right]$$

The calibration estimator $\hat{t}_{yc}$ in Eq. (1.1) can explicitly be expressed as

$$\hat{t}_{yc} = \hat{t}_{y\pi} + \widehat{B}(t_x - \hat{t}_{x\pi}) \tag{1.3}$$

where $\widehat{B} = \sum_{k=1}^{n} d_k q_k x_k y_k / \sum_{k=1}^{n} d_k q_k x_k^2$ is a weighted regression coefficient of y on x, and $\hat{t}_{x\pi}$ is Horvitz–Thompson estimator of t_x. In fact, the estimator in Eq. (1.3) is a GREG estimator (Cassel et al. 1976). Thus, calibration technique provides an alternative derivation of GREG estimator, and it can be viewed as linear weighting of y_k with weight which is sample dependent. The variance and estimator of variance of $\hat{t}_{yc}$ are given by

$$V(\hat{t}_{yc}) = \sum_{k=1}^{N} \sum_{l=1}^{N} \Delta_{kl} (d_k E_k)(d_l E_l)$$

$$\widehat{V}(\hat{t}_{yc}) = \sum_{k=1}^{n} \sum_{l=1}^{n} \left(\frac{\Delta_{kl}}{\pi_{kl}} \right) (w_k e_k)(w_l e_l)$$

where $\Delta_{kl} = \pi_{kl} - \pi_k\pi_l$, $E_k = y_k - Bx_k, e_k = y_k - \widehat{B}x_k$, $B = \sum_{k=1}^{N} d_k q_k x_k y_k / \sum_{k=1}^{N} d_k q_k x_k^2$.

The most advantage of calibration approach is that it is model free, whereas GREG is based on model-assisted approach. Deville and Särndal (1992) further argued in support of calibration approach that if $\sum_{k=1}^{n} w_k x_k$ provides exact value of the population total t_x, then $\sum_{k=1}^{n} w_k y_k$ may be more precise than $\sum_{k=1}^{n} d_k y_k$ in practice if y and x are strongly correlated either positive or negative.

Following Deville and Särndal (1992), various researchers have contributed substantially in calibration approach-based estimation. Estevao and Särndal (2006) and Särndal (2007) have dealt with survey estimates by calibration on complex auxiliary variables where they have outlined calibration estimation in two-stage sampling in one of the sections. Aditya et al. (2016) developed calibration-based regression-type estimator in two-stage sampling when auxiliary information is available at primary stage unit (psu) level. Mourya et al. (2016a, b) have also dealt with calibration estimation in two-stage sampling and cluster sampling when auxiliary information is available at secondary stage unit (ssu) level for selected psu(s). Various researchers have extended the idea of calibration approach-based estimation to stratified random sampling, where they have calibrated the stratum weights using the auxiliary information. Notably, among them are Singh et al. (1998, 1999), Tracy et al. (2003), Kim et al. (2007), and Nidhi et al. (2016). In fact, most of the large-scale sample surveys are conducted in multistage stratified random sampling.

Recently, Singh et al. (2017) dealt with calibration estimators in two-stage stratified random sampling when the auxiliary information is available at ssu level for the selected psu(s) level. The psu(s) level auxiliary information can easily be obtained rather than ssu(s) level. For example, in surveys for expenditure and income estimates, households are generally ssu(s), and villages consisting of households in a given district/state are psu(s). Tehsil (Taluka) in the districts or divisions (Mandals) of the state can be strata. The auxiliary information on several auxiliary variables at village level such as population through census, land-holding through agriculture census, etc., can easily be obtained. Obviously, when psu level auxiliary information is available, the total of auxiliary information for the population is automatically known.

In forest surveys, someone may be interested to estimate the population of active caterpillar or estimation of proportion of trees infected by the caterpillars in order to find out the density and distribution of caterpillars over the tree stand. In such surveys, forest divisions and forest ranges within forest division can be psu and ssu level units, respectively. The forest divisions can further be grouped to form the strata. The forest area/density in given forest division is generally known which can be used as known auxiliary information at psu level.

2 Notations Used

Let the population of elements $U = (1, 2, \ldots, k, \ldots, N)$ is partitioned into $U_I = (U_1, U_2, \ldots, U_i, \ldots, U_{N_I})$ psu's. N_i is the size of U_i, $U = \sum_{i=1}^{N_I} U_i$, and $N = \sum_{i=1}^{N_I} N_i$. Let the population of psu's U_I is stratified into G strata. Suppose gth stratum consists of N_g psu's such that $\sum_{g=1}^{G} N_g = N_I$. Let N_{gi} is the number of ssu of ith psu in gth stratum $(i = 1, 2, \ldots, N_g)$, such that $N_{go} = \sum_{i=1}^{N_g} N_{gi}$, the total number of elements in gth stratum. Let the population of N_g psu's in the gth stratum is denoted by $U_g = (U_{g1}, U_{g2}, \ldots, U_{gi}, \ldots, U_{gN_g})$. We further define

$\overline{N}_{go} = \frac{N_{go}}{N_g}$, average number of ssu per psu in the gth stratum
t_{ygik} = value of y corresponding to kth element of ith psu in the gth stratum
$t_{ygi} = \sum_{k=1}^{N_{gi}} t_{ygik}$, total of y in ith psu of gth stratum, $\bar{t}_{ygi} = \frac{1}{N_{gi}} \sum_{k=1}^{N_{gi}} t_{ygik}$, mean per ssu in ith psu
$t_{yg} = \sum_{i=1}^{N_g} \sum_{k=1}^{N_{gi}} t_{ygik}$, total of y in the gth stratum, $\tilde{t}_{yg} = \frac{1}{N_g} \sum_{i=1}^{N_g} t_{ygi}$, average total of y per psu
$\bar{t}_{yg} = \frac{t_{yg}}{N_g \overline{N}_{go}} = \frac{1}{N_g} \sum_{i=1}^{N_g} \frac{N_{gi}}{\overline{N}_{go}} \bar{t}_{ygi}$, mean of y per ssu in gth stratum.

3 Mean Estimator in Two-Stage Stratified Random Sampling

Suppose at the first stage, a random sample s_g of n_g psu's from N_g psu's is drawn according to sampling design $P_g(.)$ with the inclusion probabilities π_{gi} and π_{gij} at psu level in gth stratum. At second stage, we draw a random sample s_i of size n_i elements from the selected ith psu in gth stratum $(i = 1, 2, \ldots, n_g)$ according to design $P_i(\cdot)$ with inclusion probabilities $\pi_{gk/i}$ and $\pi_{gkl/i}$ of kth and lth elements. We also define

$$\Delta_{gij} = \pi_{gij} - \pi_{gi}\pi_{gj} \text{ with } \widetilde{\Delta}_{gij} = \Delta_{gij}/\pi_{gij} \text{ and}$$
$$\Delta_{gkl/i} = \pi_{gkl/i} - \pi_{gk/i}\pi_{gl/i}, \text{with } \widetilde{\Delta}_{gkl/i} = \frac{\Delta_{gkl/i}}{\pi_{gkl/i}}$$

The objective is to estimate the population mean

$$\bar{t}_y = \frac{1}{N} \sum_{g=1}^{G} \sum_{i=1}^{N_g} \sum_{k=1}^{N_{gi}} t_{ygik} = \sum_{g=1}^{G} \Omega_g \bar{t}_{yg},$$

where the stratum weight $\Omega_g = \frac{N_{go}}{N}$, such that $\sum_{g=1}^{G} \Omega_g = 1$.

Different estimators of $\bar{t}_y$ under the above sampling design have been developed as follows.

3.1 Estimator Without Using Auxiliary Information

The usual Horvitz–Thompson estimator of $\bar{t}_{yg}$ is given by

$$\hat{\bar{t}}_{yg(\mathrm{HT})} = \frac{1}{N_g\overline{N}_{go}}\sum_{i=1}^{n_g}\sum_{k=1}^{n_i} a_{gi}a_{gk/i}t_{ygik} = \frac{1}{N_g\overline{N}_{go}}\sum_{i=1}^{n_g} a_{gi}\hat{t}_{ygi(\mathrm{HT})}$$
$$= \frac{1}{N_g}\sum_{i=1}^{n_g} a_{gi}\frac{N_{gi}}{\overline{N}_{go}}\hat{\bar{t}}_{ygi(\mathrm{HT})} = \frac{\hat{t}_{yg(\mathrm{HT})}}{N_g\overline{N}_{go}} \tag{3.1}$$

where $\hat{t}_{yg(\mathrm{HT})} = \sum_{i=1}^{n_g} a_{gi}\hat{t}_{ygi(\mathrm{HT})}$ and $\hat{\bar{t}}_{ygi(\mathrm{HT})} = \frac{1}{N_{gi}}\sum_{i=1}^{n_i} a_{gk/i}t_{ygik}$ are the Horvitz–Thompson estimators of t_{yg} and $\bar{t}_{ygi}$, respectively, with $a_{gi} = \frac{1}{\pi_{gi}}$ and $a_{gk/i} = \frac{1}{\pi_{gk/i}}$.

The variance of $\hat{\bar{t}}_{yg(\mathrm{HT})}$ can be written as sum of two components as per Särndal et al. (2003)

$$V(\hat{\bar{t}}_{yg(\mathrm{HT})}) = \frac{V_{\mathrm{psu}} + V_{\mathrm{ssu}}}{N_g^2\overline{N}_{go}^2} \tag{3.2}$$

where $V_{\mathrm{psu}} = \sum\sum_{U_g}\Delta_{gij}\frac{t_{ygi}}{\pi_{gi}}\frac{t_{ygj}}{\pi_{gj}}$, $V_{\mathrm{ssu}} = \sum_{U_g}\frac{V_i}{\pi_{gi}}$, and $V_i = \sum\sum_{U_{gi}}\Delta_{gkl/i}\frac{t_{ygik}}{\pi_{gk/i}}\frac{t_{ygil}}{\pi_{gl/i}}$.

The first component V_{psu} is unbiasedly estimated by

$$\widehat{V}_{\mathrm{psu}} = \sum\sum_{s_g}\widetilde{\Delta}_{gij}\frac{\hat{t}_{ygi}}{\pi_{gi}}\frac{\hat{t}_{ygj}}{\pi_{gj}} - \sum_{s_g}\frac{1}{\pi_{gi}}\left(\frac{1}{\pi_{gi}} - 1\right)\widehat{V}_i$$

where $\widehat{V}_i = \sum\sum_{s_i}\widetilde{\Delta}_{gkl/i}\frac{t_{ygik}}{\pi_{gk/i}}\frac{t_{ygil}}{\pi_{gl/i}}$.

The second component V_{ssu} is unbiasedly estimated by

$$\widehat{V}_{\mathrm{ssu}} = \sum_{s_g}\frac{\widehat{V}_i}{\pi_{gi}^2}$$

Therefore,

$$\widehat{V}(\hat{\bar{t}}_{yg(\mathrm{HT})}) = \frac{\widehat{V}_{\mathrm{psu}} + \widehat{V}_{\mathrm{ssu}}}{N_g^2\overline{N}_{go}^2} = \frac{1}{N_g^2\overline{N}_{go}^2}\left(\sum\sum_{s_g}\widetilde{\Delta}_{gij}\frac{\hat{t}_{ygi}}{\pi_{gi}}\frac{\hat{t}_{ygj}}{\pi_{gj}} + \sum_{s_g}\frac{\widehat{V}_i}{\pi_{gi}}\right) \tag{3.3}$$

Now, the estimator of $\bar{t}_y$ and $V(\hat{\bar{t}}_y)$ using Eq. (3.2) in two-stage stratified random sampling is given by

$$\hat{\bar{t}}_y = \sum_{g=1}^{G} \Omega_g \hat{\bar{t}}_{yg(\text{HT})}$$

$$V(\hat{\bar{t}}_y) = \sum_{g=1}^{G} \Omega_g^2 V(\hat{\bar{t}}_{yg(\text{HT})})$$

Now, using Eq. (3.3), the estimator of $V(\hat{\bar{t}}_y)$ is

$$\widehat{V}(\hat{\bar{t}}_y) = \sum_{g=1}^{G} \Omega_g^2 \widehat{V}(\hat{\bar{t}}_{yg(\text{HT})}),$$

Let SRSWOR is denoted as SI. Therefore, the estimator $\hat{\bar{t}}_{yg(\text{SI})}$ under SRSWOR is given by

$$\hat{\bar{t}}_{yg(\text{SI})} = \frac{1}{n_g} \sum_{i=1}^{n_g} \frac{N_{gi}}{\overline{N}_{go}} \hat{\bar{t}}_{ygi}, \quad \text{where } \hat{\bar{t}}_{ygi} = \frac{1}{n_i} \sum_{k=1}^{n_i} t_{ygik} \tag{3.4}$$

with

$$V\left(\hat{\bar{t}}_{yg(\text{SI})}\right) = \frac{N_g - n_g}{n_g N_g} S_{byg}^2 + \frac{1}{n_g N_g} \sum_{i=1}^{N_g} \left(\frac{N_{gi}}{\overline{N}_{go}}\right)^2 \frac{(N_{gi} - n_i)}{n_i N_{gi}} S_{ygi}^2 \tag{3.5}$$

where

$$S_{byg}^2 = \frac{1}{N_g - 1} \sum_{i=1}^{N_g} \left(\frac{N_{gi}}{\overline{N}_{go}} \bar{t}_{ygi} - \bar{t}_{yg}\right)^2 \quad \text{and} \quad S_{ygi}^2 = \frac{1}{N_{gi} - 1} \sum_{k=1}^{N_{gi}} \left(t_{ygik} - \bar{t}_{ygi}\right)^2.$$

The unbiased variance estimator is given by

$$\widehat{V}\left(\hat{\bar{t}}_{yg(\text{SI})}\right) = \frac{N_g - n_g}{n_g N_g} s_{byg}^2 + \frac{1}{n_g N_g} \sum_{i=1}^{n_g} \frac{(N_{gi} - n_i)}{n_i N_{gi}} s_{ygi}^2 \tag{3.6}$$

where

$$s_{byg}^2 = \frac{1}{n_g - 1} \sum_{i=1}^{n_g} \left(\frac{N_{gi}}{\overline{N}_{go}} \hat{\bar{t}}_{ygi} - \hat{\bar{t}}_{yg}\right)^2, \quad s_{ygi}^2 = \frac{1}{n_i - 1} \sum_{k=1}^{n_i} \left(\frac{N_{gi}}{\overline{N}_{go}} t_{ygik} - \tilde{t}_{ygi}\right)^2,$$

$$\tilde{t}_{ygi} = \frac{1}{n_i} \sum_{k=1}^{n_i} \frac{N_{gi}}{\overline{N}_{go}} t_{ygik}.$$

Using Eq. (3.4), the estimator $\hat{\bar{t}}_y$ in SRSWOR can be expressed as

$$\hat{\bar{t}}_{y(\text{SI})} = \sum_{g=1}^{G} \Omega_g \hat{\bar{t}}_{yg(\text{SI})}$$

Using Eq. (3.5), we have

$$V\left(\hat{\bar{t}}_{y(\text{SI})}\right) = \sum_{g=1}^{G} \Omega_g^2 V\left(\hat{\bar{t}}_{yg(\text{SI})}\right)$$

The unbiased variance estimator using Eq. (3.6) is given by

$$\widehat{V}\left(\hat{\bar{t}}_{y(\text{SI})}\right) = \sum_{g=1}^{G} \Omega_g^2 \widehat{V}\left(\hat{\bar{t}}_{yg(\text{SI})}\right).$$

3.2 Calibration Estimator Using Auxiliary Information at psu Level

3.2.1 Calibration of Design Weight at psu Level

Consider that the auxiliary information t_{xgi} related to the study variate y is available at psu level corresponding to ith psu ($i = 1, 2, \ldots, N_g$) in gth stratum. It means $t_{xg} = \sum_{i=1}^{N_g} t_{xgi}$, the population total for x in gth stratum is known; therefore, the total for whole population, $t_x = \sum_{g=1}^{G} t_{xg}$, is also known.

Using the calibration approach, we calibrate the design weight a_{gi} in Eq. (3.1), and then, the calibration estimator of $\bar{t}_{yg}$ with the calibrated weight w_{gi} is given by

$$\hat{\bar{t}}_{yg}^{c} = \frac{1}{N_g \overline{N}_{go}} \sum_{i=1}^{n_g} w_{gi} \hat{t}_{ygi(\text{HT})} \tag{3.7}$$

To find w_{gi}, we minimize the chi-square distance measure $\sum_{i=1}^{n_g} \left(w_{gi} - a_{gi}\right)^2 / a_{gi} q_{gi}$ subject to the constraints

$$\frac{1}{N_g\overline{N}_{go}}\sum_{i=1}^{n_g} w_{gi}t_{xgi} = \frac{1}{N_g\overline{N}_{go}}\sum_{i=1}^{N_g} t_{xgi} = \bar{t}_{xg} \quad \text{and} \quad \sum_{i=1}^{n_g} a_{gi} = \sum_{i=1}^{n_g} w_{gi},$$

where q_{gi} is some scalar quantity.

Therefore, minimizing the following function with respect to w_{gi}

$$\phi(w_{gi},\lambda) = \frac{\sum_{i=1}^{n_g}\left(w_{gi}-a_{gi}\right)^2}{a_{gi}q_{gi}} - 2\lambda_1\left(\frac{1}{N_g\overline{N}_{go}}\sum_{i=1}^{n_g} w_{gi}t_{xgi} - \bar{t}_{xg}\right) - 2\lambda_2\left(\sum_{i=1}^{n_g} w_{gi} - \sum_{i=1}^{n_g} a_{gi}\right),$$

where λ_1 and λ_2 are Lagrange multipliers which give

$$w_{gi} = a_{gi} + \frac{N_g\overline{N}_{go}a_{gi}q_{gi}\left(\bar{t}_{xg} - \frac{\sum_{i=1}^{n_g} a_{gi}t_{xgi}}{N_g\overline{N}_{go}}\right)}{\sum_{i=1}^{n_g} a_{gi}q_{gi}t_{xgi}^2 - \frac{\left(\sum_{i=1}^{n_g} a_{gi}q_{gi}t_{xgi}\right)^2}{\sum_{i=1}^{n_g} a_{gi}q_{gi}}}\left[t_{xgi} - \frac{\sum_{i=1}^{n_g} a_{gi}q_{gi}t_{xgi}}{\sum_{i=1}^{n_g} a_{gi}q_{gi}}\right]$$

Putting the value of w_{gi} in Eq. (3.7), we have

$$\hat{\bar{t}}_{yg}^{c} = \hat{\bar{t}}_{yg(\mathrm{HT})} + \widehat{B}\left(\bar{t}_{xg} - \hat{\bar{t}}_{xg(\mathrm{HT})}\right)$$

where

$$\widehat{B} = \frac{\sum_{i=1}^{n_g} a_{gi}q_{gi}\hat{t}_{ygi(\mathrm{HT})}t_{xgi} - \sum_{i=1}^{n_g} a_{gi}q_{gi}\hat{t}_{ygi(\mathrm{HT})}\frac{\sum_{i=1}^{n_g} a_{gi}q_{gi}t_{xgi}}{\sum_{i=1}^{n_g} a_{gi}q_{gi}}}{\sum_{i=1}^{n_g} a_{gi}q_{gi}t_{xgi}^2 - \frac{\left(\sum_{i=1}^{n_g} a_{gi}q_{gi}t_{xgi}\right)^2}{\sum_{i=1}^{n_g} a_{gi}q_{gi}}}$$

and $\hat{\bar{t}}_{xg(\mathrm{HT})} = \frac{\sum_{i=1}^{n_g} a_{gi}t_{xgi}}{N_g\overline{N}_{go}}$ is an estimator of $\bar{t}_{xg}$.

For $q_{gi}=1$, $\hat{\bar{t}}_{yg}^{c}$ reduces to

$$\hat{\bar{t}}_{yg}^{c} = \hat{\bar{t}}_{yg(\mathrm{HT})} + \widetilde{B}\left(\bar{t}_{xg} - \hat{\bar{t}}_{xg(\mathrm{HT})}\right),$$

where

$$\widetilde{B} = \frac{\sum_{i=1}^{n_g} a_{gi}\hat{t}_{ygi(\mathrm{HT})}t_{xgi} - \sum_{i=1}^{n_g} a_{gi}\hat{t}_{ygi(\mathrm{HT})}\frac{\sum_{i=1}^{n_g} a_{gi}t_{xgi}}{\sum_{i=1}^{n_g} a_{gi}}}{\sum_{i=1}^{n_g} a_{gi}t_{xgi}^2 - \frac{\left(\sum_{i=1}^{n_g} a_{gi}t_{xgi}\right)^2}{\sum_{i=1}^{n_g} a_{gi}}} \tag{3.8}$$

Following Särndal et al. (2003) and Aditya et al. (2016), the variance of $\hat{\bar{t}}^{c}_{yg}$ for $q_{gi} = 1$ can be obtained as

$$V\left(\hat{\bar{t}}^{c}_{yg}\right) = \frac{1}{N_g^2 \overline{N}_{go}^2}\left[\sum_{i=1}^{N_g}\sum_{j=1}^{N_g} \Delta_{gij}\frac{U_{gi}U_{gj}}{\pi_{gi}\pi_{gj}} + \sum_{i=1}^{N_g}\frac{1}{\pi_{gi}}\sum_{k=1}^{N_{gi}}\sum_{l=1}^{N_{gi}} \Delta_{gkl/i}\frac{t_{ygik}t_{ygil}}{\pi_{gk/i}\pi_{gl/i}}\right] \tag{3.9}$$

where

$$U_{gi} = t_{ygi} - Bt_{xgi} \quad \text{and} \quad B = \frac{\sum_{i=1}^{N_g} a_{gi}t_{ygi}t_{xgi} - \sum_{i=1}^{N_g} a_{gi}t_{ygi}\frac{\sum_{i=1}^{N_g} a_{gi}t_{xgi}}{\sum_{i=1}^{N_g} a_{gi}}}{\sum_{i=1}^{N_g} a_{gi}t_{xgi}^2 - \frac{\left(\sum_{i=1}^{N_g} a_{gi}t_{xgi}\right)^2}{\sum_{i=1}^{N_g} a_{gi}}}$$

Similarly, the estimator of variance of $\hat{\bar{t}}^{c}_{yg}$ is given by

$$\hat{V}\left(\hat{\bar{t}}^{c}_{yg}\right) = \frac{1}{N_g^2 \bar{N}_{go}^2}\left[\frac{1}{2}\sum_{i=1}^{n_g}\sum_{j=1}^{n_g}\left(-\tilde{\Delta}_{gij}\right)\left(w_{gi}u_{gi} - w_{gj}u_{gj}\right)^2 \right.$$
$$\left. + \frac{1}{2}\sum_{i=1}^{n_g}\frac{1}{\pi_{gi}^2}\sum_{k=1}^{n_i}\sum_{l=1}^{n_i}\left(-\tilde{\Delta}_{gkl/i}\right)\left(\frac{t_{ygik}}{\pi_{gk/i}} - \frac{t_{ygil}}{\pi_{gl/i}}\right)^2\right] \tag{3.10}$$

where $u_{gi} = \hat{t}_{ygi(\mathrm{HT})} - \widetilde{B}t_{xgi}$ and $\widetilde{B}$ is given by in Eq. (3.8).

Now, the calibration estimator of $\bar{t}_y$ in two-stage stratified random sampling is given by

$$\hat{\bar{t}}^{c1}_{y} = \sum_{g=1}^{G} \Omega_g \hat{\bar{t}}^{c}_{yg}$$

Using Eq. (3.9), we have

$$V\left(\hat{\bar{t}}^{c1}_{y}\right) = \sum_{g=1}^{G} \Omega_g^2 V\left(\hat{\bar{t}}^{c}_{yg}\right)$$

Using Eq. (3.10), the estimate of $V\left(\hat{\bar{t}}^{c1}_{y}\right)$ is given by

$$\widehat{V}\left(\hat{\bar{t}}^{c1}_{y}\right) = \sum_{g=1}^{G} \Omega_g^2 \widehat{V}\left(\hat{\bar{t}}^{c}_{yg}\right)$$

If sampling design is SRSWOR, then $\hat{\bar{t}}^{c}_{yg}$ for $q_{gi} = 1$ reduces to

$$\hat{\bar{t}}^{c}_{yg(\mathrm{SI})} = \frac{1}{n_g}\sum_{i=1}^{n_g}\frac{N_{gi}}{\overline{N}_{go}}\hat{\bar{t}}_{ygi} + \widetilde{B}_{(\mathrm{SI})}\left(\bar{t}_{xg} - \frac{1}{n_g}\sum_{i=1}^{n_g}\frac{N_{gi}}{\overline{N}_{go}}\bar{t}_{xgi}\right), \quad \text{where}$$

$$\widetilde{B}_{(\mathrm{SI})} = \frac{n_g\sum_{i=1}^{n_g}\hat{t}_{ygi}t_{xgi} - \sum_{i=1}^{n_g}\hat{t}_{ygi}\sum_{i=1}^{n_g}t_{xgi}}{n_g\sum_{i=1}^{n_g}t_{xgi}^2 - \left(\sum_{i=1}^{n_g}t_{xgi}\right)^2},$$

which is the estimated regression coefficient of t_{ygi} on t_{xgi} in gth stratum.

The variance of $\hat{\bar{t}}^{c}_{yg(\mathrm{SI})}$ is given by

$$V\left(\hat{\bar{t}}^{c}_{yg(\mathrm{SI})}\right) = \frac{1}{N_g^2\bar{N}_{go}^2}\left[\frac{(N_g - n_g)}{n_g(N_g - 1)}\sum_{i=1}^{N_g}\sum_{j=1}^{N_g}\left(-U_{gi}U_{gj}\right) - \frac{N_g}{n_g}\sum_{i=1}^{N_g}\frac{(N_{gi} - n_i)}{n_i(N_{gi} - 1)}\sum_{k=1}^{N_{gi}}\sum_{l=1}^{N_{gi}}t_{ygik}t_{ygil}\right] \tag{3.11}$$

where

$$U_{gi} = t_{ygi} - B_{(\mathrm{SI})}t_{xgi} \quad \text{and} \quad B_{(\mathrm{SI})} = \frac{N_g\sum_{i=1}^{N_g}t_{ygi}t_{xgi} - \sum_{i=1}^{N_g}t_{ygi}\sum_{i=1}^{N_g}t_{xgi}}{N_g\sum_{i=1}^{N_g}t_{xgi}^2 - \left(\sum_{i=1}^{N_g}t_{xgi}\right)^2}.$$

Similarly, the estimator of $V\left(\hat{\bar{t}}^{c}_{yg(\mathrm{SI})}\right)$ is given by

$$\widehat{V}\left(\hat{\bar{t}}^{c}_{yg(\mathrm{SI})}\right) = \frac{1}{N_g^2\overline{N}_{go}^2}\left[\frac{1}{2}\frac{(N_g - n_g)}{N_g(n_g - 1)}\sum_{i=1}^{n_g}\sum_{j=1}^{n_g}\left(w_{gi}u_{gi} - w_{gj}u_{gj}\right) + \frac{1}{2}\frac{N_g^2}{n_g^2}\sum_{i=1}^{n_g}\frac{N_{gi}(N_{gi} - n_i)}{n_i^2(n_i - 1)}\sum_{k=1}^{n_i}\sum_{l=1}^{n_i}\left(t_{ygik} - t_{ygil}\right)^2\right] \tag{3.12}$$

where

$$u_{gi} = \hat{t}_{ygi} - \widehat{B}_{(\mathrm{SI})}t_{xgi} \quad \text{and} \quad \widehat{B}_{(\mathrm{SI})} = \frac{\sum_{i=1}^{n_g}\hat{t}_{ygi}t_{xgi} - \frac{\sum_{i=1}^{n_g}\hat{t}_{ygi}\sum_{i=1}^{n_g}t_{xgi}}{n_g}}{\sum_{i=1}^{n_g}t_{xgi}^2 - \frac{\left(\sum_{i=1}^{n_g}t_{xgi}\right)^2}{n_g}}.$$

Under SRSWOR and for $q_{gi} = 1$, the calibration estimator of $\bar{t}_y$ in two-stage stratified random sampling is

$$\hat{\bar{t}}^{c1}_{y(\mathrm{SI})} = \sum_{g=1}^{G} \Omega_g \hat{\bar{t}}^{c}_{yg(\mathrm{SI})}$$

Using Eq. (3.11), we have

$$V\left(\hat{\bar{t}}^{c1}_{y(\mathrm{SI})}\right) = \sum_{g=1}^{G} \Omega_g^2 V\left(\hat{\bar{t}}^{c}_{yg(\mathrm{SI})}\right).$$

Using Eq. (3.12), the estimate of $V\left(\hat{\bar{t}}^{c1}_{y(\mathrm{SI})}\right)$ is

$$\widehat{V}\left(\hat{\bar{t}}^{c1}_{y(\mathrm{SI})}\right) = \sum_{g=1}^{G} \Omega_g^2 \widehat{V}\left(\hat{\bar{t}}^{c}_{yg(\mathrm{SI})}\right).$$

3.2.2 Calibration of Stratum Weights

The estimator of the population mean $\bar{t}_y$ has been developed in two-stage stratified random sampling in Sect. 3.1 without using auxiliary information, and it is again reproduced here

$$\hat{\bar{t}}_y = \sum_{g=1}^{G} \Omega_g \hat{\bar{t}}_{yg(\mathrm{HT})},$$

where $\hat{\bar{t}}_{yg(\mathrm{HT})}$ is given in Eq. (3.1).

We do not use the suffix HT onward for simplicity of expression. Here, we wish to calibrate Ω_g, and hence, the calibrated estimator of $\hat{\bar{t}}_y$ is given by

$$\hat{\bar{t}}^{c2}_y = \sum_{g=1}^{G} \Omega'_g \hat{\bar{t}}_{yg}, \quad \text{where} \ \ \Omega'_g \text{ is the calibrated weight} \tag{3.13}$$

We find out the Ω'_g by minimizing the chi-square distance measure $\sum_{g=1}^{G} \frac{(\Omega'_g - \Omega_g)^2}{q_g \Omega_g}$ subject to the constraints

$$\sum_{g=1}^{G} \Omega'_g \hat{\bar{t}}_{xg} = \bar{t}_x = \sum_{g=1}^{G} \Omega_g \bar{t}_{xg} \quad \text{and} \quad \sum_{g=1}^{G} \Omega'_g = 1$$

where $\hat{\hat{t}}_{xg} = \frac{1}{N_g \bar{N}_{go}} \sum_{i=1}^{n_g} \sum_{k=1}^{n_i} a_{gi} a_{gk/i} t_{xgik}$, $\bar{t}_x = \frac{t_x}{N}$, and q_g is the some scalar quantity.

Therefore, minimizing the following function with respect to Ω'_g

$$\phi(\Omega'_g, \lambda) = \sum_{g=1}^{G} \frac{(\Omega'_g - \Omega_g)^2}{q_g \Omega_g} - 2\lambda_1 \left(\sum_{g=1}^{G} \Omega'_g \hat{\hat{t}}_{xg} - \sum_{g=1}^{G} \Omega_g \bar{t}_{xg} \right) - 2\lambda_2 \left(\sum_{g=1}^{G} \Omega'_g - 1 \right)$$

where λ_1 and λ_2 are Lagrange multipliers, we get

$$\Omega'_g = \Omega_g + q_g \Omega_g \left[\frac{\hat{\hat{t}}_{xg} \left(\sum_{g=1}^{G} \Omega_g \bar{t}_{xg} - \sum_{g=1}^{G} \Omega_g \hat{\hat{t}}_{xg} \right)}{\sum_{g=1}^{G} q_g \Omega_g \hat{\hat{t}}^2_{xg} - \frac{\left(\sum_{g=1}^{G} q_g \Omega_g \hat{\hat{t}}_{xg} \right)^2}{\sum_{g=1}^{G} q_g \Omega_g}} \right.$$
$$\left. - \frac{\sum_{g=1}^{G} q_g \Omega_g \hat{\hat{t}}_{xg} \left(\sum_{g=1}^{G} \Omega_g \bar{t}_{xg} - \sum_{g=1}^{G} \Omega_g \hat{\hat{t}}_{xg} \right)}{\sum_{g=1}^{G} q_g \Omega_g \left(\sum_{g=1}^{G} q_g \Omega_g \hat{\hat{t}}^2_{xg} \right) - \left(\sum_{g=1}^{G} q_g \Omega_g \hat{\hat{t}}_{xg} \right)^2} \right]$$

Putting the value of Ω'_g in Eq. (3.13), we get calibration estimator as follows:

$$\hat{\hat{t}}_y^{c2} = \sum_{g=1}^{G} \Omega'_g \hat{\hat{t}}_{yg} = \hat{\hat{t}}_y + \hat{B}_1 \left(\bar{t}_x - \hat{\hat{t}}_x \right)$$

where

$$\hat{B}_1 = \left[\frac{\left(\sum_{g=1}^{G} q_g \Omega_g \right) \left(\sum_{g=1}^{G} q_g \Omega_g \hat{\hat{t}}_{xg} \hat{\hat{t}}_{yg} \right) - \left(\sum_{g=1}^{G} q_g \Omega_g \hat{\hat{t}}_{xg} \right) \left(\sum_{g=1}^{G} q_g \Omega_g \hat{\hat{t}}_{yg} \right)}{\left(\sum_{g=1}^{G} q_g \Omega_g \right) \left(\sum_{g=1}^{G} q_g \Omega_g \hat{\hat{t}}^2_{xg} \right) - \left(\sum_{g=1}^{G} q_g \Omega_g \hat{\hat{t}}_{xg} \right)^2} \right],$$

$$\hat{\hat{t}}_y = \sum_{g=1}^{G} \Omega_g \hat{\hat{t}}_{yg}, \ \bar{t}_x = \sum_{g=1}^{G} \Omega_g \bar{t}_{xg} \quad \text{and} \quad \hat{\hat{t}}_x = \sum_{g=1}^{G} \Omega_g \hat{\hat{t}}_{xg}.$$

For $q_g = 1$, the estimator $\hat{t}_y^{c2}$ reduces to

$$\hat{\hat{t}}_y^{c2} = \hat{\hat{t}}_y + \hat{B}'_1 \left(\bar{t}_x - \hat{\hat{t}}_x \right)$$

where

$$\hat{B}'_1 = \frac{\sum_{g=1}^{G} \Omega_g \hat{\hat{t}}_{yg} \hat{\hat{t}}_{xg} - \left(\sum_{g=1}^{G} \Omega_g \hat{\hat{t}}_{yg} \right) \left(\sum_{g=1}^{G} \Omega_g \hat{\hat{t}}_{xg} \right)}{\sum_{g=1}^{G} \Omega_g \hat{\hat{t}}^2_{xg} - \left(\sum_{g=1}^{G} \Omega_g \hat{\hat{t}}_{xg} \right)^2}.$$

Under SRSWOR (say, SI), and for $q_g = 1$, the estimator $\hat{\bar{t}}^{c2}_{y(\mathrm{SI})}$ can be expressed as

$$\hat{\bar{t}}^{c2}_{y(\mathrm{SI})} = \sum_{g=1}^{G} \Omega'_{g(\mathrm{SI})} \hat{\bar{t}}_{yg(\mathrm{SI})}, \quad \text{where,} \quad \hat{\bar{t}}_{yg(\mathrm{SI})} = \frac{1}{n_g} \sum_{s_g} \frac{N_{gi}}{\overline{N}_{go}} \bar{t}_{ygi} \quad \text{and} \quad \bar{t}_{ygi} = \frac{1}{n_i} \sum_{s_i} t_{ygik}$$

$\Omega'_{g(\mathrm{SI})}$ for $q_{gi} = 1$ under SRSWOR is given by

$$\Omega'_{g(\mathrm{SI})} = \Omega_g + \Omega_g \frac{\left(\bar{t}_x - \hat{\bar{t}}_{x(\mathrm{SI})}\right)\left(\hat{\bar{t}}_{xg(\mathrm{SI})} - \hat{\bar{t}}_{x(\mathrm{SI})}\right)}{\sum_{g=1}^{G} \Omega_g \hat{\bar{t}}^2_{xg(\mathrm{SI})} - \hat{\bar{t}}^2_{x(\mathrm{SI})}}$$

where $\hat{\bar{t}}_{x(\mathrm{SI})} = \sum_{g=1}^{G} \Omega_g \hat{\bar{t}}_{xg(\mathrm{SI})}$ and $\hat{\bar{t}}_{xg(\mathrm{SI})} = \frac{1}{n_g} \sum_{s_g} \frac{N_{gi}}{\overline{N}_{go}} \bar{t}_{xgi}$, $\bar{t}_{xgi} = \frac{1}{n_i} \sum_{s_i} t_{xgik}$. Now, we have

$$\begin{aligned} \hat{\bar{t}}^{c2}_{y(\mathrm{SI})} &= \sum_{g=1}^{G} \left[\Omega_g + \Omega_g \frac{\left(\bar{t}_x - \hat{\bar{t}}_{x(\mathrm{SI})}\right)\left(\hat{\bar{t}}_{xg(\mathrm{SI})} - \hat{\bar{t}}_{x(\mathrm{SI})}\right)}{\sum_{g=1}^{G} \Omega_g \hat{\bar{t}}^2_{xg(\mathrm{SI})} - \hat{\bar{t}}^2_{x(\mathrm{SI})}} \right] \hat{\bar{t}}_{yg(\mathrm{SI})} \\ &= \hat{\bar{t}}^{c2}_{y(\mathrm{SI})} = \sum_{g=1}^{G} \Omega_g \hat{\bar{t}}_{yg(\mathrm{SI})} + \widehat{B}'_{1(\mathrm{SI})} \left(\bar{t}_x - \hat{\bar{t}}_{x(\mathrm{SI})}\right) \end{aligned}$$

where

$$\widehat{B}'_{1(\mathrm{SI})} = \frac{\sum_{g=1}^{G} \Omega_g \hat{\bar{t}}_{yg(\mathrm{SI})} \hat{\bar{t}}_{xg(\mathrm{SI})} - \left(\sum_{g=1}^{G} \Omega_g \hat{\bar{t}}_{yg(\mathrm{SI})}\right)\left(\sum_{g=1}^{G} \Omega_g \hat{\bar{t}}_{xg(\mathrm{SI})}\right)}{\sum_{g=1}^{G} \Omega_g \hat{\bar{t}}^2_{xg(\mathrm{SI})} - \left(\sum_{g=1}^{G} \Omega_g \hat{\bar{t}}_{xg(\mathrm{SI})}\right)^2}.$$

Following the procedure given by Särndal et al. (2003, Chaps. 4 and 8), the conditional variance of $\hat{\bar{t}}^{c2}_y$ for given calibrated weight Ω'_g can be written as

$$V(\hat{\bar{t}}^{c2}_y) = \sum_{g=1}^{G} \Omega'^2_g V\left(\hat{\bar{t}}_{yg}\right),$$

where $V\left(\hat{\bar{t}}_{yg}\right)$ is given in Eq. (3.2).

Using Eq. (3.3), the estimator of $V\left(\hat{\bar{t}}^{c2}_y\right)$ is given by

$$\widehat{V}(\hat{\bar{t}}^{c2}_y) = \sum_{g=1}^{G} \Omega'^2_g \widehat{V}(\hat{\bar{t}}_{yg})$$

Under SRSWOR at both stages (say, SI), the calibration estimator of $\bar{t}_y$ can also be expressed as

$$\hat{\bar{t}}^{c2}_{y(\mathrm{SI})} = \sum_{g=1}^{G} \Omega'_{g(\mathrm{SI})} \hat{\bar{t}}_{yg(\mathrm{SI})}$$

Using the Eqs. (3.5) and (3.6), respectively, we have

$$V\left(\hat{\bar{t}}^{c2}_{y(\mathrm{SI})}\right) = \sum_{g=1}^{G} \Omega'^{2}_{g(\mathrm{SI})} V\left(\hat{\bar{t}}_{yg(\mathrm{SI})}\right) \quad \text{and}$$

$$\widehat{V}\left(\hat{\bar{t}}^{c2}_{y(\mathrm{SI})}\right) = \sum_{g=1}^{G} \Omega'^{2}_{g(\mathrm{SI})} \widehat{V}\left(\hat{\bar{t}}_{yg(\mathrm{SI})}\right).$$

3.2.3 Calibration of Both Stratum Weight and Design Weight at psu Level

The calibration estimator of $\bar{t}_y$ as developed in Sect. 3.2.1 by calibrating design weight at first step is given by

$$\hat{\bar{t}}^{c1}_{y} = \sum_{g=1}^{G} \Omega_g \hat{\bar{t}}^{c}_{yg}, \quad \text{where } \Omega_g = \frac{N_{go}}{N} = \text{stratum weight}$$

Now, we want to calibrate stratum weight Ω_g at second step. Let Ω'_g be the calibrated weight o f Ω_g. Therefore, two-step calibration estimator is proposed as

$$\hat{\bar{t}}^{cc}_{y} = \sum_{g=1}^{G} \Omega'_g \hat{\bar{t}}^{c}_{yg} \tag{3.14}$$

To find out Ω'_g, we minimize the chi-square distance function $\sum_{g=1}^{G} \left(\Omega'_g - \Omega_g\right)^2 / q_g \Omega_g$ subject to the calibration constraints

$$\sum_{g=1}^{G} \Omega'_g \hat{\bar{t}}_{xg} = \bar{t}_x \quad \text{and} \quad \sum_{g=1}^{G} \Omega_g = \sum_{g=1}^{G} \Omega'_g$$

where q_g is some scalar quantity.

Minimizing the following function with respect to Ω'_g

$$\phi(\Omega_g', \lambda) = \sum_{g=1}^{G} \left(\Omega_g' - \Omega_g\right)^2 / q_g \Omega_g$$
$$+ 2\lambda_1 \left(\sum_{g=1}^{G} \Omega_g' \hat{\bar{t}}_{xg} - \bar{t}_x \right) + 2\lambda_2 \left(\sum_{g=1}^{G} \Omega_g - \sum_{g=1}^{G} \Omega_g' \right)$$

where λ_1 and λ_2 are Lagrange multipliers, we get

$$\Omega_g' = \Omega_g + \frac{q_g \Omega_g \left[\hat{\bar{t}}_{xg} - \frac{\sum_{g=1}^{G} q_g \Omega_g \hat{\bar{t}}_{xg}}{\sum_{g=1}^{G} q_g \Omega_g} \right]}{\sum_{g=1}^{G} q_g \Omega_g (\hat{\bar{t}}_{xg})^2 - \frac{\left(\sum_{g=1}^{G} q_g \Omega_g \hat{\bar{t}}_{xg}\right)^2}{\sum_{g=1}^{G} q_g \Omega_g}} \left(\bar{t}_x - \sum_{g=1}^{G} \Omega_g \hat{\bar{t}}_{xg} \right)$$

Putting the value of Ω_g' in Eq. (3.14), the estimator of $\hat{\bar{t}}_y^{cc}$ reduces to

$$\hat{\bar{t}}_y^{cc} = \hat{\bar{t}}_y^{c} + \hat{b}_1 \left(\bar{t}_x - \sum_{g=1}^{G} \Omega_g \hat{\bar{t}}_{xg} \right)$$

where

$$\hat{b}_1 = \frac{\sum_{g=1}^{G} \Omega_g q_g \hat{\bar{t}}_{yg}^{c} \hat{\bar{t}}_{xg} - \frac{\sum_{g=1}^{G} \Omega_g q_g \hat{\bar{t}}_{yg}^{c} \sum_{g=1}^{G} \Omega_g q_g \hat{\bar{t}}_{xg}}{\sum_{g=1}^{G} \Omega_g q_g}}{\sum_{g=1}^{G} \Omega_g q_g (\hat{\bar{t}}_{xg})^2 - \frac{\left(\sum_{g=1}^{G} \Omega_g q_g \hat{\bar{t}}_{xg}\right)^2}{\sum_{g=1}^{G} \Omega_g q_g}}.$$

Two-step calibration estimator for $q_g = 1$ is given by

$$\hat{\bar{t}}_y^{cc} = \hat{\bar{t}}_y^{c} + \hat{b}_1' \left(\bar{t}_x - \sum_{g=1}^{G} \Omega_g \hat{\bar{t}}_{xg} \right)$$

where

$$\hat{b}_1' = \frac{\sum_{g=1}^{G} \Omega_g \hat{\bar{t}}_{yg}^{c} \hat{\bar{t}}_{xg} - \sum_{g=1}^{G} \Omega_g \hat{\bar{t}}_{yg}^{c} \sum_{g=1}^{G} \Omega_g \hat{\bar{t}}_{xg}}{\sum_{g=1}^{G} \Omega_g (\hat{\bar{t}}_{xg})^2 - \left(\sum_{g=1}^{G} \Omega_g \hat{\bar{t}}_{xg}\right)^2}.$$

Using Eq. (3.9), the conditional approximate variance of $\hat{\bar{t}}_{y(\mathrm{SI})}^{cc}$ for given Ω_g' is

$$V(\hat{\bar{t}}_y^{cc}) = \sum_{g=1}^{G} \Omega_g'^2 V(\hat{\bar{t}}_{yg}^{c}),$$

Using Eq. (3.10), the conditional approximate estimate of $V(\hat{\hat{t}}_y^{cc})$ for given Ω'_g is

$$\widehat{V}(\hat{\hat{t}}_y^{cc}) = \sum_{g=1}^{G} \Omega'^2_g \widehat{V}(\hat{\hat{t}}_{yg}^{c}),$$

Under SRSWOR (say, SI), the two-step calibration estimator for $q_g = 1$ reduces to

$$\hat{\hat{t}}_{y(\text{SI})}^{cc} = \hat{\hat{t}}_y^{c} + \hat{b}'_1 \left(\bar{t}_x - \sum_{g=1}^{G} \Omega_g \hat{\hat{t}}_{xg} \right)$$

where

$$\hat{b}'_1 = \frac{\sum_{g=1}^{G} \Omega_g \hat{\hat{t}}_{yg(\text{SI})}^{c} \hat{\hat{t}}_{xg(\text{SI})} - \sum_{g=1}^{G} \Omega_g \hat{\hat{t}}_{yg(\text{SI})}^{c} \sum_{g=1}^{G} \Omega_g \hat{\hat{t}}_{xg(\text{SI})}}{\sum_{g=1}^{G} \Omega_g \left(\hat{\hat{t}}_{xg(\text{SI})} \right)^2 - \left(\sum_{g=1}^{G} \Omega_g \hat{\hat{t}}_{xg(\text{SI})} \right)^2}.$$

Using Eq. (3.11), the conditional approximate variance of $\hat{\hat{t}}_{y(\text{SI})}^{cc}$ for given Ω'_g is

$$V\left(\hat{\hat{t}}_{y(\text{SI})}^{cc}\right) = \sum_{g=1}^{G} \Omega'^2_g V\left(\hat{\hat{t}}_{yg(\text{SI})}^{c}\right)$$

Using Eq. (3.12), the conditional approximate estimate of variance of $\hat{\hat{t}}_y^{cc}$ for given Ω'_g is given by

$$\widehat{V}\left(\hat{\hat{t}}_{y(\text{SI})}^{cc}\right) = \sum_{g=1}^{G} \Omega'^2_g \widehat{V}\left(\hat{\hat{t}}_{yg(\text{SI})}^{c}\right).$$

4 Simulation Study

The data of MU284 given in Appendix B and Appendix C of Särndal et al. (2003) have been used for the purpose of simulation study. There are 284 municipalities. The data on various variables for each municipality are given. These municipalities have been classified into 50 psu's of varying size (See, Annexure C, cluster of MU284 population). We have considered three populations of MU284 data. The description of the population is given in Table 1.

The 50 psu's are stratified into four strata considering the value of x in ascending order. The strata I, II, III, and IV consist of 13, 14, 12 and 11 psu's, respectively. The

Table 1 Description of the populations considered for simulation study

Populations	Study variable (y)	Auxiliary variable (x)
I	P85: Population of 1985	P75: Population of 1975
II	REV84: Real estate values according to 1984 assessment (in millions of kronor)	ME84: Number of municipal employees in 1984
III	RMT85: Revenues from 1985 municipal taxation (in millions of kronor)	REV84: Real estate values according to 1984 assessment (in millions of kronor)

samples of size 4 psu's were drawn by SRSWOR independently in each stratum. This process has been repeated 300 times independently. That means we obtained 300 samples of size 4 psu's from each stratum. Subsamples of size 3 ssu's are drawn by SRSWOR from each sample of psu's in each stratum. The values of y and x in subsamples were used to compute the population mean. In this process, we get 300 estimates of $\hat{\hat{t}}_{yg(\mathrm{SI})}$, $\hat{\hat{t}}^{c1}_{yg(\mathrm{SI})}$, $\hat{\hat{t}}^{c2}_{y(\mathrm{SI})}$, and $\hat{\hat{t}}^{cc}_{y(\mathrm{SI})}$ from 300 subsamples in each stratum. We computed the values of $\hat{\overline{T}}_i$, the estimate of population mean, based on usual estimator $\hat{\hat{t}}_{y(\mathrm{SI})}$ without using auxiliary information and calibration estimators $\hat{\hat{t}}^{c1}_{y(\mathrm{SI})}$, $\hat{\hat{t}}^{c2}_{y(\mathrm{SI})}$, and $\hat{\hat{t}}^{cc}_{y(\mathrm{SI})}$ from 1200 sample estimates of strata. The true means of $y(\overline{T})$ for the populations I, II and III have been computed to be 29.363, 3,077.525, and 244.993, respectively. The following two criteria were used for assessing the relative performance of these estimators:

(i) The percent absolute relative bias,

$$\%\mathrm{ARB}(\widehat{T}) = \frac{1}{S}\left(\sum_{i=1}^{S}\left|\frac{\hat{\overline{T}}_i - \overline{T}}{\overline{T}}\right|\right) \times 100$$

(ii) The percent relative root mean square error,

$$\%\mathrm{RRMSE}(\hat{\theta}) = \sqrt{\frac{1}{S}\sum_{i=1}^{S}\left(\frac{\hat{\overline{T}}_i - \overline{T}}{\overline{T}}\right)^2} \times 100$$

where S is the number of simulation. *R* software was used for simulation study.

The coefficient of correlation between y and $x(r_{yx})$ using 284 municipal data, the %ARB and the %RRMSE of estimators have been computed for each of the population (Table 2).

It can be observed from the results of Table 2 that calibration approach for estimation of the population mean of y has drastically decreased the %RRMSE as compared to that of the usual estimator without using auxiliary information in all the populations. Among the calibration estimators, two-step calibration estimator $\hat{\hat{t}}^{cc}_{y(\mathrm{SI})}$ was

Table 2 %ARB and %RRMSE of estimators in different populations

Estimators	Population I		Population II		Population III	
	%ARB	%RRMSE	%ARB	%RRMSE	%ARB	%RRMSE
$\hat{t}_{y(\mathrm{SI})}$	*	7.013	*	9.873	*	10.509
$\hat{t}^{c1}_{y(\mathrm{SI})}$	1.461	1.812	2.581	2.766	2.961	3.118
$\hat{t}^{c2}_{y(\mathrm{SI})}$	0.630	0.956	1.746	1.922	2.312	2.826
$\hat{t}^{cc}_{y(\mathrm{SI})}$	0.578	0.798	0.915	1.647	1.672	2.081
Correlations	$r_{yx} = 0.9984$		$r_{yx} = 0.9401$		$r_{yx} = 0.9358$	

*Unbiased estimator

found to be the best in all the populations. The %ARB of calibration estimators has been found to be between 0.578 and 2.9% in all the populations. The value of r_{yx} has been found greater than 0.90 in all the populations (maximum was 0.9984 in population I followed by 0.9401 and 0.9358 in populations II and III, respectively). Table 2 clearly indicates that higher is the coefficient of correlation, higher is the precision in the estimate of the population mean as well as smaller is the %ARB. It is also interesting to note from the results that the calibration estimator based on calibrating stratum weight is preferable to the calibration estimator based on calibrating design weight in two-stage stratified random sampling. However, when stratum weights and design weights are calibrated jointly in two-step, then two-step calibration estimators have outperformed other calibration estimators including usual estimator without using auxiliary information.

5 Conclusions

When the auxiliary information is available at the psu level, then the calibration approach-based calibration estimators have brought out significant improvement in the precision of the estimate of population mean in two-stage stratified random sampling. It may, however, be mentioned that the two-step calibration estimator $\hat{t}^{cc}_{y(\mathrm{SI})}$ has been found better than the other calibration estimators. Therefore, two-step calibration estimator is recommended to use in practice for relatively high precision of the estimates of population mean.

References

Aditya, K., Sud, U. C., & Chandra, H. (2016). Calibration based regression type estimator of the population total under two-stage sampling design. *Journal of the Indian Society of Agricultural Statistics, 70*(1), 19–24.

Cassel, C. M., Särndal, C. E., & Wretman, J. H. (1976). Some results on generalized difference estimation and generalized regression estimation for finite population. *Biometrika, 63,* 615–620.

Deville, J. C., & Särndal, C. E. (1992). Calibration estimators in survey sampling. *Journal of the American Statistical Association, 87,* 376–382.

Estevao, V. M., & Särndal, C. E. (2006). Survey estimates by calibration on complex auxiliary information. *International Statistical Review, 74,* 127–147.

Horvitz, D. G., & Thompson, D. J. (1952). A generalization of sampling without replacement from a finite universe. *Journal of the American Statistical Association, 47,* 663–685.

Kim, J. M., Sunger, E. A., & Heo, T. Y. (2007). Calibration approach estimators in stratified sampling. *Statistics and Probability Letters, 77,* 99–103.

Mourya, K. K., Sisodia, B. V. S., & Chandra, H. (2016a). Calibration approach for estimating finite population parameter in two-stage sampling. *Journal of Statistical Theory and Practice, 10*(3), 550–562.

Mourya, K. K., Sisodia, B. V. S., Kumar, S., & Singh, A. (2016b). Calibrated estimators of finite population parameters in equal cluster sampling. *International Journal of Agricultural Science, 12,* 351–358.

Nidhi, Sisodia, B. V. S., Singh, S., & Singh, S. K. (2016). Calibration approach estimation of mean in stratified sampling and stratified double sampling. *Communication in Statistics-Theory and Methods, 46*(10), 4932–4942.

Särndal, C. E. (2007). The calibration approach in survey theory and practice. *Survey Methodology, 33*(2), 99–119.

Särndal, C. E., Swensson, B., & Wretman, J. (2003). *Model-assisted survey sampling*. New York: Springer.

Singh, S., Horn, S., & Yu, F. (1998). Estimation of variance of the general regression estimators: Higher level calibration approach. *Survey methodology, 24*(1), 41–50.

Singh, S., Horn, S., Chodhury, S., & Yu, F. (1999). Calibration of the estimators of variance. *Australian & New Zealand Journal of Statistics, 41*(2), 199–212.

Singh, D., Sisodia, B. V. S., Rai, V. N., & Kumar, S. (2017). A calibration approach based regression and ratio type estimators of finite population mean in two-stage stratified random sampling. *Journal of the Indian Society of Agricultural Statistics, 71*(3), 217–224.

Sukhatme, P.V., Sukhatme, B.V., Sukhatme, S. and Asok, C. (1984). Sampling theory of surveys with applications. Iowa State University Press, Ames, Iowa and Indian Society of Agricultural Statistics, New Delhi.

Tracy, D. S., Singh, S., & Arnab, R. (2003). Note on calibration estimators in stratified and double sampling. *Survey Methodology, 29,* 99–106.

A Joint Calibration Estimator of Population Total Under Minimum Entropy Distance Function Based on Dual Frame Surveys

Piyush Kant Rai, G. C. Tikkiwal and Alka

Abstract The concept of dual frame-based estimators has been already developed in sample surveys. These dual frame estimators are theoretically optimal in some cases but difficult to apply in practice, while the others are generally applicable but may have larger variances. In this chapter, we propose Joint Calibration Estimator (JCE) under minimum entropy distance function for the dual frame surveys. The proposed estimator has smaller bias and considerable decrement in variance under Lahiri–Midzuno design as compared to Simple Random Sampling Without Replacement when sample size increases. In addition, we obtain the optimal weights along with its sensitive weighing interval for the combined JCE under non-overlapping frames for which it is more efficient than individual frame-based estimator.

Keywords Dual frame survey · Distance function · Joint calibration estimator · Auxiliary information.

1 Introduction

The classical sampling theory of estimation generally makes use of a single sampling frame under the assumption that it consists of all finite population units. A sampling frame is defined as a list of population units from which the sample is selected, or a set of geographic regions, or even a sequential procedure specifying how units are to be located and selected. After deciding sampling frame, a probability sample is selected using a proper sampling design and the obtained information is applied for estimation purposes and drawing inferences for the study characteristics.

P. Kant Rai (✉)
Department of Statistics, Banaras Hindu University, Varanasi, UP, India
e-mail: raipiyush5@gmail.com

G. C. Tikkiwal
Faculty of Science, Manipal University, Jaipur, Rajasthan, India

Alka
Department of Mathematics and Statistics, Banasthali University, Banasthali, Rajasthan, India

G. Chandra et al. (eds.), *Statistical Methods and Applications in Forestry and Environmental Sciences*, Forum for Interdisciplinary Mathematics,
https://doi.org/10.1007/978-981-15-1476-0_8

The sampling frame determines the quality of the sample in terms of coverage and information associated with the drawn sample units. To obtain more precise estimates of a population parameter(s) without increasing survey costs, it is suggested to consider more than one population frame or to use additional information at the design and/or estimation stage. Because of rapid changes in the cost of survey data collection, changes in population coverage patterns, and sample unit accessibility, the use of more than one frame is important. Due to the biased estimate obtained from an incomplete or outdated single frame, it is more common to use dual frame sample surveys for improving the estimates.

In conducting sample surveys, a common problem of incomplete and outdated frame becomes apparent when dealing with a rare population such as racial/ethnic minorities, human suffering with some uncommon disease, with elusive and/or hidden populations such as homeless, illegal immigrants or drug consumers, recent births, persons over 80 years of age, academic cheating or plagiarism and in general, when treating with special or hard-to-reach populations such as households living in poverty, illegal fishing and hunting, gay men, and disabled persons. In such situations, the single frame individually does not cover the whole target populations, but the union of these available single frames will provide more efficient and reliable estimates. Thus, the dual frame methodology offers the investigator to consider various data collection procedures and/or different sampling designs, one for each frame. In the view of this popularity, many countries (e.g., Canada, Germany, Denmark) used dual frame sampling methodology when the survey uses two frames.

In dual frame surveys, two random samples are taken from two frames A and B independently. These two frames may be non-overlapping (Fig. 1), partially overlapping (Fig. 2) or completely overlapping (Fig. 3) with an assumption that they together cover the complete target population. The data obtained from the two independent surveys are then combined to estimate population parameters of interest. The non-overlapping dual frame design resembles with distinguishable strata; thus, estimation under such frames is similar to stratified random sampling design. But due to overlapping frames, direct addition of the individual estimated results under two frames increases the exact estimate of the overall population total and thus results in biased estimation conclusions. Thus, some methodology is required to combine the estimated results from the two frames in a way to get reduced biased estimate of parameters with a minimum mean squared error.

Early examples of dual frame surveys include the 1949 Survey of Retail Stores and the 1960 Survey of Agriculture. The advantages of dual frame surveys have been demonstrated in several research disciplines. Hartley (1962, 1974) shows that if one frame A (area frame) is complete but expensive to sample, and another frame B (list frame) is incomplete but inexpensive to sample, then a dual frame survey can reduce cost with comparable precision over a single frame survey by augmenting expensive data from Frame A with inexpensive data from Frame B. For instance, in many agricultural surveys, an area frame comprises land segments; enumerators visit a random sample of these segments while the list frame includes the names and addresses of agricultural producers. The area frame is complete and insensitive to changes in farm ownership and activity, but very expensive to sample, whereas

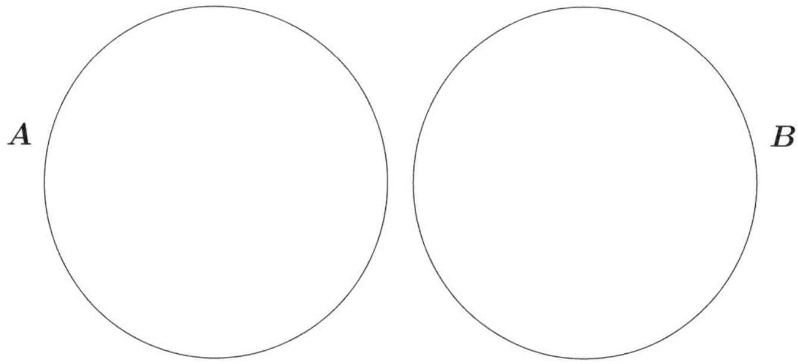

Fig. 1 Non-overlapping frames ***A*** and ***B***

Fig. 2 Partial-overlapping frames ***A*** and ***B*** with three domains ***a***, ***b***, and ***ab***

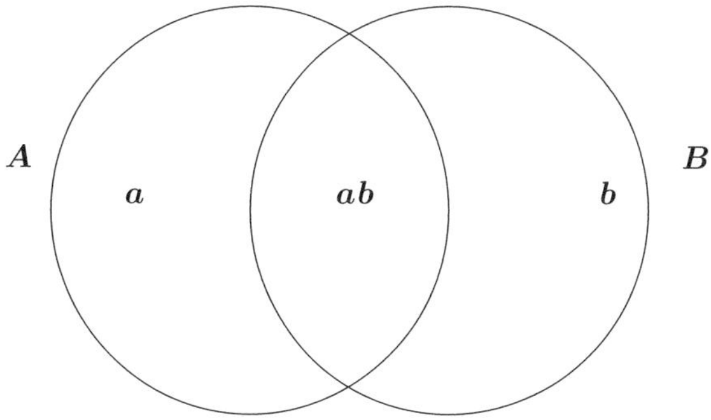

Fig. 3 Completely overlapping frames ***A*** and ***B*** with two domains ***a*** and ***ab***

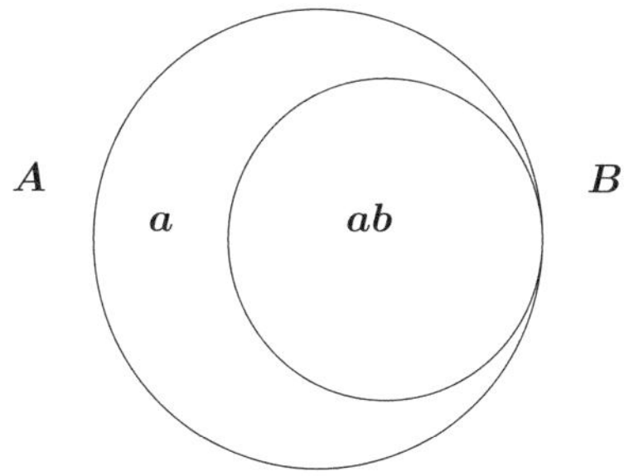

the list frame is less costly to sample as because all the names and addresses will be at one place, but the list may not include all the producers. So, both the frames were used to get the accurate estimates of parameter (National Statistics Service 1999). Similarly, accuracy of estimates can be increased with wise use of multiple frames. For example, samples taken from the single frame for general population health studies are generally not very informative for studies of human populations where the objective is to study specific features of people, such as people with certain rare diseases (e.g., HIV). Other frames, such as general hospital lists and/or separate therapy centers, often provide more information and expanded demographic coverage.

Haines and Pollock (1998) applied dual frame surveys to estimate population sizes in the cases where both frames were incomplete. Kalton and Anderson (1986) described an application to sampling rare populations, in which a screening sample is drawn from an address frame, and a sample is drawn from a small, incomplete list frame with a high proportion of rare events. Kuo (1989) and Lohr and Rao (2000) presented examples based on multiple frames with different sampling designs for different purposes. Overall, to get better estimation some desirable properties are required to satisfy for the dual frame estimators.

1.1 Desirable Properties

Lohr (2011) has given a list of desirable properties for dual frame estimators. The first property states that an estimator should be unbiased for the corresponding finite population quantity. Second and third properties are based on internal consistency and better efficiency with low mean squared error (MSE), respectively, and the fourth property is related to the analytical form of the estimator for easy computation and the last property explains that estimator should be robust to non-sampling errors. Elkasabi et al. (2015) added following three properties for dual frame estimator in addition to the above properties.

(a) Data requirements for estimator should be reasonable.
(b) An estimator should be robust to non-sampling errors in the estimator's requirements.
(c) An estimator should be applicable for dual frame and multiple frame surveys.

To work with the exhaustive desirable properties of dual frame estimator, several efforts have made to develop the estimators in sample surveys. Here, a new dual frame estimator will be discussed based on general calibration approach. The implicit potential of the developed calibration estimator under dual frame samples will also be theoretically discussed in the chapter.

1.2 Calibration Approach in Sample Survey

Calibration is the prime theme in the theory of estimation in survey sampling which provides an efficient way to integrate supplementary information in the procedure. It has become an important methodological tool in the generation of statistics on a large scale. Survey statisticians use additional information to improve the estimates of the survey in many ways. The approach to calibration contrasts with (generalized) regression estimation, which is an alternative but conceptually distinct way to take into consideration additional data.

Deville and Särndal (1992) introduced the term calibration estimation as *"a procedure of minimizing a distance measure between initial weights and final weights subject to calibration equations (constraints) using auxiliary information"*. The final weights are called calibrated weights as they have the calibration property in the sense of reproducing exactly known population quantities when applied to the sample values of the corresponding auxiliary variables. They made an intuitive argument that "weights that work well with the auxiliary variables should also work well with the study variable." The estimators developed by these calibrated weights are asymptotically design-unbiased and consistent with a lower variance than the HT estimator. Their method of calibration provides a class of estimators to which several well-known estimators belong, e.g., the classical ratio estimator among others. Särndal (2007) defined the calibration approach as: "*The calibration approach to estimate parameters of finite populations consists of*

(a) *computation of weights that incorporate specified auxiliary information and are restrained by calibration equation(s),*
(b) *the use of these weights to compute linearly weighted estimates of totals and other finite population parameters: weight times variable value, summed over a set of observed units,*
(c) *an objective to obtain nearly design-unbiased estimates as long as non-response and other non-sampling errors are absent.*"

The calibration weight(s) specified in the calibration equation(s) by minimizing a distance measure may be very large or even negative. If the weights are used to estimate the total population, it seems reasonable that there should be no individual weight less than one.

1.3 Concept of Distance Function

As discussed earlier, the calibration approach is well known for increasing the accuracy of population parameters using the known auxiliary information. This method works by minimizing the distance measure between design weights and the calibrated weights satisfying some calibration constraints. Distance measure is important as it ensures that the design weights and calibrated weights are as close as possible. A variety of such distance functions given by Deville and Särndal (1992) is shown in Table 1. They also defined the desirable properties of these functions, i.e., for every fixed design weights $d > 0$, $D(\omega, d)$ should be non-negative, strictly convex, continuous, and differentiable with respect to calibrated weights ω.

Here, q is the scale factor or tuning parameter, chosen by the statisticians. The standard choice for q is 1. The choice of q has some (but often limited) impact on the accuracy of generalized regression estimator, near-unbiasedness holds for any specification (barring outrageous choices) for the q. In practice, the choice of distance function usually depends on the approach adopted by users and the type of problem. Thus, the choice of distance function is insignificant for large samples but rather depends on the computational effort of solving the calibration equations.

Table 1 List of distance functions D(ω, d) adopted by Deville and Särndal

S. No.	Types	D(ω, d)
1	Hellinger distance function	$\frac{2(\sqrt{\omega}-\sqrt{d})^2}{q}$
2	Chi-square distance function	$\frac{(\omega-d)^2}{qd}$
3	Minimum entropy distance function	$q^{-1}(-dlog\frac{\omega}{d}+\omega-d)$
4	Modified minimum entropy distance function	$q^{-1}(-\omega log\frac{\omega}{d}-\omega-d)$
5	Modified chi-square distance function	$\frac{(\omega-d)^2}{q\omega}$

The literature shows that chi-square distance function is widely used in the calibration-based approach in survey sampling. But, there are some drawbacks of chi-square—as the sample size increases, then absolute differences become a smaller and smaller proportion of the expected value. Also, chi-square is based on the assumption that if the expected frequency is smaller than 5, it can provide misleading conclusions. One more limitation of calibration estimator based on chi-square distance function is that the weights can be extremely large or may take negative values. These unrealistic weights may occur for some rare (unlucky) samples, which becomes undesirable in a survey sampling context or unacceptable to some users. One might want to avoid a function that gives excessively extreme weights because applying these weights to make estimates for different subpopulations could produce unrealistic estimates, which would result in no use.

Deville and Särndal (1992) acknowledged this issue and showed how weights can be restricted to a certain range and encouraged the use of other available distance functions. One can ensure the positive weights by using Hellinger, minimum entropy, and modified chi-squared distance functions under calibration approach. Singh (2004) investigated a new model of assisted chi-square distance function by modifying the methodology introduced by Deville and Särndal (1992). Singh and Arnab (2011) proposed a subclass of the class of calibrated estimators, in which they set the sum of calibrated weights and sum of design weights equal to each other. Rai et al. (2018) explained the benefit of using minimum entropy distance function under calibration approach in estimating the population total. In addition, Alka et al. (2019) proposed a two-step calibration estimator under two auxiliary variables for estimating the population parameter(s) under single sampling frame. The same idea can be applied to the dual frame estimation, where strong association between the auxiliary variable and the study variable results into asymptotically unbiased dual frame estimates.

2 An Application to Forestry and Environment

To meet some of the challenges faced by single frame survey discussed earlier, multiple frame surveys are very useful where independent samples are selected from individual frames and combined sample provide an estimate of the parameter(s). The dual frame sampling method combines an incomplete but inexpensive list frame with a complete but expensive area frame to achieve full coverage of the target population by reducing the overall sample cost. This dual frame or multiple frame technique has been used for more than 60 years in the estimation of various population parameters of interest. Hartley (1962) proposed the first dual frame estimator. While many others have proposed improved dual frame estimators based on some new techniques (refer Lund 1968; Fuller and Burmeister 1972; Bankier 1986; Kalton and Anderson 1986; and Skinner 1991). A brief overview of some of its applications is given in this section particularly related to environmental research.

Haines and Pollock (1998) estimated the number of active eagle nests and the total number of successful nests in a specific region using dual frame sampling techniques. A similar idea was employed by U.S. Fish and Wildlife Service (2009) to estimate the total number of occupied eagle nests within the study area of nest density. Also, Public Opinion Poll Center (POPC) has been conducting landline telephone surveys to assess the public's opinion toward many political, social, and economic issues in Egypt since 2003 were another example of the method. Earlier they used a single list frame, but due to an increase in the number of households using cell phones the problem of non-coverage of all population units in one sampling frame is faced. Thus, to overcome this challenge, dual frame telephone surveys have been used since June 2014 with a purpose to cover the 93.3 % telephone households in Egypt having access to either communication mode. Recently, Shyvers et al. (2018) used dual frame survey methodology for estimating number of active leks, lekking males in combination with occupancy analysis to adjust for imperfect detection. The proportion of active leks and lekking males counted on annual lek surveys over three consecutive breeding seasons in a small, low-density greater sage-grouse population in northwestern Colorado, USA.

2.1 Application of Least Absolute Shrinkage and Selection Operator (LASSO) Method of Estimation for Tree Canopy Cover

A better example of the application of model-assisted tools in forest research was given by McConville et al. (2017) in their study. They have suggested a model-assisted survey regression estimation with LASSO method of estimation and extended the same to adaptive LASSO. They have used the method for estimation of tree canopy cover for a region in Utah. A simulation study is considered for the purpose. A ridge regression approximation and a calibration approach are used to develop LASSO survey regression weights. In the study, authors have considered ten different estimators of LASSO and traditional one and compared their efficiency using Utah Tree Canopy Cover Data Set. In USA and other developed countries, government agencies are increasingly interested in improving survey data products through the use of "big data" available from different sources incorporating multiple frames. Methods developed in the chapter may have future applications with such auxiliary information related to forest areas.

3 Joint Calibration Estimator (JCE) Under Dual Frame Surveys

Let $A = \{1, \ldots, i, \ldots N_A\}$ and $B = \{1, \ldots, i, \ldots N_B\}$ be two mutually non-exclusive overlapping frames of $\Omega = \{1, \ldots, k, \ldots N\}$, i.e., $A \bigcap B = ab \neq \phi$ but $A \bigcup B = \Omega$. Assuming a case where information on single auxiliary variable is available. The study and auxiliary variables are denoted as y and x, respectively, and also suppose y_k and x_k represent the values of the kth $(k \in \Omega)$ unit in the population by y and x respectively. There are two samples $s_A (s_A \subseteq A)$ and $s_B (s_B \subseteq B)$ drawn from dual frames A and B using a sample design $p(.)$ with inclusion probabilities $\pi_i^A = p(i \in s_A)$ and $\pi_i^B = p(i \in s_B)$. The weights d_i are defined as $d_i = d_i^A = 1/\pi_i^A$ for $i \in s_A$ and $d_i = d_i^B = 1/\pi_i^B$ for $i \in s_B$. Here, N_A and N_B denote the population sizes, whereas n_A and n_B denote the sample sizes for frame A and B respectively. Let $a = A \bigcap B^c$, $b = A^c \bigcap B$, and $ab = A \bigcap B$ or $B \bigcap A$, where c refers to the complement of a set, and $s_a = a \bigcap s_A$, $s_b = b \bigcap s_B$, $s_{ab}^A = ab \bigcap s_A$, and $s_{ab}^B = ab \bigcap s_B$. In either of the surveys or both surveys, the units in these overlapping domain ab can be sampled.

The standard form of dual frame estimator is given as $\hat{T} = \hat{T}_a + \hat{T}_{ab} + \hat{T}_b$ for the estimation of population total $T = T_a + T_{ab} + T_b$, where $T = \sum_{i \in \Omega} y_i$, $T_a = \sum_{i \in a} y_i$, $T_{ab} = \sum_{i \in ab} y_i$, $T_b = \sum_{i \in b} y_i$. The HT estimator of totals in domains a and b for characteristic T is $\hat{T}_a = \sum_{i \in s_a} d_i y_i$ and $\hat{T}_b = \sum_{i \in s_b} d_i y_i$, and the estimators of overlap ab are $\hat{T}_{ab}^A = \sum_{i \in s_{ab}^A} d_i y_i$ and $\hat{T}_{ab}^B = \sum_{i \in s_{ab}^B} d_i y_i$. For each sample, the estimators of population totals are unbiased for the corresponding totals T_a, T_{ab} and T_b, i.e.,

$$E[\hat{T}_a + \hat{T}_{ab}^A] = T_a + T_{ab} \tag{3.1}$$

$$\text{and} \quad E[\hat{T}_b + \hat{T}_{ab}^B] = T_b + T_{ab} \tag{3.2}$$

where $E(.)$ refers to design-based expectation. Thus, addition of above two estimators will result into biased estimate of population total

$$E[\hat{T}_a + \hat{T}_{ab}^A + \hat{T}_b + \hat{T}_{ab}^B] \approx T_a + 2T_{ab} + T_b \neq T.$$

So, an unbiased estimate for T can be achieved by considering weighted average of the estimators $\hat{T}_{ab}^A$ and $\hat{T}_{ab}^B$, given by Hartley (1962)

$$\hat{T} = \hat{T}_a + \Phi\hat{T}^A_{ab} + (1-\Phi)\hat{T}^B_{ab} + \hat{T}_b, \tag{3.3}$$

where Φ is a composite factor between 0 and 1 that combines $\hat{T}^A_{ab}$ and $\hat{T}^B_{ab}$. As each domain a, b and ab is estimated by their corresponding HT estimator, so for a given value of Φ, $\hat{T}$ is an unbiased estimator and its variance is given as

$$V(\hat{T}) = V(\hat{T}_a + \Phi\hat{T}^A_{ab}) + V((1-\Phi)\hat{T}^B_{ab} + \hat{T}_b). \tag{3.4}$$

Since frames A and B are sampled independently, so the covariance term will vanishes. The first and second components on the right-hand side are calculated under $p_A(.)$ and $p_B(.)$, respectively. Much attention has given for the choice of Φ for which the variance in Eq. (3.4) is minimum. Unknown population variances and covariances decide the value of Φ and, when estimated from the data, it depends on the values of the study variable. Thus, for every value of study variable y, there is a need to recalculate these weights that are appropriate to use for the agencies conducting sample surveys and results into consistent estimates (refer, Lohr Lohr (2009)).

3.1 Calibration Estimator

In single frame survey design, the sample s is drawn randomly from the finite population Ω with inclusion probability π_i. With the usual notation for study variable y_i, let $\boldsymbol{x}_i = (x_{i1}, \ldots, x_{ij}, \ldots, x_{iJ})'$ be the auxiliary variable vector of dimension $j = 1, 2, \ldots, J$ observed for the sample elements $i \in s$. The HT estimator for the population total is $\hat{\text{T}}_{HT} = \sum_{i \in s} d_i y_i$. With j known auxiliary totals, i.e., $\boldsymbol{X} = (X_1, \ldots, X_j, \ldots, X_J)' = \left(\sum_{i\in\Omega} x_{i1}, \ldots, \sum_{i\in\Omega} x_{ij}, \ldots, \sum_{i\in\Omega} x_{iJ}\right)'$, Deville and Särndal (1992) discovered calibration method to find new weights ω_i which minimizes a distance measure $D(\omega_i, d_i)$ subject to the constraint $\sum_{i\in s} \omega_i \boldsymbol{x}_i = \boldsymbol{X}$. This problem can be viewed as Lagrangian problem which on solving leads to final calibrated weights $\omega_i = d_i F(q_i, \boldsymbol{x}_i, \lambda)$, where λ denotes a vector of lagrange multipliers, q_i is a positive value which scales the calibrated weights, and F is the Langrangian function of $q_i, \boldsymbol{x}_i, \lambda$.

As discussed in Introduction section, several distance functions have been proposed in the literature of calibration approach and the estimators derived from these alternative distance measures have empirically small differences [see Singh and Mohl (1996), Stukel et al. (1996)]. In view of the limitations of chi-square distance function discussed earlier, the focus is given on the minimum entropy distance function with the assumption of uniform weights, i.e., $q_i = 1$. Improved calibrated weights can be obtained by minimizing the minimum entropy distance function

$$-d_i log(\omega_i/d_i) + \omega_i - d_i \tag{3.5}$$

subject to the calibration equation $\sum_{i\in s}\omega_i\boldsymbol{x}_i = \boldsymbol{X}$. The lagrangian function to minimize Eq. (3.5) under calibration constraint is given as:

$$L = \sum_{i\in s}\{-d_i log(\omega_i/d_i) + \omega_i - d_i\} - \lambda\left(\sum_{i\in s}\omega_i\boldsymbol{x}_i - \boldsymbol{X}\right). \quad (3.6)$$

Minimizing Eq. (3.6) with respect to ω_i, it is obtained as

$$\omega_i = d_i(1 - \lambda'\boldsymbol{x}_i)^{-1}. \quad (3.7)$$

And under higher-order approximations, the expression of Lagrange multiplier vector λ' is obtained as

$$\lambda' = \frac{2(\sum_{i\in\Omega}\boldsymbol{x}_i - \sum_{i\in s}d_i\boldsymbol{x}_i)'}{\sum_{i\in s}d_i\boldsymbol{x}_i\boldsymbol{x}_i'}. \quad (3.8)$$

On putting the value of λ', the final calibrated weights are obtained as

$$\omega_i = d_i\left\{1 + \frac{2(\sum_{i\in\Omega}\boldsymbol{x}_i - \sum_{i\in s}d_i\boldsymbol{x}_i)'}{\sum_{i\in s}d_i\boldsymbol{x}_i\boldsymbol{x}_i'}\boldsymbol{x}_i\right\} \quad (3.9)$$

and the calibrated estimated total is $\hat{T}_{cal} = \sum_{i\in s}\omega_i y_i$, where ω_i are given in Eq. (3.9).

Now, this method of calibration estimation can be used to combine two samples for the case of dual frame surveys to develop Joint Calibration Estimator (JCE). Let us assume $E(.)$ as design-based expectation and

$$E\left(\sum_{i\in s_A}d_i\boldsymbol{x}_i\right) = \boldsymbol{X}_A\,,\quad E\left(\sum_{i\in s_B}d_i\boldsymbol{x}_i\right) = \boldsymbol{X}_B\,, \quad (3.10)$$

where $\boldsymbol{X}_A = \left(\sum_{i\in A}x_{i1}, \ldots, \sum_{i\in A}x_{ij}, \ldots, \sum_{i\in A}x_{iJ}\right)'$ and $\boldsymbol{X}_B = \left(\sum_{i\in B}x_{i1}, \ldots, \sum_{i\in B}x_{ij}, \ldots, \sum_{i\in B}x_{iJ}\right)'$. Also, $E(\sum_{i\in s_A}d_i\boldsymbol{x}_i + \sum_{i\in s_B}d_i\boldsymbol{x}_i) = \boldsymbol{X}_A + \boldsymbol{X}_B \neq \boldsymbol{X}$, because the two frames are overlapping. The calibration constraint under dual frames is defined as

$$\sum_{i\in s_A}\omega_i\boldsymbol{x}_i + \sum_{i\in s_B}\omega_i\boldsymbol{x}_i = \boldsymbol{X}. \quad (3.11)$$

It should achieve $E\left(\sum_{i\in s_A}\omega_i\boldsymbol{x}_i + \sum_{i\in s_B}\omega_i\boldsymbol{x}_i\right) = \boldsymbol{X}$. Consequently, a powerful set of auxiliary variables, that are strong predictors for y, should result in $E(\sum_{i\in s_A}\omega_i y_i + \sum_{i\in s_B}\omega_i y_i) \approx T$.

To get weights ω_i such that $\sum_{i\in s}\omega_i\boldsymbol{x}_i = \sum_{i\in s_A}\omega_i\boldsymbol{x}_i + \sum_{i\in s_B}\omega_i\boldsymbol{x}_i = \boldsymbol{X}$ by minimizing the chi-square distance function

$$D_1(\omega_i, d_i) = \frac{1}{2}\left\{\sum_{i\in s_A}\frac{(\omega_i - d_i)^2}{d_i} + \sum_{i\in s_B}\frac{(\omega_i - d_i)^2}{d_i}\right\}.$$

The calibrated weights are obtained as

$$\omega_i = d_i\left\{1 + \frac{\left(\sum_{i\in\Omega}\boldsymbol{x}_i - \sum_{i\in s_A} d_i\boldsymbol{x}_i - \sum_{i\in s_B} d_i\boldsymbol{x}_i\right)'}{\sum_{i\in s_A} d_i\boldsymbol{x}_i\boldsymbol{x}_i' + \sum_{i\in s_B} d_i\boldsymbol{x}_i\boldsymbol{x}_i'}\boldsymbol{x}_i\right\}. \tag{3.12}$$

Thus, the JCE for population total given by Elkasabi et al. (2015) as

$$\hat{T}^*_{JCE} = \sum_{i\in s}\omega_i y_i = \sum_{i\in s_A}\omega_i y_i + \sum_{i\in s_B}\omega_i y_i, \tag{3.13}$$

where calibrated weights are given in Eq. (3.12). Now, to develop JCE under minimum entropy distance function defined as

$$D_2(\omega_i, d_i) = \sum_{i\in s_A}\{-d_i log(\omega_i/d_i) + \omega_i - d_i\} + \sum_{i\in s_B}\{-d_i log(\omega_i/d_i) + \omega_i - d_i\}.$$

This is to be minimized under calibration constraint Eq. (3.11) using Lagrange's function, the calibrated weights under higher-order approximations are obtained as

$$\omega_i = d_i\left\{1 + 2\frac{\left(\sum_{i\in\Omega}\boldsymbol{x}_i - \sum_{i\in s_A} d_i\boldsymbol{x}_i - \sum_{i\in s_B} d_i\boldsymbol{x}_i\right)'}{\sum_{i\in s_A} d_i\boldsymbol{x}_i\boldsymbol{x}_i' + \sum_{i\in s_B} d_i\boldsymbol{x}_i\boldsymbol{x}_i'}\boldsymbol{x}_i\right\}. \tag{3.14}$$

Therefore, new JCE for the population total under minimum entropy distance function is obtained as

$$\hat{T}_{JCE} = \sum_{i\in s_A}\omega_i y_i + \sum_{i\in s_B}\omega_i y_i.$$

On substituting the value of ω_i, we have

$$\begin{aligned}
\hat{T}_{JCE} &= \sum_{i\in s_A} d_i\left\{1 + 2\frac{\left(\sum_{i\in\Omega}\boldsymbol{x}_i - \sum_{i\in s_A} d_i\boldsymbol{x}_i - \sum_{i\in s_B} d_i\boldsymbol{x}_i\right)'}{\sum_{i\in s_A} d_i\boldsymbol{x}_i\boldsymbol{x}_i' + \sum_{i\in s_B} d_i\boldsymbol{x}_i\boldsymbol{x}_i'}\boldsymbol{x}_i\right\} y_i \\
&+ \sum_{i\in s_B} d_i\left\{1 + 2\frac{\left(\sum_{i\in\Omega}\boldsymbol{x}_i - \sum_{i\in s_A} d_i\boldsymbol{x}_i - \sum_{i\in s_B} d_i\boldsymbol{x}_i\right)'}{\sum_{i\in s_A} d_i\boldsymbol{x}_i\boldsymbol{x}_i' + \sum_{i\in s_B} d_i\boldsymbol{x}_i\boldsymbol{x}_i'}\boldsymbol{x}_i\right\} y_i \\
&= \sum_{i\in s_A} d_i y_i + \sum_{i\in s_B} d_i y_i \\
&+ 2\frac{\left(\sum_{i\in\Omega}\boldsymbol{x}_i - \sum_{i\in s_A} d_i\boldsymbol{x}_i - \sum_{i\in s_B} d_i\boldsymbol{x}_i\right)'}{\sum_{i\in s_A} d_i\boldsymbol{x}_i\boldsymbol{x}_i' + \sum_{i\in s_B} d_i\boldsymbol{x}_i\boldsymbol{x}_i'}\left(\sum_{i\in s_A} d_i\boldsymbol{x}_i y_i + \sum_{i\in s_B} d_i\boldsymbol{x}_i y_i\right)
\end{aligned}$$

$$\hat{T}_{JCE} = \sum_{i\in s_A} d_i y_i + \sum_{i\in s_B} d_i y_i + 2\frac{\sum_{i\in s_A} d_i \boldsymbol{x}_i y_i + \sum_{i\in s_B} d_i \boldsymbol{x}_i y_i}{\sum_{i\in s_A} d_i \boldsymbol{x}_i \boldsymbol{x}_i' + \sum_{i\in s_B} d_i \boldsymbol{x}_i \boldsymbol{x}_i'} \left\{\sum_{i\in\Omega} \boldsymbol{x}_i - \left(\sum_{i\in s_A} d_i \boldsymbol{x}_i + \sum_{i\in s_B} d_i \boldsymbol{x}_i\right)\right\}'. \tag{3.15}$$

The JCE can be written as,

$$\hat{T}_{JCE} = \hat{T}_{HT}^{A} + \hat{T}_{HT}^{B} + \left\{\sum_{i\in\Omega} \boldsymbol{x}_i - \left(\sum_{i\in s_A} d_i \boldsymbol{x}_i + \sum_{i\in s_B} d_i \boldsymbol{x}_i\right)\right\}' \hat{B}_s^{A,B}, \tag{3.16}$$

where

$$\hat{T}_{HT}^{A} = \sum_{i\in s_A} d_i y_i\,,\quad \hat{T}_{HT}^{B} = \sum_{i\in s_B} d_i y_i \text{and}\quad \hat{B}_s^{A,B} = 2\left(\frac{\sum_{i\in s_A} d_i \boldsymbol{x}_i y_i + \sum_{i\in s_B} d_i \boldsymbol{x}_i y_i}{\sum_{i\in s_A} d_i \boldsymbol{x}_i \boldsymbol{x}_i' + \sum_{i\in s_B} d_i \boldsymbol{x}_i \boldsymbol{x}_i'}\right).$$

Similarly, the above results can also be generalized for the group mean model and group ratio model based on the different choices of auxiliary variable $\boldsymbol{x}_i$. In the next section, the expressions for bias and variance of proposed JCE are derived.

4 Bias and Variance of JCE

Here, an approximate bias expression for JCE in Eq. (3.15) is derived which makes it easier to understand the joint calibration approach-based mechanism in combining the dual frame design-based samples under minimum entropy distance function. Also, it highlights the JCE as a model-assisted design-based estimator that affects the design-based bias properties by the association between the survey variable and the auxiliary variable vector. From Eq. (3.16),

$$\begin{aligned}
\hat{T}_{JCE} &= \sum_{i\in s_A} d_i y_i + \sum_{i\in s_B} d_i y_i + \left\{\sum_{i\in\Omega} \boldsymbol{x}_i' - \left(\sum_{i\in s_A} d_i \boldsymbol{x}_i' + \sum_{i\in s_B} d_i \boldsymbol{x}_i'\right)\right\} \hat{B}_s^{A,B} \\
&= \sum_{i\in s_A} d_i y_i + \sum_{i\in s_B} d_i y_i + \sum_{i\in\Omega} \boldsymbol{x}_i' \hat{B}_s^{A,B} - \sum_{i\in s_A} d_i \boldsymbol{x}_i' \hat{B}_s^{A,B} - \sum_{i\in s_B} d_i \boldsymbol{x}_i' \hat{B}_s^{A,B} \\
&= \sum_{i\in s_A} d_i y_i + \sum_{i\in s_B} d_i y_i + \sum_{i\in\Omega} y_i - \sum_{i\in s_A} d_i y_i - \sum_{i\in s_B} d_i y_i \\
&= \sum_{i\in\Omega} y_i + \sum_{i\in s_A} d_i (y_i - y_i) + \sum_{i\in s_B} d_i (y_i - y_i)\,,
\end{aligned} \tag{4.1}$$

where $y_i = \boldsymbol{x}_i^{'}\hat{B}_s^{A,B}$. On rearranging the terms in Eq. (4.1), we have

$$\hat{T}_{JCE} = \sum_{i\in\Omega} \boldsymbol{x}_i^{'}\hat{B}_s^{A,B} + \sum_{i\in s_A} d_i y_i - \sum_{i\in s_A} d_i\boldsymbol{x}_i^{'}\hat{B}_s^{A,B} + \sum_{i\in s_B} d_i y_i - \sum_{i\in s_B} d_i\boldsymbol{x}_i^{'}\hat{B}_s^{A,B}. \quad (4.2)$$

On subtracting T both sides, we get

$$\hat{T}_{JCE} - T = \sum_{i\in\Omega} \boldsymbol{x}_i^{'}\hat{B}_s^{A,B} + \sum_{i\in s_A} d_i y_i - \sum_{i\in s_A} d_i\boldsymbol{x}_i^{'}\hat{B}_s^{A,B} + \sum_{i\in s_B} d_i y_i - \sum_{i\in s_B} d_i\boldsymbol{x}_i^{'}\hat{B}_s^{A,B} - \sum_{i\in\Omega} y_i.$$

To obtain bias of JCE, the terms $\sum_{i\in\Omega}\boldsymbol{x}_i^{'}B_\Omega$, $\sum_{i\in s_A} d_i\boldsymbol{x}_i^{'}B_\Omega$, and $\sum_{i\in s_B} d_i\boldsymbol{x}_i^{'}B_\Omega$ are to be added and subtracted on right-hand side as

$$\begin{aligned}\hat{T}_{JCE} - T = &\sum_{i\in\Omega} \boldsymbol{x}_i^{'}\hat{B}_s^{A,B} + \sum_{i\in s_A} d_i y_i - \sum_{i\in s_A} d_i\boldsymbol{x}_i^{'}\hat{B}_s^{A,B} + \sum_{i\in s_B} d_i y_i \\ &- \sum_{i\in s_B} d_i\boldsymbol{x}_i^{'}\hat{B}_s^{A,B} - \sum_{i\in\Omega} y_i - \sum_{i\in\Omega}\boldsymbol{x}_i^{'}B_\Omega - \sum_{i\in s_A} d_i\boldsymbol{x}_i^{'}B_\Omega \\ &- \sum_{i\in s_B} d_i\boldsymbol{x}_i^{'}B_\Omega + \sum_{i\in\Omega}\boldsymbol{x}_i^{'}B_\Omega + \sum_{i\in s_A} d_i\boldsymbol{x}_i^{'}B_\Omega + \sum_{i\in s_B} d_i\boldsymbol{x}_i^{'}B_\Omega,\end{aligned}$$

where $e_i = y_i - \boldsymbol{x}_i^{'}B_\Omega$ and $B_\Omega = 2\dfrac{\sum_{i\in\Omega}\boldsymbol{x}_i y_i}{\sum_{i\in\Omega}\boldsymbol{x}_i\boldsymbol{x}_i^{'}}$.

$$\begin{aligned}\hat{T}_{JCE} - T = &-\sum_{i\in\Omega} y_i + \sum_{i\in\Omega}\boldsymbol{x}_i^{'}B_\Omega + \sum_{i\in s_A} d_i y_i - \sum_{i\in s_A} d_i\boldsymbol{x}_i^{'}B_\Omega + \sum_{i\in s_B} d_i y_i - \sum_{i\in s_B} d_i\boldsymbol{x}_i^{'}B_\Omega \\ &+ \sum_{i\in\Omega}\boldsymbol{x}_i^{'}\hat{B}_s^{A,B} - \sum_{i\in s_A} d_i\boldsymbol{x}_i^{'}\hat{B}_s^{A,B} - \sum_{i\in s_B} d_i\boldsymbol{x}_i^{'}\hat{B}_s^{A,B} - \sum_{i\in\Omega}\boldsymbol{x}_i^{'}B_\Omega \\ &+ \sum_{i\in s_A} d_i\boldsymbol{x}_i^{'}B_\Omega + \sum_{i\in s_B} d_i\boldsymbol{x}_i^{'}B_\Omega \\ = &-\sum_{i\in\Omega}(y_i - \boldsymbol{x}_i^{'}B_\Omega) + \sum_{i\in s_A} d_i(y_i - \boldsymbol{x}_i^{'}B_\Omega) + \sum_{i\in s_B} d_i(y_i - \boldsymbol{x}_i^{'}B_\Omega) \\ &+ \left(\sum_{i\in\Omega}\boldsymbol{x}_i^{'} - \sum_{i\in s_A} d_i\boldsymbol{x}_i^{'} - \sum_{i\in s_B} d_i\boldsymbol{x}_i^{'}\right)(\hat{B}_s^{A,B} - B_\Omega) \\ = &-\sum_{i\in\Omega} e_i + \sum_{i\in s_A} d_i e_i + \sum_{i\in s_B} d_i e_i + \left(\sum_{i\in\Omega}\boldsymbol{x}_i^{'} - \sum_{i\in s_A} d_i\boldsymbol{x}_i^{'} - \sum_{i\in s_B} d_i\boldsymbol{x}_i^{'}\right)(\hat{B}_s^{A,B} - B_\Omega) \quad = P + Q,\end{aligned} \quad (4.3)$$

where
$$P = \sum_{i\in s_A} d_i e_i + \sum_{i\in s_B} d_i e_i - \sum_{i\in\Omega} e_i \tag{4.4}$$

$$Q = \left(\sum_{i\in\Omega} \boldsymbol{x}_i^{'} - \sum_{i\in s_A} d_i \boldsymbol{x}_i^{'} - \sum_{i\in s_B} d_i \boldsymbol{x}_i^{'}\right)\left(\hat{B}_s^{A,B} - B_\Omega\right). \tag{4.5}$$

On taking expectation of Eq. (4.3), the bias will be obtained as

$$E(\hat{T}_{JCE} - T) = E(P+Q) = E(P) + E(Q). \tag{4.6}$$

Finding individual expectations of P and Q as

$$E(P) = E\left\{\sum_{i\in s_A} d_i e_i + \sum_{i\in s_B} d_i e_i - \sum_{i\in\Omega} e_i\right\} = E\left\{\sum_{i\in\Omega_A} e_i + \sum_{i\in\Omega_B} e_i - \sum_{i\in\Omega} e_i\right\} = \sum_{i\in\Omega_{ab}} e_i, \tag{4.7}$$

and
$$\begin{aligned} E(Q) &= E\left\{\left(\sum_{i\in\Omega} \boldsymbol{x}_i^{'} - \sum_{i\in s_A} d_i \boldsymbol{x}_i^{'} - \sum_{i\in s_B} d_i \boldsymbol{x}_i^{'}\right)\left(\hat{B}_s^{A,B} - B_\Omega\right)\right\} \\ &= E\left(\sum_{i\in\Omega} \boldsymbol{x}_i^{'} - \sum_{i\in s_A} d_i \boldsymbol{x}_i^{'} - \sum_{i\in s_B} d_i \boldsymbol{x}_i^{'}\right) E\left(\hat{B}_s^{A,B} - B_\Omega\right) \\ &= E\left(\sum_{i\in\Omega} \boldsymbol{x}_i^{'} - \sum_{i\in\Omega_A} \boldsymbol{x}_i^{'} - \sum_{i\in\Omega_B} \boldsymbol{x}_i^{'}\right) E\left(\hat{B}_s^{A,B} - B_\Omega\right) \\ &= -\sum_{i\in\Omega_{ab}} \boldsymbol{x}_i^{'} E\left(\hat{B}_s^{A,B} - B_\Omega\right). \end{aligned} \tag{4.8}$$

Using linearization via Taylor Series, the estimator $\hat{B}_s^{A,B}$ may be written as

$$\hat{B}_s^{A,B} = B_\Omega^{A,B} + \frac{\sum_{i\in s'} d_i \boldsymbol{x}_i y_i - \sum_{i\in\Omega'} \boldsymbol{x}_i y_i}{\sum_{i\in\Omega'} \boldsymbol{x}_i \boldsymbol{x}_i^{'}} - \sum_{i\in\Omega'} \boldsymbol{x}_i y_i \frac{\sum_{i\in s'} d_i \boldsymbol{x}_i \boldsymbol{x}_i^{'} - \sum_{i\in\Omega'} \boldsymbol{x}_i \boldsymbol{x}_i^{'}}{(\sum_{i\in\Omega'} \boldsymbol{x}_i \boldsymbol{x}_i^{'})^2},$$

where
$$\sum_{i\in s'} d_i \boldsymbol{x}_i y_i = \sum_{i\in s_A} d_i \boldsymbol{x}_i y_i + \sum_{i\in s_B} d_i \boldsymbol{x}_i y_i,$$

$$\sum_{i\in\Omega'} \boldsymbol{x}_i y_i = \sum_{i\in\Omega_A} \boldsymbol{x}_i y_i + \sum_{i\in\Omega_B} \boldsymbol{x}_i y_i = \sum_{i\in\Omega} \boldsymbol{x}_i y_i + \sum_{i\in\Omega_{ab}} \boldsymbol{x}_i y_i,$$

$$\sum_{i\in s'} d_i \boldsymbol{x}_i \boldsymbol{x}_i^{'} = \sum_{i\in s_A} d_i \boldsymbol{x}_i \boldsymbol{x}_i^{'} + \sum_{i\in s_B} d_i \boldsymbol{x}_i \boldsymbol{x}_i^{'},$$

$$\sum_{i\in\Omega'} \boldsymbol{x}_i \boldsymbol{x}_i^{'} = \sum_{i\in\Omega_A} \boldsymbol{x}_i \boldsymbol{x}_i^{'} + \sum_{i\in\Omega_B} \boldsymbol{x}_i \boldsymbol{x}_i^{'} = \sum_{i\in\Omega} \boldsymbol{x}_i \boldsymbol{x}_i^{'} + \sum_{i\in\Omega_{ab}} \boldsymbol{x}_i \boldsymbol{x}_i^{'} \tag{4.9}$$

and $\hat{B}_s^{A,B} = 2\frac{\sum_{i\in s'} d_i \boldsymbol{x}_i y_i}{\sum_{i\in s'} d_i \boldsymbol{x}_i \boldsymbol{x}_i^{'}}, \quad B_\Omega^{A,B} = 2\frac{\sum_{i\in \Omega'} \boldsymbol{x}_i y_i}{\sum_{i\in \Omega'} \boldsymbol{x}_i \boldsymbol{x}_i^{'}}.$

$$\begin{aligned} \text{It is known that} \quad E(\hat{B}_s^{A,B}) &= B_\Omega^{A,B} \\ &= B_\Omega + B_\Omega^{A,B} - B_\Omega \text{ (on adding and subtracting} B_\Omega) \\ E(\hat{B}_s^{A,B} - B_\Omega) &= B_\Omega^{A,B} - B_\Omega. \end{aligned} \tag{4.10}$$

Thus, the expression of $E(Q)$ from Eqs. (4.8) and (4.10) will be

$$E(Q) = -\sum_{i\in\Omega_{ab}} \boldsymbol{x}_i^{'}\left(B_\Omega^{A,B} - B_\Omega\right). \tag{4.11}$$

Using Eqs. (4.6), (4.7), and (4.11),

$$\begin{aligned} E(\hat{T}_{JCE} - T) &= \sum_{i\in\Omega_{ab}} e_i - \sum_{i\in\Omega_{ab}} \boldsymbol{x}_i^{'}\left(B_\Omega^{A,B} - B_\Omega\right) \\ &= \sum_{i\in\Omega_{ab}} \left(e_i - \boldsymbol{x}_i^{'} B_\Omega^{A,B} + \boldsymbol{x}_i^{'} B_\Omega\right) \qquad \text{on putting the value of } e_i \\ &= \sum_{i\in\Omega_{ab}} \left\{(y_i - \boldsymbol{x}_i^{'} B_\Omega) - \boldsymbol{x}_i^{'} B_\Omega^{A,B} + \boldsymbol{x}_i^{'} B_\Omega\right\} \\ &= \sum_{i\in\Omega_{ab}} \left(y_i - \boldsymbol{x}_i^{'} B_\Omega^{A,B}\right) \end{aligned} \tag{4.12}$$

$$\text{Therefore,} \quad B(\hat{T}_{JCE}) = \sum_{i\in\Omega_{ab}} e_i^{A,B}, \tag{4.13}$$

where $e_i^{A,B} = (y_i - \boldsymbol{x}_i^{'} B_\Omega^{A,B})$.

Also, under dual frame design, approximate variance of $\hat{T}_{JCE}$ is given as

$$\begin{aligned} V(\hat{T}_{JCE}) &= E(\hat{T}_{JCE} - T)^2 = E(\hat{T}_{JCE})^2 + T^2 - 2E(\hat{T}_{JCE}T) \\ &= E\left(\sum_{i\in s_A} d_i y_i + \sum_{i\in s_B} d_i y_i + \sum_{i\in\Omega} y_i - \sum_{i\in s_A} d_i y_i - \sum_{i\in s_B} d_i y_i\right)^2 + T^2 - 2E(\hat{T}_{JCE}T). \end{aligned}$$

then after making rearrangement and approximations to higher-order term of $o(n)^2$, the variance is obtained as

$$V(\hat{T}_{JCE}) = \sum\sum_{i,j\in\Omega_A} \Theta_{ij}^{A}\left(\frac{e_i^A}{\pi_i^A}\right)\left(\frac{e_j^A}{\pi_j^A}\right) + \sum\sum_{i,j\in\Omega_B} \Theta_{ij}^{B}\left(\frac{e_i^B}{\pi_i^B}\right)\left(\frac{e_j^B}{\pi_j^B}\right)$$
$$+ \sum\sum_{i,j\in\Omega_{ab}} \Theta_{ij}^{ab}\left(\frac{e_i^{ab}}{\pi_i^{ab}}\right)\left(\frac{e_j^{ab}}{\pi_j^{ab}}\right), \quad (4.14)$$

where $s_{ab} = s_A \bigcap s_B$ and for $D = (A, B, ab),\ \ k = i, j,$

$$\Theta_{ij}^{D} = \left(\pi_{ij}^{D} - \pi_i^{D}\pi_j^{D}\right),\ \ \pi_{ij}^{D} = Pr\left(i\ \&j \in s_D\right),\ \ \pi_i^{D} = Pr(i \in s_D),$$
$$\pi_j^{D} = Pr(j \in s_D),\ \ e_k^{D} = y_k - \boldsymbol{x}_k^{'} B_{\Omega_D},\ \ B_{\Omega_D} = 2\frac{\sum_{i\in\Omega_D} \boldsymbol{x}_i y_i}{\sum_{i\in\Omega_D} \boldsymbol{x}_i \boldsymbol{x}_i^{'}}.$$

Assuming negligibly small values for π_i^{ab}, π_{ij}^{ab}, π_j^{ab}, the estimated variance is given by

$$\hat{V}(\hat{T}_{JCE}) = \sum\sum_{i,j\in s_A} \frac{\Theta_{ij}^{A}}{\pi_{ij}}\left(\omega_i\hat{e}_i^{A}\right)\left(w_j\hat{e}_j^{A}\right) + \sum\sum_{i,j\in s_B} \frac{\Theta_{ij}^{B}}{\pi_{ij}}\left(\omega_i\hat{e}_i^{B}\right)\left(w_j\hat{e}_j^{B}\right), \quad (4.15)$$

where $\hat{e}_k^{D} = y_k - \boldsymbol{x}_k^{'}\hat{B}_{ws_D}$ and $\hat{B}_{ws_D} = 2\frac{\sum_{i\in s_D} \omega_i \boldsymbol{x}_i y_i}{\sum_{i\in s_D} \omega_i \boldsymbol{x}_i \boldsymbol{x}_i^{'}}$.

Since the variance estimation does not provide an unbiased result, a simulation study to evaluate the performance of the proposed estimator is also required. In the next section, the performance of JCE under minimum entropy distance function is compared theoretically with JCE under the chi-square distance function.

5 Performance of Proposed JCE

Theoretically, the performance of the JCE under minimum entropy distance function and chi-square distance function under SRSWOR and Lahiri–Midzuno (L-M) sampling designs is shown. This study focuses on estimating the population total of study variable under dual frame surveys.

Under chi-square distance function, the expression for $B(\hat{T}_{JCE})$ is given as:

$$B_1(\hat{T}_{JCE}) = \sum_{i\in\Omega_{ab}} \left(y_i - \boldsymbol{x}_i^{'} B_{\Omega}^{*,A,B}\right), \quad (5.1)$$

$$\text{where}\quad B_{\Omega}^{*,A,B} = \frac{\sum_{i\in\Omega} \boldsymbol{x}_i y_i}{\sum_{i\in\Omega} \boldsymbol{x}_i \boldsymbol{x}_i^{'}} \quad (5.2)$$

and under minimum entropy distance function, the expression of bias of proposed JCE is given by

$$B_2(\hat{T}_{JCE}) = \sum_{i \in \Omega_{ab}} \left(y_i - \boldsymbol{x}_i^{'} B_{\Omega}^{A,B} \right), \tag{5.3}$$

$$\text{where} \quad B_{\Omega}^{A,B} = 2 \frac{\sum_{i \in \Omega} \boldsymbol{x}_i y_i}{\sum_{i \in \Omega} \boldsymbol{x}_i \boldsymbol{x}_i^{'}} = 2 B_{\Omega}^{*,A,B}. \tag{5.4}$$

Now, on taking the difference

$$\begin{aligned} B_1(\hat{T}_{JCE}) - B_2(\hat{T}_{JCE}) &= \sum_{i \in \Omega_{ab}} \left(y_i - \boldsymbol{x}_i^{'} B_{\Omega}^{*,A,B} \right) - \sum_{i \in \Omega_{ab}} \left(y_i - \boldsymbol{x}_i^{'} B_{\Omega}^{A,B} \right) \\ &= \sum_{i \in \Omega_{ab}} \left(y_i - \boldsymbol{x}_i^{'} B_{\Omega}^{*,A,B} \right) - \sum_{i \in \Omega_{ab}} \left(y_i - 2\boldsymbol{x}_i^{'} B_{\Omega}^{*,A,B} \right) \\ &= \sum_{i \in \Omega_{ab}} \boldsymbol{x}_i^{'} B_{\Omega}^{*,A,B} \\ &= \sum_{i \in \Omega_{ab}} \boldsymbol{x}_i^{'} \frac{\sum_{i \in \Omega} \boldsymbol{x}_i y_i}{\sum_{i \in \Omega} \boldsymbol{x}_i \boldsymbol{x}_i^{'}} \end{aligned}$$

Therefore,

$$B_1(\hat{T}_{JCE}) = B_2(\hat{T}_{JCE}) + \sum_{i \in \Omega_{ab}} \boldsymbol{x}_i^{'} \frac{\sum_{i \in \Omega} \boldsymbol{x}_i y_i}{\sum_{i \in \Omega} \boldsymbol{x}_i \boldsymbol{x}_i^{'}}. \tag{5.5}$$

The second term on right-hand side of Eq. (5.5) decides the difference between the bias of the two estimators. Both calibration estimators will be same in terms of bias if $\sum_{i \in \Omega_{ab}} \boldsymbol{x}_i^{'} \frac{\sum_{i \in \Omega} \boldsymbol{x}_i y_i}{\sum_{i \in \Omega} \boldsymbol{x}_i \boldsymbol{x}_i^{'}} = 0$.

The expressions of approximated variance estimator of JCE under chi-square distance function can be obtained as

$$\begin{aligned} \hat{V}_1(\hat{T}_{JCE}) &= \sum\sum_{i,j \in s_A} \frac{\Theta_{ij}^{A}}{\pi_{ij}} \left(\omega_i \hat{e}_i^{A} \right) \left(w_j \hat{e}_j^{A} \right) + \sum\sum_{i,j \in s_B} \frac{\Theta_{ij}^{B}}{\pi_{ij}} \left(\omega_i \hat{e}_i^{B} \right) \left(w_j \hat{e}_j^{B} \right) \\ &= \sum\sum_{i,j \in s_A} \frac{\Theta_{ij}^{A}}{\pi_{ij}} \left\{ \omega_i \left(y_i - \boldsymbol{x}_i^{'} \hat{B}_{ws_A} \right) \right\} \left\{ w_j \left(y_j - \boldsymbol{x}_j^{'} \hat{B}_{ws_A} \right) \right\} \\ &+ \sum\sum_{i,j \in s_B} \frac{\Theta_{ij}^{B}}{\pi_{ij}} \left\{ \omega_i \left(y_i - \boldsymbol{x}_i^{'} \hat{B}_{ws_B} \right) \right\} \left\{ w_j \left(y_j - \boldsymbol{x}_j^{'} \hat{B}_{ws_B} \right) \right\}, \end{aligned} \tag{5.6}$$

where $\hat{e}_k^{D} = y_k - \boldsymbol{x}_k^{'} \hat{B}_{ws_D}, \quad \hat{B}_{ws_D} = \frac{\sum_{i \in s_D} \omega_i \boldsymbol{x}_i y_i}{\sum_{i \in s_D} \omega_i \boldsymbol{x}_i \boldsymbol{x}_i^{'}}$.

and under minimum entropy distance function,

$$\hat{V}_2(\hat{T}_{JCE}) = \sum\sum_{i,j\in s_A} \frac{\Theta_{ij}^A}{\pi_{ij}} \left(\omega_i \hat{e}_i^A\right)\left(w_j \hat{e}_j^A\right) + \sum\sum_{i,j\in s_B} \frac{\Theta_{ij}^B}{\pi_{ij}} \left(\omega_i \hat{e}_i^B\right)\left(w_j \hat{e}_j^B\right)$$

$$= \sum\sum_{i,j\in s_A} \frac{\Theta_{ij}^A}{\pi_{ij}} \left\{\omega_i \left(y_i - 2\boldsymbol{x}_i^{'}\hat{B}_{ws_A}\right)\right\}\left\{w_j \left(y_j - 2\boldsymbol{x}_j^{'}\hat{B}_{ws_A}\right)\right\}$$

$$+ \sum\sum_{i,j\in s_B} \frac{\Theta_{ij}^B}{\pi_{ij}} \left\{\omega_i \left(y_i - 2\boldsymbol{x}_i^{'}\hat{B}_{ws_B}\right)\right\}\left\{w_j \left(y_j - 2\boldsymbol{x}_j^{'}\hat{B}_{ws_B}\right)\right\}. \quad (5.7)$$

To compare the performance of these estimators, the difference of the variance estimators can be obtained as

$$\hat{V}_1(\hat{T}_{JCE}) - \hat{V}_2(\hat{T}_{JCE}) = \sum\sum_{i,j\in s_A} \frac{\Theta_{ij}^A}{\pi_{ij}} \left\{\omega_i \left(y_i - \boldsymbol{x}_i^{'}\hat{B}_{ws_A}\right)\right\}\left\{w_j \left(y_j - \boldsymbol{x}_j^{'}\hat{B}_{ws_A}\right)\right\}$$

$$+ \sum\sum_{i,j\in s_B} \frac{\Theta_{ij}^B}{\pi_{ij}} \left\{\omega_i \left(y_i - \boldsymbol{x}_i^{'}\hat{B}_{ws_B}\right)\right\}\left\{w_j \left(y_j - \boldsymbol{x}_j^{'}\hat{B}_{ws_B}\right)\right\}$$

$$- \sum\sum_{i,j\in s_A} \frac{\Theta_{ij}^A}{\pi_{ij}} \left\{\omega_i \left(y_i - 2\boldsymbol{x}_i^{'}\hat{B}_{ws_A}\right)\right\}\left\{w_j \left(y_j - 2\boldsymbol{x}_j^{'}\hat{B}_{ws_A}\right)\right\}$$

$$- \sum\sum_{i,j\in s_B} \frac{\Theta_{ij}^B}{\pi_{ij}} \left\{\omega_i \left(y_i - 2\boldsymbol{x}_i^{'}\hat{B}_{ws_B}\right)\right\}\left\{w_j \left(y_j - 2\boldsymbol{x}_j^{'}\hat{B}_{ws_B}\right)\right\}$$

On further simplification, $\hat{V}_1(\hat{T}_{JCE}) - \hat{V}_2(\hat{T}_{JCE})$ is obtained as

$$= \sum\sum_{i,j\in s_A} \frac{\Theta_{ij}^A}{\pi_{ij}} \Big\{\omega_i w_j \Big(y_i y_j - y_i \boldsymbol{x}_j^{'}\hat{B}_{ws_A} + \boldsymbol{x}_i^{'}\boldsymbol{x}_j^{'}\hat{B}_{ws_A}^2 - y_j \boldsymbol{x}_i^{'}\hat{B}_{ws_A} - y_i y_j$$

$$+ 2y_i \boldsymbol{x}_j^{'}\hat{B}_{ws_A} - 4\boldsymbol{x}_i^{'}\boldsymbol{x}_j^{'}\hat{B}_{ws_A}^2 + 2y_j \boldsymbol{x}_i^{'}\hat{B}_{ws_A}\Big)\Big\} + \sum\sum_{i,j\in s_B} \frac{\Theta_{ij}^B}{\pi_{ij}} \Big\{\omega_i w_j \Big(y_i y_j - y_i \boldsymbol{x}_j^{'}\hat{B}_{ws_B}$$

$$+ \boldsymbol{x}_i^{'}\boldsymbol{x}_j^{'}\hat{B}_{ws_B}^2 - y_j \boldsymbol{x}_i^{'}\hat{B}_{ws_B} - y_i y_j + 2y_i \boldsymbol{x}_j^{'}\hat{B}_{ws_B} - 4\boldsymbol{x}_i^{'}\boldsymbol{x}_j^{'}\hat{B}_{ws_B}^2 + 2y_j \boldsymbol{x}_i^{'}\hat{B}_{ws_B}\Big)\Big\}.$$

On simplifying the expressions, it is obtained as

$$\hat{V}_1(\hat{T}_{JCE}) - \hat{V}_2(\hat{T}_{JCE}) = \sum\sum_{i,j\in s_A} \frac{\Theta_{ij}^A}{\pi_{ij}} \left\{\omega_i w_j \left(y_i \boldsymbol{x}_j^{'} - 3\boldsymbol{x}_i^{'}\boldsymbol{x}_j^{'}\hat{B}_{ws_A} + y_j \boldsymbol{x}_i^{'}\right)\hat{B}_{ws_A}\right\}$$

$$+ \sum\sum_{i,j\in s_B} \frac{\Theta_{ij}^B}{\pi_{ij}} \left\{\omega_i w_j \left(y_i \boldsymbol{x}_j^{'} - 3\boldsymbol{x}_i^{'}\boldsymbol{x}_j^{'}\hat{B}_{ws_B} + y_j \boldsymbol{x}_i^{'}\right)\hat{B}_{ws_B}\right\}. \quad (5.8)$$

The above Eq. (5.8) does not provide any theoretical support about the superiority of $\hat{T}_{JCE}$ under the two considered distance functions. Thus, a numerical or simulation study is required to analyze the efficiency of both estimators under different sampling designs.

6 Higher-Order Calibration for Variance Estimation of JCE

Singh et al. (1998) were the first to use a higher-order calibration approach to estimate the variance of GREG estimator using the known variance of auxiliary variable. Using this concept, the precision of proposed estimator can be improved. Many authors, Das and Tripathi (1978), Singh et al. (1999), gave the estimators of variance of estimators of population total with known auxiliary variable variance. Théberge (1999) also discussed calibration-based approach to variance estimation. In this view, the weights Θ_{ij}^A and Θ_{ij}^B are revised by some new weights Ψ_{ij}^A and Ψ_{ij}^B in such a way that the chi-square distance function is minimized subject to some calibration constraint.

The estimator of variance of JCE is given as:

$$\hat{V}_1(\hat{T}_{JCE}) = \sum\sum_{i,j\in s_A} \Theta_{ij}^A(\omega_i\hat{e}_i^A)(w_j\hat{e}_j^A) + \sum\sum_{i,j\in s_B} \Theta_{ij}^B(\omega_i\hat{e}_i^B)(w_j\hat{e}_j^B) \tag{6.1}$$

$$\text{where} \quad \Theta_{ij}^D = \frac{\pi_{ij}^D - \pi_i^D\pi_j^D}{\pi_{ij}}, \quad \hat{e}_k^D = y_k - \boldsymbol{x}_k^{'}\hat{B}_{ws_D} \text{ and } \hat{B}_{ws_D} = \frac{\sum_{i\in s_D}\omega_i\boldsymbol{x}_i y_i}{\sum_{i\in s_D}\omega_i\boldsymbol{x}_i\boldsymbol{x}_i^{'}}.$$

An improved variance estimator is provided as

$$\hat{V}_2(\hat{T}_{JCE}) = \sum\sum_{i,j\in s_A} \Psi_{ij}^A(\omega_i\hat{e}_i^A)(w_j\hat{e}_j^A) + \sum\sum_{i,j\in s_B} \Psi_{ij}^B(\omega_i\hat{e}_i^B)(w_j\hat{e}_j^B), \tag{6.2}$$

where Ψ_{ij} are calibrated weights such that the chi-square distance function

$$D_3(\Psi_{ij}, \Theta_{ij}) = \sum\sum_{i,j\in s_A} \frac{(\Psi_{ij}^A - \Theta_{ij}^A)^2}{\Theta_{ij}^A} + \sum\sum_{i,j\in s_B} \frac{(\Psi_{ij}^B - \Theta_{ij}^B)^2}{\Theta_{ij}^B} \tag{6.3}$$

is minimum subject to the constraint

$$\sum\sum_{i,j\in s_A} \Psi_{ij}^A(d_ix_i^A)(d_jx_j^A) + \sum\sum_{i,j\in s_B} \Psi_{ij}^B(d_ix_i^B)(d_jx_j^B) = V(\hat{X}_{JCE}). \tag{6.4}$$

Here, the information on auxiliary variable is assumed to be known. Thus, like its mean, the variance of auxiliary variable is also incorporated to improve the variance estimator for study variable. The Lagrangian function will be

$$L = \sum\sum_{i,j\in s_A} \frac{(\Psi_{ij}^A - \Theta_{ij}^A)^2}{\Theta_{ij}^A} + \sum\sum_{i,j\in s_B} \frac{(\Psi_{ij}^B - \Theta_{ij}^B)^2}{\Theta_{ij}^B}$$
$$- 2\lambda \left\{ \sum\sum_{i,j\in s_A} \Psi_{ij}^A (d_i x_i^A)(d_j x_j^A) + \sum\sum_{i,j\in s_B} \Psi_{ij}^B (d_i x_i^B)(d_j x_j^B) - V(\hat{X}_{JCE}) \right\}. \tag{6.5}$$

On partial differentiating the Lagrangian function with respect to Ψ_{ij}^A and Ψ_{ij}^B and setting equal to zero, the value of new calibrated weights will be obtained as

$$\Psi_{ij}^A = \Theta_{ij}^A + \lambda\Theta_{ij}^A (d_i x_i^A)(d_j x_j^A) \tag{6.6}$$

and $$\Psi_{ij}^B = \Theta_{ij}^B + \lambda\Theta_{ij}^B (d_i x_i^B)(d_j x_j^B). \tag{6.7}$$

On putting the values of $(\Psi_{ij}^A - \Theta_{ij}^A)$ and $(\Psi_{ij}^B - \Theta_{ij}^B)$ in Eq. (6.5), we have

$$\sum\sum_{i,j\in s_A} \lambda^2\Theta_{ij}^A (d_i x_i^A)^2 (d_j x_j^A)^2 + \sum\sum_{i,j\in s_B} \lambda^2\Theta_{ij}^B (d_i x_i^B)^2 (d_j x_j^B)^2$$
$$- 2\lambda \left\{ \sum\sum_{i,j\in s_A} \left(\Theta_{ij}^A + \lambda\Theta_{ij}^A (d_i x_i^A)(d_j x_j^A)\right) (d_i x_i^A)(d_j x_j^A) \right.$$
$$\left. + \sum\sum_{i,j\in s_B} \left(\Theta_{ij}^B + \lambda\Theta_{ij}^B (d_i x_i^B)(d_j x_j^B)\right) (d_i x_i^B)(d_j x_j^B) - V(\hat{X}_{JCE}) \right\} = 0$$
$$\Rightarrow \lambda \left\{ \sum\sum_{i,j\in s_A} \Theta_{ij}^A (d_i x_i^A)^2 (d_j x_j^A)^2 + \sum\sum_{i,j\in s_B} \Theta_{ij}^B (d_i x_i^B)^2 (d_j x_j^B)^2 \right\}$$
$$= -2 \left\{ \sum\sum_{i,j\in s_A} \Theta_{ij}^A (d_i x_i^A)(d_j x_j^A) + \sum\sum_{i,j\in s_B} \Theta_{ij}^B (d_i x_i^B)(d_j x_j^B) - V(\hat{X}_{JCE}) \right\}$$
$$\Rightarrow \lambda \left\{ \sum\sum_{i,j\in s_A} \Theta_{ij}^A (d_i x_i^A)^2 (d_j x_j^A)^2 + \sum\sum_{i,j\in s_B} \Theta_{ij}^B (d_i x_i^B)^2 (d_j x_j^B)^2 \right\} = 2\left\{ V(\hat{X}_{JCE}) - \hat{V}(\hat{X}_{JCE}) \right\}$$
$$\Rightarrow \lambda = \frac{2\left\{ V(\hat{X}_{JCE}) - \hat{V}(\hat{X}_{JCE}) \right\}}{\sum\sum_{i,j\in s_A} \Theta_{ij}^A (d_i x_i^A)^2 (d_j x_j^A)^2 + \sum\sum_{i,j\in s_B} \Theta_{ij}^B (d_i x_i^B)^2 (d_j x_j^B)^2}, \tag{6.8}$$

where $\hat{V}(\hat{X}_{JCE}) = \sum\sum_{i,j\in s_A} \lambda\Theta_{ij}^A (d_i x_i^A)(d_j x_j^A) + \sum\sum_{i,j\in s_B} \lambda\Theta_{ij}^B (d_i x_i^B)(d_j x_j^B)$. On putting the value of λ from Eq. (6.8) in Eq. (6.6)

$$\begin{aligned}\Psi_{ij}^{A} &= \Theta_{ij}^{A} + \frac{\left\{V(\hat{X}_{JCE}) - \hat{V}(\hat{X}_{JCE})\right\}\Theta_{ij}^{A}(d_i x_i^A)(d_j x_j^A)}{\sum\sum_{i,j\in s_A}\Theta_{ij}^{A}(d_i x_i^A)^2(d_j x_j^A)^2 + \sum\sum_{i,j\in s_B}\Theta_{ij}^{B}(d_i x_i^B)^2(d_j x_j^B)^2} \\ &= \Theta_{ij}^{A} + \hat{B}^{A}\left\{V(\hat{X}_{JCE}) - \hat{V}(\hat{X}_{JCE})\right\},\end{aligned} \tag{6.9}$$

where

$$\hat{B}^{A} = \frac{\Theta_{ij}^{A}(d_i x_i^A)(d_j x_j^A)}{\sum\sum_{i,j\in s_A}\Theta_{ij}^{A}(d_i x_i^A)^2(d_j x_j^A)^2 + \sum\sum_{i,j\in s_B}\Theta_{ij}^{B}(d_i x_i^B)^2(d_j x_j^B)^2}. \tag{6.10}$$

Similarly,

$$\Psi_{ij}^{B} = \Theta_{ij}^{B} + \hat{B}^{B}\left\{V(\hat{X}_{JCE}) - \hat{V}(\hat{X}_{JCE})\right\}, \tag{6.11}$$

where $$\hat{B}^{B} = \frac{\Theta_{ij}^{B}(d_i x_i^B)(d_j x_j^B)}{\sum\sum_{i,j\in s_A}\Theta_{ij}^{A}(d_i x_i^A)^2(d_j x_j^A)^2 + \sum\sum_{i,j\in s_B}\Theta_{ij}^{B}(d_i x_i^B)^2(d_j x_j^B)^2}. \tag{6.12}$$

On substituting the values of Ψ_{ij}^{A} and Ψ_{ij}^{B} from Eqs. (6.9) and (6.11) in Eq. (6.2), an improved estimator of variance is obtained as:

$$\begin{aligned}\hat{V}_2(\hat{T}_{JCE}) &= \sum\sum_{i,j\in s_A}\Theta_{ij}^{A}(\omega_i\hat{e}_i^A)(w_j\hat{e}_j^A) + \sum\sum_{i,j\in s_B}\Theta_{ij}^{B}(\omega_i\hat{e}_i^B)(w_j\hat{e}_j^B) \\ &+ \left\{\sum\sum_{i,j\in s_A}\hat{B}^{A}(\omega_i\hat{e}_i^A)(w_j\hat{e}_j^A) + \sum\sum_{i,j\in s_B}\hat{B}^{B}(\omega_i\hat{e}_i^B)(w_j\hat{e}_j^B)\right\}\left\{V(\hat{X}_{JCE}) - \hat{V}(\hat{X}_{JCE})\right\}.\end{aligned} \tag{6.13}$$

Now using Eqs. (6.1) and (6.13), we have

$$\hat{V}_2(\hat{T}_{JCE}) = \hat{V}_1(\hat{T}_{JCE}) + \hat{B}^{*}\left\{V(\hat{X}_{JCE}) - \hat{V}(\hat{X}_{JCE})\right\} \tag{6.14}$$

where $$\hat{B}^{*} = \left\{\sum\sum_{i,j\in s_A}\hat{B}^{A}(\omega_i\hat{e}_i^A)(w_j\hat{e}_j^A) + \sum\sum_{i,j\in s_B}\hat{B}^{B}(\omega_i\hat{e}_i^B)(w_j\hat{e}_j^B)\right\}.$$

7 Combining the Individual Frame Estimators

First, consider the case of non-overlapping dual frames, and the two estimates of frame A and B can be combined using a convex combination of weights to obtain a composite estimator as,

$$\hat{T}_{combined} = \Phi\hat{T}_A + (1-\Phi)\hat{T}_B.$$

Variance expression (after ignoring the covariance term) will be given as

$$V(\hat{T}_{combined}) = \Phi^2 V(\hat{T}_A) + (1-\Phi)^2 V(\hat{T}_B). \tag{7.1}$$

To compute optimum weights, the above expression is minimized with respect to Φ as

$$\frac{\delta\left(V(\hat{T}_{combined})\right)}{\delta\Phi} = 0 \implies 2\Phi V(\hat{T}_A) + 2(1-\Phi)V(\hat{T}_B) = 0,$$

$$\text{or} \quad \Phi[V(\hat{T}_A) + V(\hat{T}_B)] - V(\hat{T}_B) = 0,$$

$$\therefore \quad \Phi_{opt} = \frac{V(\hat{T}_B)}{V(\hat{T}_A) + V(\hat{T}_B)}. \tag{7.2}$$

Now, another way to get a sensitive weighing interval for weights Φ when $\hat{T}_{combined}$ is more efficient than $\hat{T}_A$, i.e., $V(\hat{T}_{combined}) \leq V(\hat{T}_A)$.
On substituting the value of $V(\hat{T}_{combined})$ from Eq. (7.1) and solving further

$$\Phi^2 V(\hat{T}_A) + (1-\Phi)^2 V(\hat{T}_B) \leq V(\hat{T}_A),$$

$$\text{or} \quad \Phi = \frac{2V(\hat{T}_B) \pm \sqrt{4V^2(\hat{T}_B) - 4(V^2(\hat{T}_B) - V^2(\hat{T}_A))}}{2(V(\hat{T}_A) + V(\hat{T}_B))}$$

$$\therefore \quad \Phi = \frac{V(\hat{T}_A) \pm V(\hat{T}_B)}{V(\hat{T}_A) + V(\hat{T}_B)}. \tag{7.3}$$

$$\text{Thus,} \qquad \Phi_{opt} \in \left(1 - \frac{2V(\hat{T}_A)}{V(\hat{T}_A) + V(\hat{T}_B)},\ 1\right). \tag{7.4}$$

Similarly, the weighing interval for weights Φ when $\hat{T}_{combined}$ is more efficient than $\hat{T}_B$ is given by

$$\left(1 - \frac{2V(\hat{T}_B)}{V(\hat{T}_A) + V(\hat{T}_B)},\ 1\right). \tag{7.5}$$

Now, in case of overlapping frames, simply adding the two estimates results in the biased estimator. So, an unbiased dual frames estimator for $\hat{T}_{combined}$ can be obtained by the weighted average of the overlap frame estimator, i.e., $\hat{T}^A_{ab}$ and $\hat{T}^B_{ab}$, $\hat{T}_{combined} = \hat{T}_A + \Phi\hat{T}^A_{ab} + (1-\Phi)\hat{T}^B_{ab} + \hat{T}_B$, where $\Phi \in [0, 1]$ is a composite factor combining the estimators of overlapping frames which is already discussed earlier. Now, to examine the performance of JCE under minimum entropy distance function, an empirical study is carried out in the next section.

8 A Simulation Study

Here, the MU284 population data of Sweden among its 284 municipalities from Särndal et al. (1992) book are considered. It contains different variables that describe their municipalities. Typically, a municipality consists of a town and the surrounding area that vary considerably in size and other characteristics. Few variables that describe the municipality in some way are considered. In our study, y, the study variable denotes the revenue for the year 1984 while, x, an auxiliary variable is the number of municipal employees in 1984, is considered. The whole population (units from first to eighth geographic regions) is divided into two overlapping frames A (populations from first to fifth geographic regions) and B (populations from third to eighth geographic regions), i.e., overlapping of third, fourth, and fifth regions and samples of proportion 5, 10, 15, 20, 25, and 30% of total population are drawn under two sampling designs, i.e., SRSWOR and L-M scheme. The values of the estimators are simulated M = 5000 using R-software.

The first-order inclusion probability π_i under Midzuno–Sen sampling scheme is given by

$$\pi_i = \left(\frac{N-n}{N-1}\right) P_i + \frac{n-1}{N-1}, \tag{8.1}$$

$$\text{where} \quad P_i = \frac{x_i}{\sum_{i \in \Omega} x_i}.$$

Here, P_i is the initial probability of selecting the ith unit with x as the size measure. Now, the two dual frame estimators, i.e., one under the chi-square distance function (CSDF) and another under minimum entropy distance function (MEDF) for population total, are analyzed in terms of absolute relative bias (ARB) and simulated relative squared error (SRSE) of the estimators, defined as:

$$ARB(\hat{T}_{JCE}) = \frac{\frac{1}{M}\sum_{i=1}^{M} \hat{T}_{JCE} - Y}{Y} \times 100 \tag{8.2}$$

$$\text{and} \quad SRSE(\hat{T}_{JCE}) = \frac{\sqrt{ASE(\hat{T}_{JCE})}}{E(\hat{T}_{JCE})} \times 100 \tag{8.3}$$

where $ASE(\hat{T}_{JCE}) = \frac{1}{M}\sum_{i=1}^{M}(\hat{T}_{JCE} - T)^2$ and $E(\hat{T}_{JCE}) = \frac{1}{M}\sum_{i=1}^{M}\hat{T}_{JCE}$.

On analyzing the dual frame estimators under two distance functions, the ARB values of JCE under chi-square distance function vary from 12.95 to 13.55 in the samples of proportions 5 to 30% using SRSWOR design while it varies from 4.37 to 13.39 for minimum entropy distance function under same sampling scheme showing significant reduction in ARB values throughout the samples; see Table 2. So, it is better to use JCE under minimum entropy distance function when complete information is not available as per ARB is concerned. But for SRSE, JCE under minimum

Table 2 ARB and SRSE for JCE under SRSWOR design for the dataset on MU284 population

Sample proportion (%)	ARB(EDF)	ARB(CSDF)	SRSE(EDF)	SRSE(CSDF)
5	4.37	13.23	51.87	17.77
10	10.88	13.21	36.10	15.31
15	12.75	12.95	29.39	14.30
20	13.57	13.24	25.49	14.17
25	13.29	13.55	22.68	14.24
30	13.39	13.41	21.00	13.95

Table 3 ARB and SRSE for JCE under L-M Design for MU284 population

Sample proportion (%)	ARB(EDF)	ARB(CSDF)	SRSE (EDF)	SRSE (CSDF)
5	7.95	14.34	32.8	18.55
10	2.35	13.92	24.1	15.9
15	5.65	13.33	21.00	14.61
20	8.13	13.55	19.47	14.43
25	9.71	13.80	18.14	14.44
30	10.21	13.57	17.31	14.10

entropy distance function does not perform better; as seen under chi-square distance function, the SRSE values are continuously smaller for varying sample proportions, under SRSWOR. A positive side for EDF is that under L-M, the variation among SRSE values seems less as compared to SRSWOR design.

Here, under L-M design, the performance of JCE regarding SRSE values under different sample proportions varies 17.31–32.8 which is less dispersed as compared to SRSE values 21–51.87 under minimum entropy distance function; see Table 3. The situation is little opposite for performance values of SRSE under chi-square distance function as it varies 13.95–17.77 and 14.10–18.55 under L-M and SRSWOR design, respectively.

Using different covariates among National Land Cover Database tree canopy cover estimates can be obtained using Compound Topographic Index and Digital Elevation Model or others. Under different sampling designs as mentioned above, the efficiency of Joint Calibration Estimator for tree canopy cover estimates can be obtained in the similar line.

9 Conclusion

In classical design-based sampling theory, a random sample with some inclusion probabilities is drawn from the single sampling frame to make inferences about the population. The HT estimator produces unbiased results in case of full coverage of the target population in the sampling frame, full response of all sampled units, and absence of measurement error. But, the practical challenges of single frame surveys faced by many government agencies (e.g., US Government) are incomplete frames, decreasing response rates, risen data collection costs, and increasing demand for small area statistics. There is a need to use multiple frames with an assumption of complete coverage of the targeted population. An empirical study is carried out to compare the performance of the proposed estimator with the existing one under SRSWOR and L-M sampling designs.

Acknowledgements We are very thankful to editor and the two learned referees for their fruitful suggestions to improve the quality and insightful content of the present chapter.

References

Alka, Rai, P. K., & Qasim, M. (2019). Two-step calibration of design weights under two auxiliary variables in sample survey. *Journal of Statistical Computation and Simulation*, *89*, 2316–2327.

Bankier, M. D. (1986). Estimators based on several stratified samples with applications to multiple frame surveys. *Journal of the American Statistical Association*, *81*, 1074–1079.

Das, A. K., & Tripathi, T. P. (1978). Use of auxiliary information in estimating the finite population variance. *Sankhya Series C*, *40*, 139–148.

Deville, J. C., & Särndal, C. E. (1992). Calibration estimators in survey sampling. *Journal of the American Statistical Association*, *87*, 376–382.

Elkasabi, M. A., Heeringa, S. G., & Lepkowski, J. M. (2015). Joint calibration estimator for dual frame surveys. *Statistics in Transition, New Series*, *16*, 7–36.

Fuller, W. A., & Burmeister, L. F. (1972). Estimators for samples selected from two overlapping frames. In *Proceedings of the Social Statistics Section, American Statistical Association*. Fort Collins, CO, pp. 245–249.

Haines, D. E., & Pollock, K. H. (1998). Estimating the number of active and successful bald eagle nests: An application of the dual frame method. *Environmental and Ecological Statistics*, *5*, 245–256.

Hartley, H. O. (1962). Multiple Frame Surveys. In *Proceedings of the Social Statistics Section, American Statistical Association*. Alexandria, VA, pp. 203–206.

Hartley, H. O. (1974). Multiple frame methodology and selected applications. *Sankhya, Series C*, *36*, 99–118.

Kalton, G., & Anderson, D. W. (1986). Sampling Rare Populations. *Journal of the Royal Statistical Society: Series A (Statistics in Society)*, *149*, 65–82.

Kuo, L. (1989). Composite estimation of totals for livestock surveys. *Journal of the American Statistical Association, 84*, 421–429.

Lohr, S. L. (2009). *Multiple-frame surveys* (Vol. 29). Handbook of Statistics: Elsevier.

Lohr, S. L. (2011). Alternative survey sample designs: Sampling with multiple over-lapping frames. *Survey Methodology 37*, 197–213.

Lohr, S. L. & Rao, J. N. K. (2000). Inference from dual frame surveys. *Journal of the American Statistical Association, 95*, 271–280.
Lund, R. E. (1968). Estimators in Multiple Frame Surveys. In *Proceedings of the social statistics section, american statistical association,*. Pittsburgh, PA, pp. 282–286.
McConville, K. S., Breidt, F. J., Lee, T. C. M., & Moisen, G. G. (2017). Model-assisted survey regression estimation with the lasso. *Journal of Survey Statistics and Methodology, 5*, 131–158.
Rai, P. K., Tikkiwal, G. C., Alka & Singh, S., (2018). Calibration approach-based estimator under minimum entropy distance Function VIS-A-VIS T-2 class of estimator. *International Journal of Applied Engineering Research, 13*, 15329–15342.
Shyvers, J., Walker, B., & Noon, B. (2018). Dual-frame lek surveys for estimating greater sage-grouse populations. *The Journal of Wildlife Management, 82*, 1689–1700.
Singh, A. C., & Mohl, C. A. (1996). Understanding calibration estimators in survey sampling. *Survey Methodology, 22*, 107–115.
Singh, S. (2004). Golden and Silver Jubilee Year-2003 of the Linear Regression Estimators. In *Proceedings of the American Statistical Association, Survey Method Section*. pp. 4382–4389.
Singh, S., & Arnab, R. (2011). On calibration of design weights. *Metron, 69*, 185–205.
Singh, S., Horn, S., Chowdhury, S., & Yu, F. (1999). Calibration of the estimators of variance. *Australian and New Zealand Journal of Statistics, 41*, 199–212.
Singh, S., Horn, S., & Yu, F. (1998). Estimation of variance of general regression estimator: Higher level calibration approach. *Survey methodology, 24*, 41–50.
Skinner, C. J. (1991). On the efficiency of raking ratio estimation for multiple frame surveys. *Journal of the American Statistical Association, 86*, 779–784.
Stukel, D., Hidiroglou, M., & Särndal, C. E. (1996). Variance estimation for calibration estimators: a comparison of jackknifing versus taylor linearization. *Survey Methodology, 22*, 117–125.
Särndal, C. E. (2007). The calibration approach in survey theory and practice. *Survey Methodology, 33*, 99–119.
Särndal, C. E., Swensson, B., & Wretman, J. (1992). *Model assisted survey sampling*. New York: Springer-Verlag.
Théberge, A. (1999). Extensions of calibration estimators in survey sampling. *Journal of the American Statistical Association, 94*, 635–644.

Fusing Classical Theories and Biomechanics into Forest Modelling

S. Suresh Ramanan, T. K. Kunhamu, Deskyong Namgyal and S. K. Gupta

Abstract There is a renaissance in forest modelling due to the application of mathematics and physics. There are some classical theories including pipe model theory, Metzger's theory, self-thinning rule, Da Vinci's tree form concept, Logarithmic spiral technique, etc., which have greater significance in forestry science. With advanced computational tools from the IT revolution at disposal, a better understanding of the above-mentioned theories is now possible. In meantime, plants' architecture and design have been a source of inspiration for biomechanists, as mechanics is an inseparable part of the abiotic realm. It is based on one important principle "all structures, whether engineered or natural, must obey the laws of physics". Trees grow, adapt and acclimate to maintain their stability. This demands a trade-off between their mechanical stability and other physiological functions. It is also vivid that mechanical forces can manipulate the tree architecture and root architecture and influence thigomorphogensis. For this reason, it is important to understand the impact of mechanical forces on tree growth. Forest modelling can take a leap forward by infusing these theories and biomechanics. The present chapter narrates some of the classical theories in forestry and simultaneously showcases the relevance of these theories based on research work done. Furthermore, it deliberates on the utilization of modelling to provide greater impetus in forest science in order to explore prudent silvicultural practice for enhancing forest productivity and product quality.

Keywords Biomechanics · Forests · Modelling · Theories

S. S. Ramanan (✉) · D. Namgyal · S. K. Gupta
Division of Agroforestry, Sher-E-Kashmir University of Agricultural Sciences and Technology of Jammu, Jammu, India
e-mail: sureshramanan01@gmail.com

T. K. Kunhamu
Department of Silviculture and Agroforestry, Kerala Agricultural University, Kerala, India

G. Chandra et al. (eds.), *Statistical Methods and Applications in Forestry and Environmental Sciences*, Forum for Interdisciplinary Mathematics,
https://doi.org/10.1007/978-981-15-1476-0_9

1 Introduction

Science is the knowledge gathered through systematic observation of the physical world with well-defined reasons and strategies by which general principles and laws are deduced. Ultimately, it involves measurement, analysis and interpretation for clarity of the concept. The general procedure adopted is to formulate theories or hypothesis to support the interpretation of the complex persona of nature (Torres and Santos 2015). In the real world, it may or may not be possible to explain all the phenomena and it is where models—a simplified representation of a system—do aid in understanding and quantification. Models are the abstraction of reality, and it serves as a means for systematizing the information pertaining to a phenomenon (Bézivin 2005). Models can be simple as a verbal statement or an arrow between two boxes or complex and extremely complicated as that of nitrogen transformation pathways in an ecosystem or intercellular signalling (Jackson et al. 2000). There are many classical theories in forestry and modelling that can be used to validate theories. Sometimes, these theories are debatable and they are also subjected to the retesting, time-to-time. In this chapter, some of the classical theories are briefly discussed. It also attempts to discuss future prospects in forest modelling as well as shows the real-time application.

Development and advancement in a scientific field correlate positively with levels of usage of models. In the twenty-first century, mathematical models have become more prevalent in biology. The advent of computers has sorted many of the problems, and there is ample evidence that modelling is the key to understand the nature(Hübner et al. 2011; Lanza et al. 2012). Forestry is the most unique branch of science because of the dynamic and complexity attribute by the long life, plasticity in morphology and adaptation to diverse habitat. This makes it very challenging for model developers for construction, validation and usage. The modelling can incorporate time as one of the variables thereby making temporal analysis feasible (Wainwright and Mulligan, 2005). This chapter deliberates on some of the classical theories in forestry, and it also attempts to highlight the relevance of these theories.

1.1 Classical Theories and Biomechanics

There are many theories in forestry science which are based on ecological, mathematical and mechanical applications, stand productivity, mortality, etc. For instance, the climax theory is an ecological concept. So from the wide array of theories, our focus will be on theories based on mathematics and mechanics for understanding the tree architecture. One of the oldest contributions to tree architecture is Leonardo's rule of area conservation by Leonardo da Vinci, the great painter and scientist. It states that branches of a tree at every stage of its height when put together are equal in thickness to the trunk (Eloy 2011). In a much simple word, the ratio between the sum cross-sectional areas of daughter branches to the cross-sectional area of the mother branch should be constant. In terms of formula, the area conservation ratio is given by the formula, $\frac{(d_1+d_2)}{m_0}$, where d_1 and d_2 are the cross-sectional areas of daughter branches and m_0 is the area of the mother branch. This ratio computed for temperate species like balsa, maple, oak, pinyon and ponderosa pine ranged from 0.90 to 1.05 (Bentley et al. 2013). The same concept has been applied to understand the root architecture (Oppelt et al. 2001).

One of the mathematical tools used to study the root architecture is logarithmic spiral. The methodology proves to be very handy and cost-effective. A detailed description is provided in the real-time application section.

One more famous theory—Metgzer's theory—on tree form was published in 1893 which states that the wind-induced stresses should be constant along the tree trunk. It is based on the axiom of uniform stress (Morgan and Cannell 1994, Mattheck 2006). This theory has paved the way in understanding plant architecture to a larger extent, and this forms a greater part under the concept of thigmomorphogenesis (response by plants to the mechanical sensation by altering their growth patterns) (McMahon and Kronauer 1976, King and Loucks 1978, Ennos 1997). The concept of thigmomorphogenesis is also widely used in modelling (Fournier et al. 2006). Similar to Metgzer's tree form theory, the concept of elastic similarity also describes the mechanical design of trees. It is an allometric law which relates to branch radii and lengths, such that the deflection of the branch tip under self-weight is proportional to its length. This law has been applied to understand the biomass allocation pattern (Enquist and Niklas 2002).

Understanding the influence of mechanical forces on tree architecture has been pestering the scientists. This is nothing but an attempt to understand the response of a living organism to the mechanical force—Biomechanics. This emerging field has helped us to understand and reshape many ideologies in the plant as well as animal science (Humphrey and O'Rourke 2015).

The development in this field can be attributed to the statement "there is no result or change in nature without a cause"—Leonardo da Vinci (Bronzino and Peterson 2014). Plants have been fascinating biomechanists for quite a long duration and also mechanics an inseparable component of abiotic interaction experienced by a plant. It can be gravity, wind, soil, aquatic current or biotic interaction with other plant or animal or microorganisms. On the whole, plant biomechanics is defined as "study of structures and functions of the biological system of plant phylum, on the concept and foundation of mechanics" (Boudaoud 2010). It encompasses continuum mechanics, solid and fluid mechanics, kinematics, statistics, dynamics, mechanism and structural analysis, the strength of the material and modelling, and of course, it combines scientific information from many biological disciplines under concern such as genomics, botany, biochemistry, ecophysiology, ecology as well as palaeobotany. Some of the prominent scientists are Galileo Galilei who illustrated that it was the peripheral material rather than the central construction material that resisted external bending using a hollow stalk of a grass. Some more examples from the great artist and inventor, Leonardo da Vinci who took nature as inspiration for many of his inventions- the auto gyroscopic propeller was out of the dandelion pappus and maple samara and list does go on. It was Simon Schwendener's book "Das mechanisheP-rinizipimanatomischenBau der Monocoltyledonen (The Mechanical Principles of the Anatomy of Monocotyledons)" which was the first comprehensive work published in 1874. Subsequently, there were many other publications from many scholars. With present-day technological tools, many of the initial theories are being evaluated and reformulated and simultaneously the application of biomechanical concept took into various dimensions in plant sciences from molecular cell structure to complex life forms. The field of biomechanics is based on one underlying important principle "all structures, whether engineered or natural must obey the laws of physics" (Niklas and Spatz 2012).

2 Pioneer Work Done

The concept of integrating both biomechanics and modelling has successfully done by the AMAPstudio which is developed as a collaborative effort of University of Montpellier, CIRAD and INRA (institut national de la recherche agronomique) and AMAP (botanique et bioinformatique de l'architecture des plantes). The AMAP-studio has two main modules, Xplo and Simeo. The Xplo is specifically designed for individual plant level, whereas Simeo is developed for stand level. Java-based software has the ability for integration of several tools such as radiative balance, biomechanics and realistic visualization. A biomechanical model called AMAPpara is an outcome. The AMAPpara has been inbuilt with the provision of finite element method (FEM) (Fourcaud and Lac 2003).

Sellier et al. (2006) investigated the tree aerial architecture changes due to tree oscillations. The experiment revealed that the foliage governs the damping of the stem and it is also found that A_3 (third branch order) has a greater influence on the damping. Thus, architectural third-order branching plays a significant role in dissipating the mechanical force. The result of the experiment clearly indicates that branching pattern has a significant role in the tree oscillations due to wind thereby effectively influencing the tree architecture. This is an outcome due to the incorporation of modelling with biomechanics. One more significant work by the University of Calgary is "*Incorporating Biomechanics into Architectural Tree Models*" instead of replicating real-time observation. The authors have made an attempt to model it in the real-time shaping as it happens in the natural environment. For this attempt, they have used the L-systems which are very commonly used in plant architect designing. Jirasek et al. (2000) work was the first of its sort to incorporate both L-system and biomechanical aspects. It was continued taking care of the biomechanical aspect of branch modelling. The primary assumptions in this concept are that the entire branch is divided into a number of small segments. The influence of load or any other biomechanical parameter is reflected by infinitesimal rotations that are expressed as three vectors aligned perpendicular to each other. This has enabled them to create a simulation of the tree much closer to the real-world scenario. Both AMAP and L-systems are prominent and promising means in modelling growth with due concern for parameters such as architecture, geometry and biomechanics (Fourcaud et al. 2008). Hence, they can be regarded as stepping stones in modelling the dynamic response of the branches and trees as a whole.

3 Real-Time Applications

Our team attempted two of the classical theories in real-time scenario. One was Leonardo's rule of area conservation, and another was logarithmic spiral trenching.

Leonardo's rule of area conservation

A simple diameter measurement was carried out at the branches and the main leader of the trees. We assessed the validity of Leonardo's rule in two conditions (i) *Acacia mangium* plantation in Kerala (ii) different tree species planted as avenue trees in Sher-e-Kashmir University Agricultural Sciences and Technology of Jammu, Jammu.

Through our field measurements where at two different sites, Leonardo's rule of area conservation holds good in both avenue trees and *Acacia* plantations. The area conservation ratio ranged from 0.92 to 1.42 in the case of avenue trees. For the *Acacia mangium*, inside the plantation, the ratio was 1.16 (Fig. 1). This value is in the agreeable limit and closely resembles the values reported by Sone et al. (2009). The work also shows the real-time application of Leonardo's rule combined with the pipe model theory.

Logarithmic spiral trench

Methods for studying the root architecture and distribution are numerous, each having their own advantages and disadvantages. Root excavation (destructive sampling) is a widely adopted method which is capable of giving more reliable results. In spite of it, tree excavation is considered a standard for coarse root biomass estimation and has been widely used (Snowdon et al. 2001, Thakur et al. 2015). However, destructive sampling is laborious and time-consuming and it is very tedious to carry out on a large scale. Furthermore, it fails to determine the actual area to which fine roots and its activity are confined which is of much relevance in agroforestry context. Nevertheless, most of the root studies reported in the literature concerning plantation forestry and agroforestry systems are based on the excavation method. In order to overcome the drawbacks of destructive sampling, a new innovative method was proposed by Huguet (1973)—known as the logarithmic spiral trenching. It was implemented by Fernández et al. (1991) for understanding the effect of drip irrigation in olive trees followed by Tomlinson et al. (1998) in *Parkiabiglobosa*. This method has been very innovative and was adapted by other researchers with slight modifications (Bhimappa 2014, Divakara et al. 2001, Srinivasan et al. 2004, Thakur et al. 2015).

A typical logarithmic spiral is shown in Fig. 2. The general equation of this spiral is given by the polar equation (Eq. 1).

Fig. 1 Validation of Leonardo's rule of area conservation

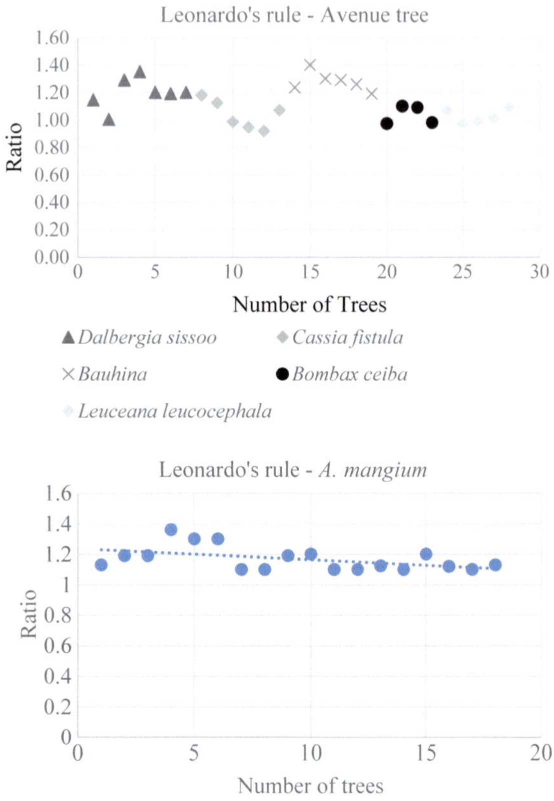

Fig. 2 Typical logarithmic spiral

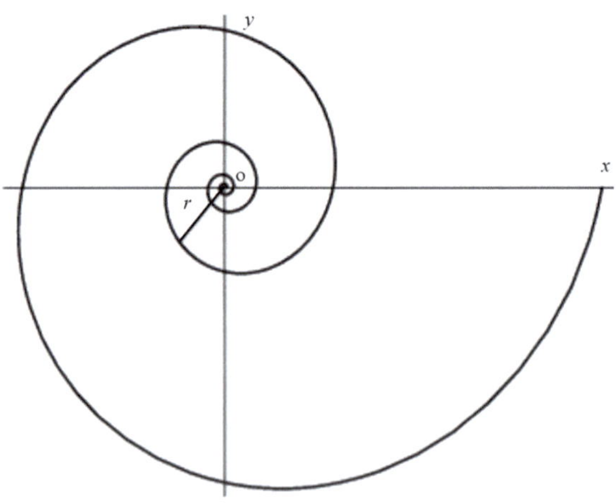

$$z = a \times e^{b\theta} \tag{3.1}$$

where z is the distance from the origin, θ is the angle from the x-axis while a and b are arbitrary constants and are both positive integers. The value of b determines the closeness of the spiral. This is the mathematical equation for logarithmic spiral (Lockwood 1967).

To employ this equation in the field, the value of the arbitrary constants has to be determined. Even though, this technique for root studies was employed by Fernández et al. (1991) without making any statement on the arbitrary values. Tomlinson et al. (1998) made a small modification to the equations by assigning values to both a and b for a study in *Parkiabiglobosa*, based on the correlation between the crown diameters, diameter at breast height (dbh) and lateral root spread. Precisely, the modification is,

$$a = 1.5 \times d \tag{3.2}$$

$$b = \frac{\ln\left(\frac{r}{d}\right)}{\pi} \tag{3.3}$$

Combining Eqs. (1), (2) and (3), we get,

$$z = 1.5 \times d \times e^{\frac{\ln\left(\frac{r}{d}\right)}{\pi} \times \theta} \tag{3.4}$$

where d is the tree diameter in metres (m), r is the average of the crown radius at four cardinal points (in m), a is the distance of the starting point of the spiral from the tree (in m), b is the natural logarithmic of the ratio of crown radius to the diameter of the tree divided by π, z is the distance of any point on the spiral from the tree base (in m) and $\theta = 0°$, 22.5°, 45°, 67.5°, 90°, 112.5°, 135°, 157.5° and 180°. A 20-year-old *Acacia mangium* tree was selected inside the plantation that was planted at the spacing of 2.5 × 2.5 m. The trajectory of the trench was laid down on the field using plastic ropes by calculating the distance on the north side of the tree which will be the origin, and further extension is done in the spiral clockwise direction with θ taking values 0°, 22.5°, 45°, 67.5°, 90°, 112.5°, 135°, 157.5° and 180°. The trench was dug to a depth of 60 cm and a width of 60 cm and care was taken so that the sides remain intact. Severed roots (living) on the internal and external trench walls were counted by placing a 50 cm × 50 cm quadrat (subdivided into 10 cm depth intervals). Roots were classified into less than 2 mm, 2–5 mm, 5–2 cm and >2 cm diameter classes at the time of counting by placing the quadrats (50 cm × 50 cm) at fixed distances from the trunk. Root counts were converted into rooting intensity (number of roots m^{-2}) (Bohm 1979).

Here, our aim was not to discuss the significant result of the study but as such the significance of the methodology. For instance, there is a need to understand the fine root distribution of a tree species to decide its suitability for agroforestry. In this regard, Table 1 and Fig. 3 can provide the relevant data.

Table 1 Mean root intensity (m^2) of 20-year-old *Acacia mangium* plantation at various distances from the tree base

Root diameter class	Distance from the tree (m)				
	0.35	1.00	1.38	2.08	2.73
<2 mm	309.60	432.20	218.60	501.00	368.40
2–5 mm	18.60	9.20	4.40	6.20	5.60
5–20 mm	0.60	0.80	0.40	3.00	3.80
>2 cm	0.00	1.20	0.00	0.00	0.60

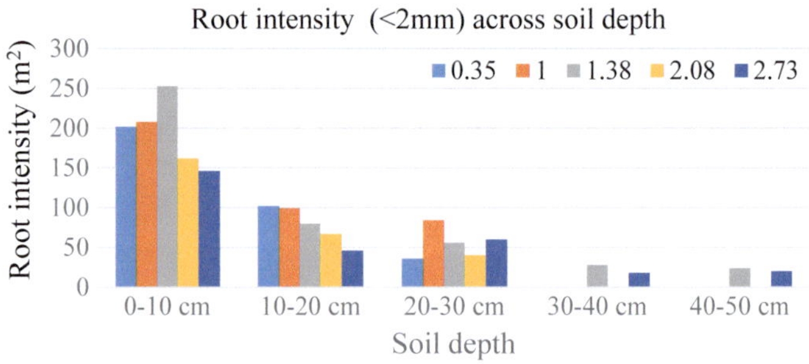

Fig. 3 Fine root intensity (m^2) of 20-year-old Acacia mangium at a different soil depth

The results from Fig. 3 clearly showed that *Acacia mangium* is surface feeder as highest root intensity is recorded in the 0-10 cm soil layer which clearly reveals the viability of incorporating the tree in agroforestry system. There are other studies using the high-end methodology to show the result on par with our study; i.e. *Acacia mangium* has profuse surface roots which even make them idea species for slope stabilization (Voottipruex et al. 2008, Ali 2010, Lateh et al. 2012, 2013, Avani et al. 2015). With sufficient samples across different sites, the simple mathematical model can be created, assess the root intensity with respect to distance from tree base; even for a different age.

4 Conclusions

The scope of classical theories and biomechanics in forestry has been discussed in this chapter. We were able to test and validate Leonardo's rule and logarithmic spiral trenching. The results can be used full-fledged modelling with furthermore studies. For instance, the work of Eloy et al. (2017) who developed a realistic tree model–MECHATREE, to show that wind load and competition for light brings the self-similarity in a tree shape. He has used many published works based on the pipe model theory, Leonardo's rule and strahler rank concept on temperate trees. It is time to develop theories suited for the tropical conditions. Such works, incorporating theories and biomechanics into modelling, can provide greater impetus to unravel the mysteries in forest science. This may help to explore prudent silvicultural practice for enhancing productivity and product quality.

References

Ali, F. (2010). Use of vegetation for slope protection: Root mechanical properties of some tropical plants. *International Journal of Physical Sciencies, 5*(5), 496–506.

Avani, N., Lateh, H., & Bibalani, G. H. (2015). Root distribution of *Acacia mangium* Willd. and Macaranga tanarius L. of rainforest. *Bangladesh Journal of Botany, 43,* 141–145.

Bentley, L. P., Stegen, J. C., Savage, V. M., Smith, D. D., Allmen, E. I., Sperry, J. S., et al. (2013). An empirical assessment of tree branching networks and implications for plant allometric scaling models. *Ecological Letters, 16,* 1069–1078.

Bézivin, J. (2005). On the unification power of models. *Software and Systematic Modelling, 4*(2), 171–188.

Bhimappa, K. (2014). Bamboo based agroforestry systems in Kerala, India: performance of turmeric (*Curcumalonga* L.) in the subcanopy of differentially spaced seven year-old bamboo stand. *Agroforestry Systems, 90*(2), 237–250.

Bohm, W. (1979). *Methods of studying root systems* (p. 188p). Berlim: Springer-Verlag.

Boudaoud, A. (2010). An introduction to the mechanics of morphogenesis for plant biologists. *Trends in Plant Science, 15,* 353–360.

Bronzino, J. D. & Peterson, D. R. (2014). Biomechanics: Principles and practices (pp. 352). CRC Press.

Divakara, B. N., Kumar, B. M., Balachandran and P. V, Kamalam, N. V. (2001). Bamboo hedgerow systems in Kerala, India: Root distribution and competition with trees for phosphorus. *Agroforestry Systems*, *51*, 189–200.

Eloy, C. (2011). Leonardo's rule, self-similarity, and wind-induced stresses in trees. *Physical Review Letters, 107*(25), 258101.

Eloy, C., Fournier, M., Lacointe, A., & Moulia, B. (2017). Wind loads and competition for light sculpt trees into self-similar structures. *Nature Communications, 8,* 1014.

Ennos, A. (1997). Wind as an ecological factor. *Trends in Ecology & Evolution, 12,* 108–111.

Enquist, B. J., & Niklas, K. J. (2002). Global allocation rules for patterns of biomass partitioning in seed plants. *Science, 80*(295), 1517–1520.

Fernández, J. E., Moreno, F., Cabrera, F., Arrue, J. L., & Martín-Aranda, J. (1991). Drip irrigation, soil characteristics and the root distribution and root activity of olive trees. *Plant and Soil, 133,* 239–251.

Fourcaud, T., & Lac, P. (2003). Numerical modelling of shape regulation and growth stresses in trees. *Trees, 17,* 23–30.

Fourcaud, T., Zhang, X., Stokes, A., Lambers, H., & Körner, C. (2008). Plant growth modelling and applications: the increasing importance of plant architecture in growth models. *Annals of Botany, 101,* 1053–1063.

Fournier, M., Stokes, A., Coutand, C., Fourcaud, T., & Moulia, B. (2006). Tree biomechanics and growth strategies in the context of forest functional ecology. *Ecology and Biomechanics—a mechancial approach to ecology of animals and plants* (pp. 1–33). Boca Raton: CRC Press.

Hübner, K., Sahle, S., & Kummer, U. (2011). Applications and trends in systems biology in biochemistry. *The FEBS Journal, 278,* 2767–2857.

Huguet, J. G. (1973). Nouvelle methode d'etude de l'enracinement des vegetaux perennes a partir d'une trachee spirale. *Annales Agronomiques, 24*(6), 707–731.

Humphrey, J. D., & O'Rourke, S. L. (2015). *An introduction to biomechanics* (p. 692). New York: Springer New York.
Jackson, L. J., Trebitz, A. S., & Cottingham, K. L. (2000). An introduction to the practice of ecological modeling. *BioScience, 50*(8), 694–706.
Jirasek, C., Prusinkiewicz, P., & Moulia, B. (2000). Integrating biomechanics into developmental plant models expressed using L-systems. *Plant Biomechanics, 19,* 615–624.
King, D., & Loucks, O. L. (1978). The theory of tree bole and branch form. *Radiation and Environmental Biophysics, 15,* 141–165.
Lanza, A. M., Crook, N. C., & Alper, H. S. (2012). Innovation at the intersection of synthetic and systems biology. *Current Opinion in Biotechnology, 23,* 712–717.
Lateh, H., Avani, N., & Bibalani, G. H. (2012). Investigation of root distribution and tensile strength of *Acaciamangium* willd (Fabaceae) in the rainforest. *Greener Journal of Biological Sciences, 4,* 45–52.
Lateh, H., Avani, N., & Bibalani, H. (2013). Effect of *Acaciamangium* root properties on shallow landslide and slope stability. *Journal of Life Science and Technology, 1,* 127–131.
Lockwood, E. H. (1967). *A book of curves* (p. 690). Cambridge University Press.
Mattheck, C. (2006). Teacher tree: The evolution of notch shape optimization from complex to simple. *Engineering Fracture Mechanics, 73,* 1732–1742.
McMahon, T. A., & Kronauer, R. E. (1976). Tree structures: Deducing the principle of mechanical design. *Journal of Theoretical Biology, 59,* 443–466.
Morgan, J., & Cannell, M. G. R. (1994). Shape of tree stems—a re-examination of the uniform stress hypothesis. *Tree Physiology, 14,* 49–62.
Niklas, K. J., & Spatz, H. C. (2012). *Plant physics* (p. 426) University of Chicago Press.
Oppelt, A. L., Kurth, W., & Godbold, D. L. (2001). Topology, scaling relations and Leonardo's rule in root systems from African tree species. *Tree Physiology, 21,* 117–128.
Sellier, D., Fourcaud, T., & Lac, P. (2006). A finite element model for investigating effects of aerial architecture on tree oscillations. *Tree Physiology, 26,* 799–806.
Snowdon, P., Keith, H., & Raison, R. J. (2001). Protocol for sampling tree and stand biomass. Technical Report No. 31, Australian Greenhouse Office, Canberra.
Sone, K., Suzuki, A. A., Miyazawa, S. I., Noguchi, K., & Terashima, I. (2009). Maintenance mechanisms of the pipe model relationship and Leonardo da Vinci's rule in the branching architecture of Acer rufinerve trees. *Journal of Plant Research, 122,* 41–52.
Srinivasan, K., Kunhamu, T. K., & Mohankumar, B. (2004). Root excavation studies in a mature rubber (*Heveabrasiliensis* Muell. Arg.) plantation. *Natural Rubber Research, 17,* 18–22.
Thakur, S., Kumar, B. M., & Kunhamu, T. K. (2015). Coarse root biomass, carbon, and nutrient stock dynamics of different stem and crown classes of silver oak (*Grevillearobusta* A. Cunn. ex. R. Br.) plantation in central Kerala, India. *Agroforestry Systems, 89,* 869–883.
Tomlinson, H., Traore, A., & Teklehaimanot, Z. (1998). An investigation of the root distribution of Parkia biglobosa in Burkina Faso, West Africa, using a logarithmic spiral trench. *Forest and Ecology Management, 107,* 173–182.
Torres, N. V., & Santos, G. (2015). The mathematical modeling process in biosciences. *Frontiers in Genetics, 6,* 354.
Voottipruex, P., Bergado, D. T., Mairaeng, W., Chucheepsakul, S., & Modmoltin, C. (2008). Soil reinforcement with combination roots system: a case study of Vetiver grass and Acacia Mangium Wild. *Lowland Technology International, 10,* 56–67.
Wainwright, J., & Mulligan, M. (2005). *Environmental modelling: finding simplicity in complexity* (p. 494). John Wiley and Sons.

Investigating Selection Criteria of Constrained Cluster Analysis: Applications in Forestry

Gavin Richard Corral

Abstract Forest measurements are inherently spatial. Soil productivity varies spatially at fine scales, and tree growth responds by changes in growth-age trajectories. Measuring spatial variability is a perquisite to more effective analysis and statistical testing. In this chapter, current techniques of partial redundancy analysis and constrained cluster analysis are used to explore how spatial variables determine structure in a managed regular spaced plantation. The objectives of this research are to measure, test, and map spatial variability in simulated forest plots. We test for spatial relationships in the data and then explore how those spatial relationships are manifested into spatially recognizable structures. Partial redundancy analysis was found to be a good method for detecting the presence or absence of spatial relationships (~95% accuracy). We found that the Calinski-Harabasz method consistently performed better at detecting the correct number of clusters when compared to several other methods. Constrained cluster analysis has promising applications in forestry and that the Calinski-Harabasz criterion will be most useful.

Keywords Calinski-Harabasz · Constrained cluster · Partial redundancy

1 Introduction

Forests are an invaluable part of our environment. They provide ecosystem services like carbon sequestration, clear air and water, fuel, food and have aesthetic value and recreational use. Timber products and livestock grazing, for example, are economic commodities. Managing forests is complex and requires advanced knowledge of tree growth and stand dynamics. Understanding stand dynamics requires knowledge of how the stands vary through time and space. Many studies in forestry have looked at how trees grow and stands develop through time (Oliver and Larson 1996). Statistical models and methods for examining the effect of time are commonplace in ecological

G. R. Corral (✉)
USDA NASS Research and Development Division, 1400 Independence Ave SW, Room 6408B, 20250 Washington, DC, USA
e-mail: Gavin.Corral@nass.usda.gov

G. Chandra et al. (eds.), *Statistical Methods and Applications in Forestry and Environmental Sciences*, Forum for Interdisciplinary Mathematics,
https://doi.org/10.1007/978-981-15-1476-0_10

studies. Studied to a much lesser extent, but equally as important is the study of how space is related to tree growth and stand development. Spatial heterogeneity is a fundamental concept to many ecological theories including patterns of succession, stand disturbance, habitat structure, and site productivity.

Environmental scientists have a major interest in spatial analysis (Legendre and Legendre 1998). Forest cover and tree distributions are in part determined by spatial variation, in particular, soil patterns within forests are seen to mostly affect tree distributions (Leduc et al. 1992). This phenomenon is prevalent in most environmental studies because ecological measurements taken by sampling geographic space are affected by spatial components (terBraakand Prentice 1988; Leduc et al. 1992; Legendre and Legendre 1998; Peres-Neto et al. 2006). Legendre and Legendre (1998) noted that ecological processes create spatially recognizable structures, which may display spatial patterns and be the subject of spatial analysis. Hulbert (1984) acknowledges that through mensurative experiments, which include measurements made at multiple points in space, will allow one to test hypotheses about patterns in space.

A need for an investigation into fine-scale spatial effects on tree development exists. Discontinuities in soil productivity often happen at fine scales. The patches formed by these discontinuities are called microsites. These microsites cause trees to grow along different height-age trajectories (Oliver and Larson 1996). Weber (1983) recognizes that the height-age curves for these microsites can be polymorphic (i.e., curves are not proportional). Polymorphism occurs when one microsite may favor rapid early growth and another may favor rapid late growth (Oliver and Larson 1996). Unlike anamorphic differences where curves develop proportionally, polymorphic patterns imply that structural differences may develop at different points in time and not necessarily be exhibited evenly through time. Oliver and Larson (1996) explain that trees on good microsites are more likely to become dominant if they grow next to trees on poor microsites. This gives rise to spatially recognizable structures, which develop spatial patterns that may be spatially analyzed (Legendre and Legendre 1998).

Testing for spatial relationships, measuring spatial variance and mapping spatial structures lead to a rich understanding of underlying ecological processes. Testing for spatial heterogeneity of growth will allow practitioners to make more informed decisions into site preparation and treatments by refining delineation of treatment areas. Many interrelated processes such as depth and composition of soil layers, water and nutrient availability, decomposition rates, and macro- and microfauna composition govern site and microsite productivity. All these processes affect the growth response of trees. Recognizing and understanding these spatial patterns is integral in being able to model tree growth and understand stand dynamics. A reasonable approach to investigate the phenomena of spatially recognizable structures in tree growth would be to first test if spatial relationships exist between growth and space. The methods of canonical analysis and ordination are well suited to analyze spatial relationships.

In this chapter, we investigate current techniques of partial redundancy analysis (pRDA) and constrained cluster analysis to explore how spatial variables explain structure in a managed regular spaced plantation. We test for spatial relationships in the data and then explore how those spatial relationships are manifested into spatially recognizable structures. The objectives of this research are to measure, test, and map spatial variability in forest plots. This entails:

1. Simulate stands with varying levels of within plot heterogeneity.
   a. Test effectiveness of pRDA to detect no spatial correlations in a control plot with no microsites. This will be measured by success or failure to detect spatial relationship.
      i. Simulate 100 times at each level of variability
      ii. Estimate the probability of detecting no spatial pattern
   b. Test effectiveness of constrained cluster analysis to detect number of microsites (k) of our specified spatial patterns with known k. This will be measured as a success or failure to detect correct number of groups.
      i. Simulate 100 times for each combination of variability and difference in mean.
      ii. Estimate the probability of successful detection by each criterion.
   c. Perform a misclassification test to estimate percent of trees correctly allocated to its each microsite when using a specified criterion.
      i. Estimate the probability of misclassification by each criterion.

2 Literature Review

2.1 Ordination and Redundancy Analysis

Ordination, simply the arrangement of units into some order, is used in canonical analysis (Legendre and Legendre 1998; Peres-Neto et al. 2006; Borcard et al. 2011). In ordination, one orders objects which are explained by values across multiple variables such that object that are near each other are more similar than object that are further from each other. Bray (1957) observed that a rise in the use and application of ordination techniques began in the 1950s. Work such as Motyka et al. (1950), Curtis and McIntosh (1951), Brown and Curtis (1952), Vries (1952), Webb (1954), and Poore (1955) began applying and building quantitative techniques for plant community classification. Through the years, ordination has been used in forestry in studying the variance of some response, generally vegetation composition, across a landscape (see Greig-Smith 1967; Peet 1981; Lahti 1987; Martel et al. 2007; Grimaldi et al. 2014).

Redundancy analysis (RDA) provides the means for conducting direct explanatory analysis in which the association among growth measures can be studied according to their shared relationship with environmental and spatial variables (Peres-Neto et al. 2006). Rao (1964) was the first to describe RDA, and it was later rediscovered and presented by Van den Wollenberg (1977). Since then, RDA has been applied widely in many different aspects of ecology. Peres-Neto et al. (2006) noted that over

1500 studies have been published applying canonical correspondence analysis (an extension of RDA) and RDA in modeling species environment relationships.

Borcard et al. (1992) introduced variation partitioning using two sets of explanatory variables. One set of variables was called environmental and described as descriptors that are not spatially structured. The other set of explanatory variables was spatially structured. They described applications of pRDA that could be used to partition the variation of observed responses into components of variation. The partitioning included pure environmental, pure spatial, environmental spatial, and undetermined components of the variation. This method allows a partition of explained variation relative to the total amount of variation. This will also allow for significance testing.

Hypothesis testing is possible with RDA, because the canonical axes (eigenvectors) are orthogonal to each other (Borcard et al. 2011). The details of hypothesis testing go beyond the scope of this research, but it is an important concept to know because it allows for testing of spatial correlation in the data set and for testing of fractions in pRDA. Permutation F-tests are commonly used in ecological data sets to deal with the complex nature of the data (Legendre and Legendre 1998, Borcard et al. 2011). terBraak and Šmilauer (2002) constructed an F statistic that can examine fractional analysis via pRDA. This analysis is done in the presence of a second explanatory matrix of spatial coordinates (Legendre et al. 2011). These methods can bring insight into whether or not spatial variables are related growth variables of trees. Once it is determined that spatial relationships exist it would be logical to address how they are manifested. This insight can be gained by cluster analysis.

2.2 Cluster Analysis

Cluster analysis is the discovery of groups in data (Everitt et al. 2011). This is a very simple way to think about an incredibly diverse subject. Legendre and Legendre (1998) explain that clustering is a family of techniques that are undergoing rapid development. This development is in large part due to that fact that clustering is not a traditional statistical method in that there are no formal procedures to follow or hypotheses to test. There is a lack of literature investigating the use of cluster analysis on regular spaced, managed forests. In the 1980s, a pulse of forestry literature utilizing cluster analysis emerged. Lorimer (1985) described how cluster analysis could be used to improve his sampling design to better understand disturbance history in the forest. Guevara et al. (1986) used cluster analysis to examine the spatial analysis of forest succession and habitat patterns for bird species. Applications of cluster analysis stayed consistent through the years with focus on forest patch diversity, canopy patterns, or disturbance patterns (see Fraver 1994, Oliveira-Filho et al. 2000, Plotkin et al. 2002, Steane et al. 2006). These applications are appropriate as spatial heterogeneity of populations, and communities play a central role in many ecological theories, in particular theories of succession, adaption, maintenance of species diversity, community stability, competition, parasitism, epidemics, and natural catastrophes (Legendre and Fortin 1989).

2.2.1 Clustering Algorithms

There are great deals of algorithms and methods of clustering so, for brevity, we describe our general approach to clustering by introducing some terms:

Sequential—This algorithm works by applying a recurrent sequence of operations to the objects (trees).
Agglomerative—Agglomerative procedures begin with all objects being considered separate from one and another. This method successively groups the objects into larger and larger clusters until a single all encompassing clusters are obtained.
Hierarchical—Hierarchical methods allocate members of inferior ranking clusters to larger, higher ranking clusters. Most of the time and in our case, this method will produce non-overlapping clusters.
Non-Probabilistic—Non-probabilistic methods do not use parametric or nonparametric methods for estimating density functions in multivariate space. There are no probabilities linked with the association matrices.

The Lance and Williams general model (Borcard et al. 2011) for clustering encompasses many agglomerative methods and is easy translatable to many computer packages (Legendre and Legendre 1998). The Lance and Williams algorithm is also appropriate for constrained cluster analysis when the objects in a cluster are contiguous (Legendre and Legendre 1998). This method is based on a similarity matrix whose elements are values that describe how related two objects are. The Lance and Williams algorithm is known as a combinatorial agglomerative method because it utilizes a similarity matrix (Legendre and Legendre 1998). This clustering procedure is used widely for classification of landscape and cover type (Drewa et al. 2002; Urban et al. 2002; Snelder et al. 2004; Perrin et al. 2006; Divíšek et al. 2014). Through the clustering process and analysis, we will be able to develop maps of spatial correlation (Legendre et al. 2009, Legendre and Legendre 2012).

2.2.2 Determining the Number of Clusters

Currently, there are several criteria to determine the appropriate number of clusters. Among these are the adjusted R^2, AIC, Calinski-Harabasz (CH) and cross-validation residual error (CVRE). In the following sections, the simulations used to test the accuracy of these criteria are described.

3 Simulation

Sites representing 625 trees (25 × 25) were simulated. Within each site, patterns of microsites were embedded. The simulations replicated 5 sites with distinct spatial patterns (Fig. 1), each with different microsites. Response and explanatory variables are generated randomly. The response variables are diameter at breast height (DBH) and total tree height (H). The DBH values are generated separately for each microsite.

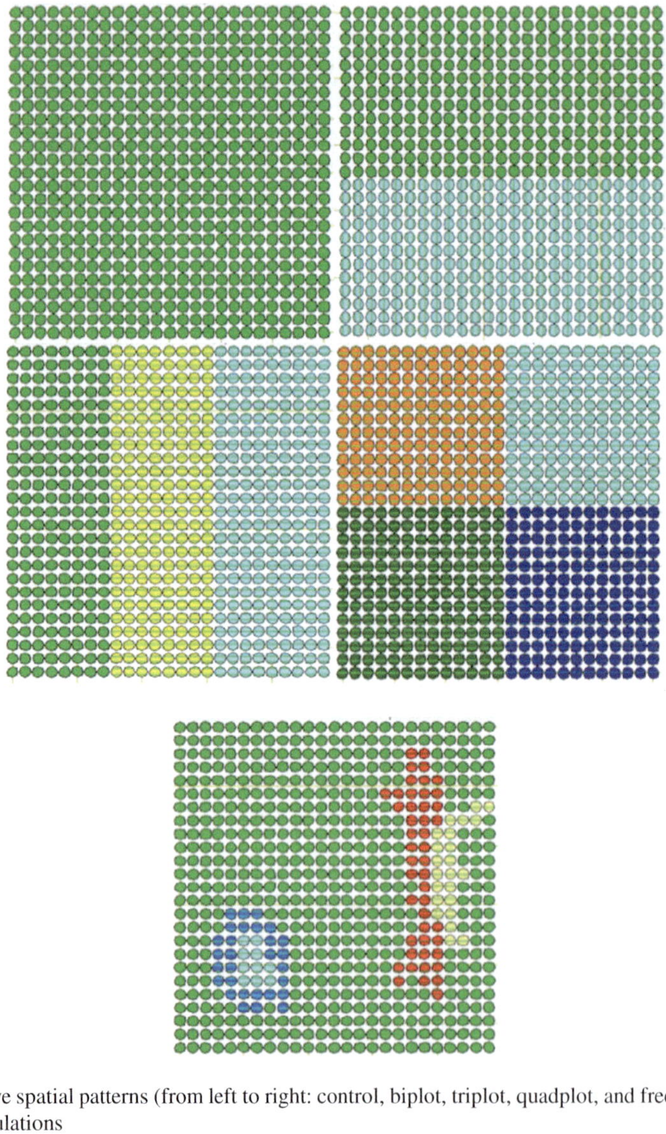

Fig. 1 Five spatial patterns (from left to right: control, biplot, triplot, quadplot, and free plot) used in the simulations

R software is used to draw randomly from a normal distribution of a specified mean DBH and coefficient of variation (CV) for each microsite. For *k* microsites, we have $\mu_1 < \mu_2 < \cdots < \mu_k$ where μ_k is the mean DBH for the *j*th microsites (subplot). For each simulation, the CV was equal among all microsites. That is, $\mathrm{DBH}_{ijk} \sim \left(\mu_j, \sigma^2\right)$, where σ^2 is chosen to achieve a pre-specified value of conditions $\mathrm{CV} = (\sigma \div \mu) * 100$. Heights are calculated based on known diameter–height relationships. Equation (3.1) from Sabatia and Burkhart (2013) can be used to model heights. The parameter estimates from Sabatia and Burkhart (2013) are from a loblolly pine study of similar planting conditions.

$$H_{ijk} = \beta_0 \mathrm{e}^{\beta_1 \mathrm{DBH}_{ijk}^{-1}} \tag{3.1}$$

where H_{ijk} and DBH_{ijk} are respectively the total height and DBH of the *k*th tree in the *j*th microsite of the *i*th simulation, in our case $i = 1 \ldots 100$, $j = 1 \ldots 5$ and $k = 1 \ldots 625$. β_0 is the upper asymptote parameter and β_1 is the rate parameter and θ_{ijk} is the random error due to the *k*th tree $[\theta_{ijk} \sim (0, \sigma_\theta^2)]$ (Sabatia and Burkhart 2013). The estimated equation from Sabatia and Burkhart (2013) is then given at Eq. (3.2).

$$H_{ijk} = 20.382\mathrm{e}^{-7.309 \mathrm{DBH}_{ijk}^{-1}} + \theta_{ijk}, \theta_{ijk} \sim (0, 0.482) \tag{3.2}$$

We now have our response matrix $\boldsymbol{Y}_{ijk} = \left[\mathrm{DBH}_{ijk}, H_{ijk}\right]$. Next, the environmental variables were determined.

We used site index (SI) as a proxy variable to measure soil productivity. Each microsite has a unique SI value that relates to its productivity. The less productive microsites have smaller SI values. We assigned SI values to each microsite within each spatial pattern. The SI values remained fixed for each microsite throughout the simulation process. The SI values were generated randomly from R, $SI_{ijk} \sim N\left(\mu_j, \sigma_j^2\right)$, where $j\,(= 1 \ldots k)$ represents the microsite and σ_j^2 satisfies the conditions $CV_j = \left(\sigma_j \div \mu_j\right) * 100$. The mean SI for each microsite is positively related to the mean DBH of the microsite. Increasing values of mean SI result in increased values of mean DBH. This is because increased values of SI represent higher levels of productivity that produce larger trees. Assigning mean SI values was based on known relationships between H and SI. The specific mean SI is less consequential for this analysis. What is important is that the SI values are higher for more productive sites. These SI values represent our matrix of environmental variables (**X**). For each microsite, we simulate unique SI distributions.

A second explanatory matrix of spatial variables is built, where $\mathbf{W} = [X_{\text{cord},}\ Y_{\text{cord}}]$, where $X_{\text{cord},}\ Y_{\text{cord}}$ are grid points measured in feet. We are simulating a managed plantation stand so values of **W** remain fixed grid points. The next step is to allocate value to spatial patterns. The designed spatial patterns are:

1. *Control*: Stand structures are spatially homogenous. Response variables are generated from the same distribution.

Table 1 Difference in mean DBH and CV for microsites

	CV				
		5%	10%	15%	20%
Difference in mean DBH	1	(1, 5%)	(1, 10%)	(1, 15%)	(1, 20%)
	2	(2, 5%)	(2, 10%)	(2, 15%)	(2, 20%)

2. *Biplot*: The whole plot is divided into two equal microsites. Response variables and explanatory variables are drawn from distinct distributions to reflect the productivity of each microsite.
3. *Triplot*: The whole plot is divided into three equal microsites. Response variables and explanatory variables are drawn from distinct distributions to reflect the productivity of each microsite.
4. *Quadplot*: The whole plot is divided into four equal microsites. Response variables and explanatory variables are drawn from distinct distributions to reflect the productivity of each microsite.
5. *Free plot*: Five microsites are created by hand. Microsites are irregular sizes and shapes. Response variables and explanatory variables are drawn from distinct distributions to reflect the productivity of each microsite.

For each spatial pattern (Fig. 1), we varied differences in mean DBH among the microsites and plot CV. The reason for this is to study how our spatial analysis will handle different spatial structures. All four spatial patterns (excluding the control) were simulated following the parameters of Table 1. This means that for each spatial pattern eight different parameters were used for the simulation. Each cell in Table 1 represents assigned differences among microsites.

Using the triplot as an example, the value of (mean difference DBH, CV) in row two and column two is (2, 10%). So for this simulation, microsites differed in mean DBH values successively by 2 inches and each microsite had a CV of 10%. Three sets of DBH values, viz. $DBH_1 \sim N\ (5, 0.25)$, $DBH_2 \sim N\ (7, 0.49)$, and $DBH_3 \sim N\ (9, 0.81)$ were then generated randomly using *R*. Using Eq. (3.2), each calculated *H* was assigned to a spatial coordinate within its appropriate microsite. The appropriate microsite was determined by its mean SI value, so the matches were SI = 70 with $DBH_1 \sim N(5, 0.25)$, SI = 75 with $DBH_2 \sim N(7, 0.49)$, and SI = 80 with $DBH_3 \sim N\ (9, 0.81)$.

Microsite number was assigned based on when it first appeared in the tree order. Tree 1 is located on the upper left corner of the plot, tree 2 and 3 follow from left to right of the row, then down each row successively. This continues down to tree 625 that is located in the bottom right corner of the plot. Using the free plot as an example, microsite 1 is green (tree 1), microsite 2 is red (starts at tree 69), microsite 3 is yellow (starts at tree 174), up to microsite 5 that is light blue (starts at tree 406). Each spatial pattern with $k > 1$ underwent 100 simulations for each cell in Table 1. The control plot ($k = 1$) underwent 1000 simulations for the four levels of CV. The control plot was tested with partial redundancy analysis and not clustering analysis so it was much more time-efficient.

The simulations were all run using the R software. We used the "vegan" and the "const.clust" (Legendre 2011) packages in R to analyze the data and "nb2listw(tri2nb())" function to apply spatial constraints. Constrained clustering methods take into account more information than other types of clustering. In our analysis, spatial information is used to build clusters and to make results more interpretable. For spatial contiguity, the only admissible clusters are those that obey a contiguity relationship (Legendre and Legendre 1998). Spatial contiguity is described by a connection scheme. We spatially constrained the cluster analysis by Delaunay triangulation methods (discussed later). The spatial constraint forces the clusters to be restricted the same way a microsite is restricted-spatially. This is a more interpretable form of cluster analysis in our case because we want our cluster map to mirror our spatial patterns (map of known microsites). A microsite is a spatially constrained patch, containing two or more trees, that has dissimilar site productivity from the surrounding area. Two microsites that have the same productivity but are not spatially connected are considered to be two distinct microsites, and therefore, we would expect two distinct clusters to identify them.

For each run of the constrained cluster simulation, we examined the selection criteria for $c = 2, \ldots, 10$ clusters in the data. The sites can be investigated for $c = 2, \ldots, n - 1$ clusters in the data, but for our analysis, $c = 2, \ldots, 10$ is sufficient. We then saved the number of clusters estimated by each criterion. For example, when using the CH statistic as a criterion, we choose the number of clusters with the highest CH value (higher values are better). The number of clusters picked by the CH was then stored into a new matrix with the other "best" selections by PT, AIC, R^2, and CVRE. We are using our cluster maps to estimate locations of microsites (our known spatial pattern). Since the microsites are known and generated by us, we can compare the effectiveness of the cluster maps to estimate or number of microsites.

After completing the constrained cluster simulations, we computed the probability of successful (POS) detection of known microsites for each spatial pattern by each criterion. This was done by counting the number of times each criterion correctly identified the known number of microsites for each simulation. Each time the criterion was correct we counted 1, otherwise we counted zero. This was done a hundred times for each simulation ($n = 100$).

For all POS values that exceeded 80%, we ran misclassification simulations. The misclassification simulations would examine how each tree is allocated based on the criterion. For a single run of "const.clust" trees were all assigned a cluster based on the criterion. If the assigned cluster of each tree matched the known microsite, then it was a successful grouping. If the assigned cluster did not match the known microsite, then the tree was misclassified. Ideally, cluster arranges mirror microsite arrangements. Each misclassification simulation was run 50 times. At the end of the simulation, the number of misclassified trees were summed and divided by the total number of trees involved. The resulting value was the probability of misclassification by criterion. The misclassification simulations estimated the probability of misclassification by criterion.

The CH statistic and PT were the only two criteria to successfully detect the correct number of microsites. The CH statistic is an F statistic comparing the among cluster sum of squares to the within cluster sum of squares (Borcard et al. 2011). The PT is a cross-validation procedure that determines the best number of groups based on a relative error ratio of the dispersion unexplained by the cluster tree divided by the overall dispersion of the response data (Borcard et al. 2011).

4 Analysis

All trees in the control plot come from the same distribution; therefore, the spatial structure should be homogenous. To test spatial homogeneity, we used the redundancy Eq. (4.1).

$$\left(\boldsymbol{S}_{\hat{y}'\hat{y}} - \boldsymbol{\lambda I}\right)\mathbf{u} = 0 \tag{4.1}$$

where $\boldsymbol{S}_{\hat{y}'\hat{y}} = \left[\frac{1}{n-1}\right]\boldsymbol{Y}'\boldsymbol{X}\left(\boldsymbol{X}'\boldsymbol{X}\right)^{-1}\boldsymbol{X}'\boldsymbol{Y}$ is the covariance matrix corresponding to fitted values $\boldsymbol{Y}$, where $\boldsymbol{Y}$ is the centered matrix of the estimated response, $\boldsymbol{\lambda}$ and $\boldsymbol{u}$ are respectively the vector of eigenvalues and the matrix of eigenvectors of the covariance matrix, and **I** is an identity matrix. The steps below use our generated data **X** and **W**. We want to partition the variation in **Y** such that we can explain the total variation through the sum of different fractions. These different four fractions are as: (a) The variation explained by the environmental variables; (b) the variation explained by the confounded variation of the environmental variables and spatial variables; (c) the variation explained by the spatial variables, and (d) the residual variation not explained by the other components. (modified from Legendre and Legendre 1998). We run the following steps to partition the variation:

1. Run an RDA of the response data **Y** by **X**. This yields the first fraction $[a + b]$.
2. Run an RDA of the response data **Y** by **W**. This yields fraction $[b + c]$.
3. Run an RDA of the response data **Y** by **X** and **W** together. This gives fraction $[a + b + c]$.
4. Compute the R^2_{adj} of the three RDA's above:

$$R^2_{\text{adj}} = 1 - \frac{n-1}{n-m-1}\left(1 - R^2\right) \tag{4.2}$$

   where n is the number of objects and m is the number of explanatory variables.
5. Compute the fractions of adjusted variation by subtraction:
   For example, fraction $[a] = [a + b + c] - [b + c]$. Repeat for fraction $[b]$ and $[c]$ (Fig. 2).

Next, the effectiveness of pRDA to detect spatial relationships in a control plot was tested. We expect to accept the null hypothesis of no spatial relationship. We iterated this process 1000 times and measured its success rate. In our case, we considered it a success to accept the null hypothesis. The permutation F test (Borcard et al. 2011) was used to test significance of explanatory and spatial components. Since we

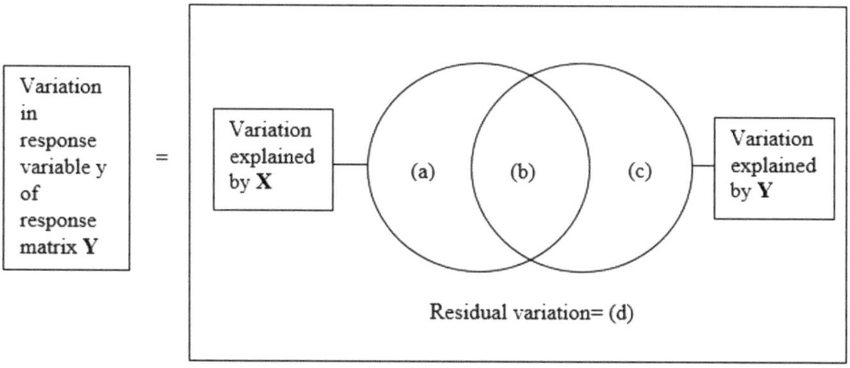

Fig. 2 Illustration of how partitioning method via RDA. Modified from Legendre and Legendre (1998)

are most interested in fraction [c], we will show the process for it here. We use an ANOVA like F test Eq. (4.3) to investigate the effectiveness of $\boldsymbol{W}$ on explaining the variation in $\boldsymbol{Y}$.

$$F = \frac{\sum_1^p \delta_i}{\text{RSS}/(\text{DF})} \tag{4.3}$$

The numerator (fraction [c]) is the contribution to the variance of **Y** from **W** after removing the contribution of **X**. The denominator, residual sum of squares (RSS) is the sum of the unconstrained eigenvalues (fraction [d]) and df are the degrees of freedom. The next step is to test the success of constrained clustering on maps with spatial heterogeneity.

The spatially constrained cluster analysis was implemented to identify spatially recognizable structures in tree growth of our known maps using the R package "const.clust". For this procedure, we had to specify which distance metric we would use and which connection network.

Distance metrics are used to measure the association between two objects (trees). The smaller a distance value or closer it is to zero, the more related objects are structural. In our data, two trees that are identical would have a distance value of 0. The common metric measure Euclidean distances Eq. (4.4) among objects using non-geographic information to create our dissimilarity matrix (**D**) were used.

$$\mathrm{D}(y_r y_{r+1}) = \sqrt{\sum\nolimits_{c=1}^{p} \left(y_{rc} - y_{(r+1)c}\right)^2} \tag{4.4}$$

where r = row of matrix **Y**, c = column of **Y**, and p is the number of variables in **Y**. For one of our simulations, $r = 1 \ldots 625$, and $p = 2$. This step is typical in many clustering algorithms, but in the next steps, we impose spatial constraints on

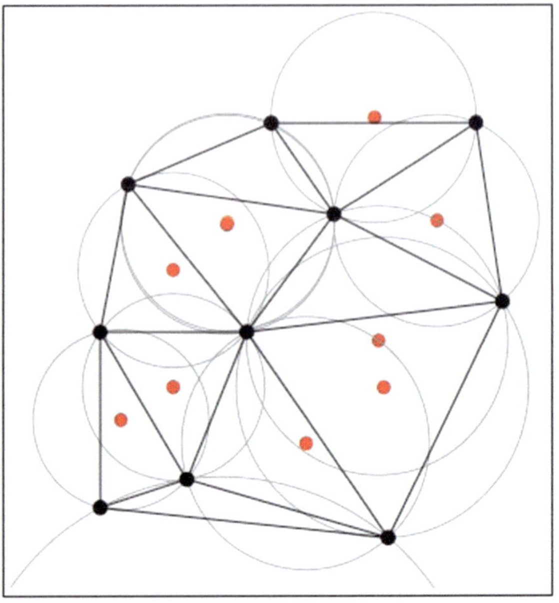

Fig. 3 Illustration of Delaunay triangulation. Black dots—the objects, red dots—the center of each circle used to circumscribe three points. Thick black lines connect "neighbors" (Wikipedia, November 3rd, 2014)

the dissimilarity matrix (Fig. 3), which is information typically not incorporated into clustering analysis.

Prior to performing spatially constrained clustering, it is important to state which trees are neighbors in space. In order for a tree to enter a cluster, it has to be a neighbor to it in space. The only admissible clusters in a spatially constrained analysis are those that obey the contiguity scheme. We relate clusters to microsites by constraining clusters so they are spatially defined in the same way as a microsite. A cluster is then a contiguous patch or group of trees that are structurally unlike the rest of the trees. Microsites create spatially recognizable structures in tree growth due to differences in productivity. This is the link from clusters to microsite, and because of this, we expect cluster location to parallel microsite location. Microsites create the structural differences which clustering recognizes.

The Delaunay triangulation uses spatial coordinates to identify neighbors. This is how we define contiguity. To determine neighbors, we produce a list of connection edges to create a contiguity matrix containing 1's for connected and 0's elsewhere (based on spatial coordinates of plot map). The contiguity matrix is how we spatially constrain our cluster analysis. The 1's and 0's are how we define neighbors and create connections among the trees. The Delaunay triangulation method states that for any triplet of non-collinear points A, B, and C, the three edges connecting these points are included if and only if the circles passing through these points (Fig. 3) include no other point (Legendre and Legendre 1998). This criterion is a robust method for defining contiguity. This connection scheme works well with regular grids and is adaptable to various patterns of planting grids and will transfer well to real plots that are slightly irregular.

The spatial constraint allows only connected trees to be clustered together. This prevents a scattering of cluster assignments on the map. Instead, clusters form distinct clumps. The cluster analysis results can be mapped with the spatial coordinates of the trees. The resulting map shows cluster assignment of each tree. When compared to our map of known microsites, we expect clusters to form over microsites and for trees within a microsite to be assigned the same cluster number.

Figure 4 illustrates the general framework for how a dissimilarity matrix interacts with the contiguity matrix to create a spatially constrained dissimilarity matrix suitable for constrained clustering. The Hadamard product between the dissimilarity matrix and contiguity matrix creates a constrained dissimilarity matrix where distance values exist only where neighbors were previously defined by the contiguity matrix. Our data file consists of growth information on 625 trees. A 625×625 dissimilarity matrix is created from Eq. (4.4). The more akin any two trees are in structure, the closer to 0 is their dissimilarity value. We then create a 625×625 contiguity matrix of 1's and 0's where 1's mark neighbors as defined by Delaunay triangulation and 0's elsewhere. The Hadamard product for our data is the dissimilarity in growth among neighboring trees.

We iterate this process 100 times for each combination of spatial pattern and parameter values. After the iteration process is completed, we can examine how successful the criteria were at detecting spatial patterns. We explored the POS allocating a tree to the correct microsite. The cluster map should reflect the microsite map if each tree is assigned correctly from the clustering algorithm. This is a simple process where we count 1 if tree i is correctly grouped into a cluster and 0 otherwise. Probability of success is p, and $n = 625$ is the number of trees per simulation.

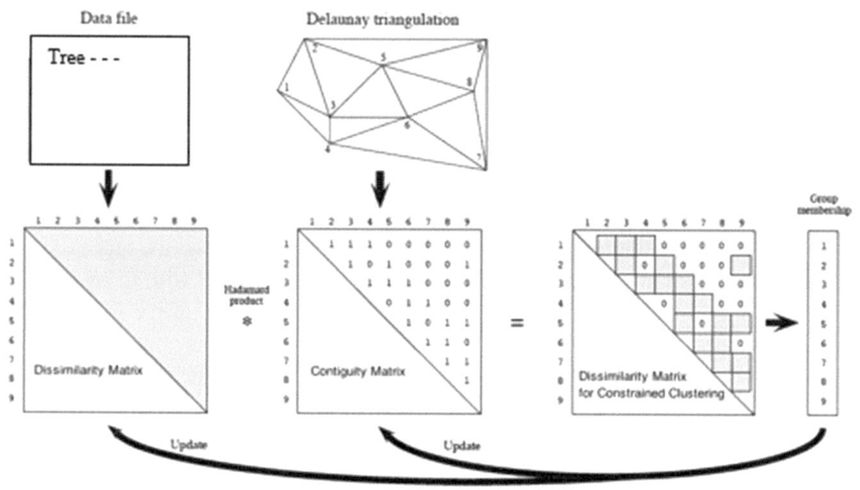

Fig. 4 Spatial constraints imposed in the clustering process. Modified from Legendre and Legendre (1998)

By the central limit theorem

$$x_i = \begin{cases} 1 \text{ if correctly identified by criterion} \\ 0 \text{ otherwise} \end{cases} \quad \text{and} \quad \hat{p} = \frac{\sum_{i=1}^{n} x_i}{n}$$

Therefore, $\hat{p} \sim (p, \frac{p(1-p)}{n})$ which gives the confidence interval $\hat{p} \pm z_{\alpha/2}\sqrt{\frac{\hat{p}(1-\hat{p})}{n}}$.

5 Results

Simulations for the control plot indicated that pRDA is an effective method for testing of spatial homogeneity. For these simulations, a *success* was considered to be those situations where the F test Eq. (4.3) failed to reject the null hypothesis of no linear relationship between the response and spatial coordinates. We found a high success rate (Table 2) at each level of CV.

We examined 2 through 10 clusters for each simulation. For each number of clusters, we would get a value from each criterion. The number of clusters corresponding to the highest CH statistic is best, or in the case of the PT, it was the cluster number with the lowest relative error ratio. The CH statistic was the most successful in that it detected the correct number of microsites most often and when the PT detection rate was high (>80%). With this in mind, we are reporting POS detection of both the CH and PT, but only the misclassification rates of the CH statistic.

There were two main expectations. The first is that less complex spatial patterns (biplot being the least and free plot the most complex) would be correctly detected more often than the complex patterns. The second was that for both the 1-inch and 2-inch difference in mean DBH, probability of detection would drop with increased amounts of variation.

As expected, the biplot was consistently detected the most by the CH statistic and the free plot the least (Figs. 5 and 6). Interestingly, the triplot and quadplot alternate in their relative success between Figs. 5 and 6. The low POS detection using the PT was unexpected (Figs. 7 and 8). The more complex patterns were generally detected more often than the less complex spatial patterns. The PT criterion was not successful in detecting the biplot in all scenarios, but was relatively successful with the quadplot.

Table 3 shows the 95% confidence intervals for each estimate of the POS for the correct number of clusters. The values of p range from 0 to 1. It is apparent that CH values are consistently more useful. This can most easily be noticed by examining successive rows. The rows of the table alternate from the CH to the PT.

Table 2 Probability for detection of spatial homogeneity for the control plot

CV	5%	10%	15%	20%	25%
POS	0.75	0.96	0.96	0.94	0.95

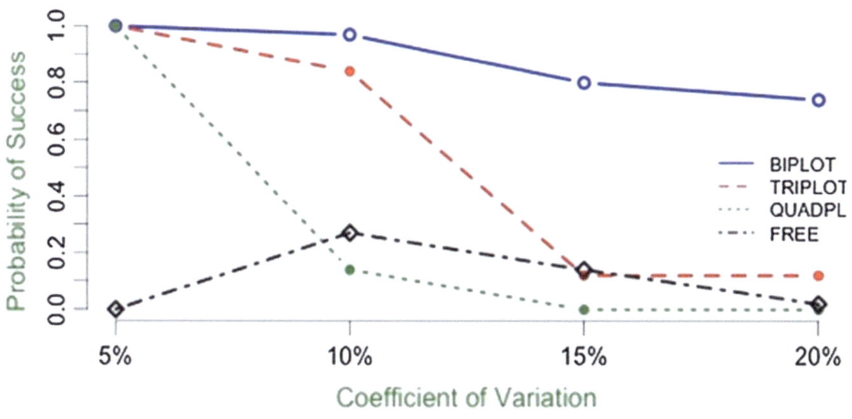

Fig. 5 POS detection by the CH statistic when the difference in successive mean DBH values between k groups is 1 inch

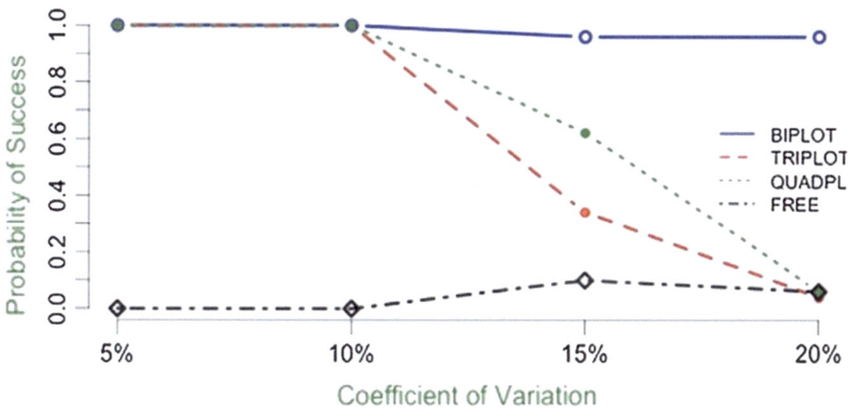

Fig. 6 POS detection by the CH statistic when the difference in successive mean DBH values between k groups is 2 inch

The interpretation of the values in Tables 4 and 5 is the probability that a tree will be correctly allocated when using the CH statistic. This is not a conditional probability (i.e., given CH identified 2 microsites). The misclassification rates were generally expected with the probability of misclassification generally increasing from left to right and from top to bottom. Range of misclassification was 0% for the biplot, triplot, and quadplot at 5% CV and 2-inch difference to 16% for the biplot at 15% CV and 1-inch difference.

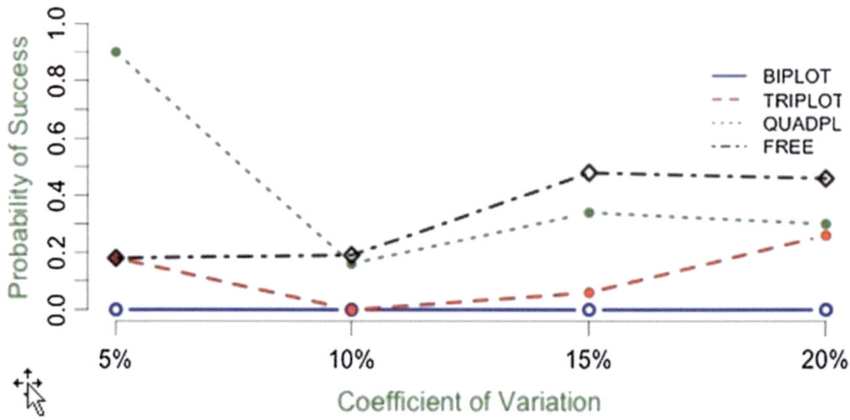

Fig. 7 POS detection by the PT when the difference in successive mean DBH values between k groups is 1 inch

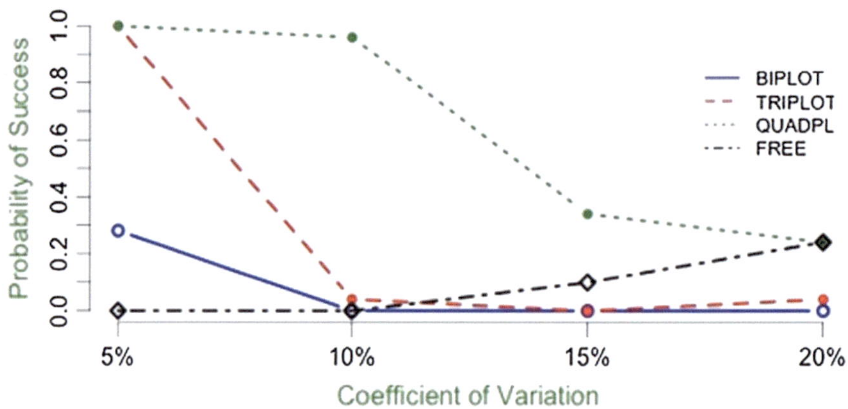

Fig. 8 POS detection by the PT when the difference in mean DBH values between successive k groups is 2 inch

6 Discussion

The POS for the free draw pattern is anomalous. For both the PT and CH statistics, the POS of the free draw pattern seems to increase with increasing values of CV. Like the other patterns, we expected a decrease in POS from low to high values of CV. There are three distinct features about the free plot that are plausible explanations for this occurrence. First is that the free plot has irregular shaped microsites. The irregular pattern of the microsites can influence which trees are usurped into a cluster. Connection schemes are carefully chosen before clustering is done in order to mitigate for possible influences from how objects (trees) and groups (microsites) are spatially dispersed. As described earlier, we choose Delaunay triangulation for our connection

Table 3 95% confidence intervals for the POS detection of the number of microsites

CV (%)	Biplot	Triplot	Quadplot	Free	Criterion 5	Mean 5 Difference
5	1,1	1,1	1,1	0,0	CH	1
5	0,0	0.10,0.26	0.84,0.96	0.1,0.26	PT	1
5	1,1	1,1	1,1	0,0	CH	2
5	0.19,0.37	1,1	1,1	0,0	PT	2
10	0.94,1	0.77,0.91	0.07,0.21	0.18,0.36	CH	1
10	0,0	0,0	0.09,0.23	0.11,0.27	PT	1
10	1,1	1,1	1,1	0,0	CH	2
10	0,0	0,0.08	0.92,0.99	0,0	PT	2
15	0.72,0.88	0.06,0.13	0,0	0.07,0.21	CH	1
15	0,0	0.01,0.11	0.25,0.43	0.38,0.58	PT	1
15	0.92,0.99	0.25,0.43	0.52,0.72	0.04,0.16	CH	2
15	0,0	0,0	0.52,0.72	0.04,0.16	PT	2
20	0.65,0.83	0.06,0.18	0,0	0,0.05	CH	1
20	0,0	0.17,0.35	0.21,0.39	0.36,0.56	PT	1
20	0.90,1	0,0.08	0.01,0.11	0.01,0.11	CH	2
20	0,0	0,0.08	0.16,0.32	0.16,0.32	PT	2

Table 4 Probability of tree misclassification when using the CH statistic at a mean difference in successive DBH values at 1 inch

CV (%)	Biplot	Triplot	Quadplot
5	0.004	0.012	0.009
10	0.080	0.276	*
15	0.159	*	*
20	*	*	*

Table 5 Probability of tree misclassification when using the CH statistic at a mean difference in successive DBH values at 2 inch

CV (%)	Biplot	Triplot	Quadplot
5	0.000	0.000	0.000
10	0.007	0.013	0.036
15	0.084	*	*
20	0.136	*	*

scheme. Based on review of our contiguity matrix, this seems an unlikely cause of free plot POS behavior. Even for the smallest microsites, trees were neighbors (spatially constrained) with other trees in their microsite. Second, the free plot has a fifth microsite and is the most complex stand we simulate. This is a cause for change in the POS, but not a factor that will cause the POS to rise from low to high CV values. Like the other patterns, as we add an additional microsite we see a general decrease in POS compared to the previous less complex pattern. This too is an unlikely candidate. Third, the five microsites are all different sizes. Size is not always equal among microsites due to the odd number of rows and columns in our stand, but up to the free plot, they were as close as possible. The number of trees per microsite in the free plot ranged from 10 trees to 506 trees. This means that the five distributions for each microsite have a varying number of trees, as well as different means and variances. It is likely that to minimize the CH statistic and the CVRE (the statistic minimized for the PT) trees from other distributions were simply split or taken into other distributions. It is most likely that the erratic POS across the CV values for the free plot was due to the large range in sizes of each microsite. This could have likely caused an over plotting effect that creates uncertainty in cluster assignment due to ranges in microsite values. The erratic behavior of the POS for the free plot is subject for further investigation and may require additional simulations and analysis.

Cluster analysis is difficult to validate. In some instances, mostly with the biplot, we measured a success rate of 100%. Although we do not know the spatial patterns in practice, we are still able to apply our algorithm. Through our simulations, we were able to mimic real situations and gather the information that will allow us to make more informed decisions. Given our findings, the CH statistic in our opinion is the best choice. There are a few interesting topics to mention pertaining to our results. These topics are applications to real data, validating with multiple methods, and extensions of this work.

Forest data is inherently complex. There are a plethora of variables that can alter growth. These variables include soil chemical reactions to stochastic weather including ice storms and lightning. We have known that productivity varies at very fine scales as a result of many processes. When we examine a forest plot, we do not know how many groups are in the data so we can follow two procedures. First, we would want to know if there is a spatial component to our data, we could use partial redundancy analysis pRDA to test for this. If there is a significant spatial relationship, we can proceed to use cluster analysis. Spatially constrained cluster analysis is well suited for describing its structure, and the CH statistic is the best criterion to follow. Once the microsites are located, it is important to verify the microsites by other means. This reduces the chance of error. Once we locate the microsites, there are a variety of methods to check if we have reliable results.

In order to check the validity of clustering results, we can implement different types of clustering. For example, K means clustering and constrained clustering can be used together. If the constrained clustering indicates that 2 microsites are present in the data then you would expect another clustering method to come to similar results if there are in fact 2 microsites in the data.

Clustering analysis has some promising applications in forestry. Even close approximations of structural differences may give foresters better ideas of how to apply expensive fertilizer and herbicidal treatments. It is fully expected that refinements in clustering techniques will improve management by foresters on the ground. Further investigation will need to ultimately be done. This includes modeling trees with software that includes complex competition interactions in estimating growth. Also, these methods will need to be compared with fine-scale soil maps that measure productivity between trees.

There is further work that needs to be done, but this is certainly a first step in developing a richer understanding of the growth dynamics of forest plots. For the first time, we have information on how to more effectively measure clusters in a forest plot. Our investigation indicates that the CH statistic is best suited for cluster analysis in forestry applications.

References

Borcard, D., Gillet, F., & Legendre, P. (2011). *Numerical ecology with R*. New York: Springer.

Borcard, D., Legendre, P., & Drapeau, P. (1992). Partialling out the spatial component of ecological variation. *Ecology, 73*(3), 1045–1055.

Bray, J. R., & Curtis, J. T. (1957). An ordination of the upland forest communities of southern Wisconsin. *Ecological Monographs, 27*(4), 325–349.

Brown, R. T. & Curtis, J. T. (1952). The upland conifer-hardwood forests of northern Wisconsin. *Ecological Monographs*, 217–234.

Curtis, J. T., & McIntosh, R. P. (1951). An upland forest continuum in the prairie-forest border region of Wisconsin. *Ecology, 32*(3), 476–496.

Delaunay Triangulation. (2014, September 19). Retrieved November 3, 2014, from http://upload.wikimedia.org/wikipedia/commons/1/1f/Delaunay_circumcircles_centers.svg.

Divíšek, J., Chytrý, M., Grulich, V., & Poláková, L. (2014). Landscape classification of the Czech Republic based on the distribution of natural habitats. *Preslia, 86,* 209–231.

Drewa, P. B., Platt, W. J., & Moser, E. B. (2002). Community structure along elevation gradients in headwater regions of longleaf pine savannas. *Plant Ecology, 160*(1), 61–78. Everitt, B., Landau, S., Leese, M., & Stahl, D. (2011). Cluster Analysis. Wiley Series in Probability and Statistics.

Everitt, B. S., Landau, S., Leese, M., & Stahl, D. (2011). An introduction to classification and clustering. *Cluster Analysis* (pp. 1–13). John Wiley & Sons, Ltd.

Fraver, S. (1994). Vegetation responses along edge-to-interior gradients in the mixed hardwood forests of the Roanoke River Basin, North Carolina. *Conservation Biology, 8*(3), 822–832.

Greig-Smith, P., Austin, M. P., & Whitmore, T. C. (1967). The Application of Quantitative Methods to Vegetation Survey: I. Association-Analysis and Principal Component Ordination of Rain Forest. *The Journal of Ecology*, 483–503.

Grimaldi, M., Oszwald, J., Dolédec, S., Hurtado, M. P., Miranda, I. S., de Sartre, X. A., et al. (2014). Ecosystem services of regulation and support in Amazonian pioneer fronts: searching for landscape drivers. *Landscape Ecology, 29*(2), 311–328.

Guevara, S., Purata, S. E., & Van der Maarel, E. (1986). The role of remnant forest trees in tropical secondary succession. *Vegetatio, 66*(2), 77–84.

Hardwood Forests of the Roanoke River Basin, North Carolina. *Conservation Biology*, *8*(3), 822–832.

Hurlbert, S. H. (1984). Pseudoreplication and the design of ecological field experiments. *Ecological Monographs, 54*(2), 187–211.

Lahti, T., & Väisänen, R. A. (1987). Ecological gradients of boreal forests in South Finland: an ordination test of Cajander's forest site type theory. *Vegetatio, 68*(3), 145156.

Leduc, A., Drapeau, P., Bergeron, Y., & Legendre, P. (1992). Study of spatial components of forest cover using partial Mantel tests and path analysis. *Journal of Vegetation Science, 3*(1), 69–78.

Legendre, P. (2011). Const.clust. Space and time constrained clustering package. R package version 1.2. http://numericalecology.com/rcode/.

Legendre, P., & Fortin, M. J. (1989). Spatial pattern and ecological analysis. *Vegetatio, 80*(2), 107–138.
Legendre, P. and Legendre, L. (1998). Numerical ecology: second English edition. *Developments in Environmental Modelling, 20*.
Legendre, P. and Legendre, L. F. (2012). *Numerical Ecology* (Vol. 20). Elsevier.
Legendre, P., Mi, X., Ren, H., Ma, K., Yu, M., Sun, I. F., et al. (2009). Partitioning beta diversity in a subtropical broad-leaved forest of China. *Ecology, 90*(3), 663–674.
Legendre, P., Oksanen, J., & terBraak, C. J. (2011). Testing the significance of canonical axes in redundancy analysis. *Methods in Ecology and Evolution, 2*(3), 269–277.
Lorimer, C. G. (1985). Methodological considerations in the analysis of forest disturbance history. *Canadian Journal of Forest Research, 15*(1), 200–213.
Martel, N., Rodriguez, M. A., & Berube, P. (2007). Multi-scale analysis of responses of stream macrobenthos to forestry activities and environmental context. *Freshwater Biology, 52*(1), 85–97.
Motyka, J., Dobrzanski, B. & Zawadzki, S. (1950). Preliminary studies on meadows in the south-east of Lublin province. *Ann. Univ. Mariae Curie-Sklodowska5E.*, 367–447.
Oliveira-Filho, A. T., & Fontes, M. A. L. (2000). Patterns of floristic differentiation among Atlantic Forests in Southeastern Brazil and the influence of climate 1. *Biotropica, 32*(4b), 793–810.
Oliver, C. D. & Larson, B. C. (1996). Forest stand dynamics: updated edition. New York: John Wiley & Sons, Inc.
Peet, R. K. (1981). Forest vegetation of the Colorado front range. *Vegetatio, 45*(1), 3–75.
Peres-Neto, P. R., Legendre, P., Dray, S., & Borcard, D. (2006). Variation partitioning of species data matrices: estimation and comparison of fractions. *Ecology, 87*(10), 2614–2625.
Perrin, P. M., Martin, J. R., Barron, S. J., & Roche, J. R. (2006). A cluster analysis approach to classifying Irish native woodlands. *Biology and Environment: Proceedings of the Royal Irish Academy, 106*(3), 261–275.
Plotkin, J. B., Chave, J., & Ashton, P. S. (2002). Cluster analysis of spatial patterns in Malaysian tree species. *The American Naturalist, 160*(5), 629–644.
Poore, M. E. D. (1955). The use of phytosociological methods in ecological investigations: III. Practical application. *The Journal of Ecology*, 606–651.
Rao, C. R. (1964). The use and interpretation of principal component analysis in applied research. *Sankhyā, A*, 329–358.
Sabatia, C. O., & Burkhart, H. E. (2013). Height and diameter relationships and distributions in loblolly pine stands of enhanced genetic material. *Forest Science, 59*(3), 278–289.
Snelder, T. H., Cattanéo, F., Suren, A. M., & Biggs, B. J. (2004). Is the river environment classification an improved landscape-scale classification of rivers? *Journal of the North American Benthological Society, 23*(3), 580–598.
Steane, D. A., Conod, N., Jones, R. C., Vaillancourt, R. E., & Potts, B. M. (2006). A comparative analysis of population structure of a forest tree, Eucalyptus globulus (Myrtaceae), using microsatellite markers and quantitative traits. *Tree Genetics and Genomes, 2*(1), 30–38.
terBraak, C. J., & Prentice, I. C. (1988). A theory of gradient analysis. *Advances in Ecological Research, 18*, 271–317.
terBraak, C. T., & Šmilauer, P. (2002). CANOCO reference manual and CanoDraw for Windows user's guide: software for canonical community ordination (version 4.5). *Section on Permutation Methods. Microcomputer Power, Ithaca, New York.*
Urban, D., Goslee, S., Pierce, K., & Lookingbill, T. (2002). Extending community ecology to landscapes. *Ecoscience, 9*(2), 200–212.
Van Den Wollenberg, A. L. (1977). Redundancy analysis an alternative for canonical correlation analysis. *Psychometrika, 42*(2), 207–219.
Vries, D. D. (1952). Objective combinations of species. *Acta Botanica Neerlandica, 1*(4), 497–499.
Webb, D. A. (1954). Is the classification of plant communities either possible or desirable. *Bot. Tidsskr, 51*, 362–370.
Weber, C. D. (1983). Height growth patterns in a juvenile Douglas-fir stand, effects of planting site, microtopography and lammas occurrence (Doctoral dissertation, University of Washington).

Ridge Regression Model for the Estimation of Total Carbon Sequestered by Forest Species

Manish Sharma, Banti Kumar, Vishal Mahajan and M. I. J. Bhat

Abstract In this chapter, the study of total carbon sequestration has been taken for *Acacia catechu*, because of its highest relative dominance, relative density, relative frequency and importance value index in forest land-use system. The ridge regression methods have been used to estimate the total carbon sequestration (dependent variable) of this species in *Kandi* belt of Jammu region of Jammu and Kashmir state located in the foothill zone of Jammu Shivaliks. The explanatory variables diameter at breast height (DBH), height, stem biomass (SB), branch biomass, leaf biomass, below-ground biomass and total above-ground biomass (TB) were chosen for the study. It has been observed that the explanatory variables were highly correlated and indicates the presence of multicollinearity and hence estimates based on the classical ordinary least square method are not precise. Ridge regression method was attempted to deal with the problem of multicollinearity. The results show that the optimum value of ridge constant is 0.02 and the variables DBH, SB and TB are significant variables to increase the total carbon content sequestered by the species *Acacia catechu*. The estimates of parameters through ridge regression technique were more stable and more reliable than ordinary least square on the basis of size, sign and significance of the regression parameters.

Keywords Carbon sequestration · Multicollinearity · Ordinary least square · Ridge constant · Ridge regression · Variance inflation factor

---

M. Sharma (✉) · B. Kumar · M. I. J. Bhat
Division of Statistics and Computer Science, Faculty of Basic Sciences, Sher-E-Kashmir University of Agricultural Sciences and Technology of Jammu, Jammu, India
e-mail: manshstat@gmail.com

V. Mahajan
Division of Agroforestry, Faculty of Agriculture, Sher-E-Kashmir University of Agricultural Sciences and Technology of Jammu, Jammu, India

G. Chandra et al. (eds.), *Statistical Methods and Applications in Forestry and Environmental Sciences*, Forum for Interdisciplinary Mathematics,
https://doi.org/10.1007/978-981-15-1476-0_11

1 Introduction

In multiple regression analysis, the method of ordinary least square (OLS) breaks down when the predictors are linearly related to each other. This linear dependency of predictors on each other is known as multicollinearity (Bowerman and O'Connell 1990). In the presence of multicollinearity in the data, two or more predictors become redundant; i.e., they give the same information which affects the efficiency of the regression model in estimation of regression coefficients. The variances of the regression coefficients become large, and hence the estimates are less precise in the presence of multicollinearity which may affect the sign, size, significance and confidence interval of the regression parameters. There are several methods available in the literature for detection of the presence of multicollinearity.

Variance inflation factor (VIF) is the one of most commonly used method which provides the index that measures how much the variance of the estimates of regression coefficient is increased because of colinearity (Alin 2010). Many different methods have been developed in order to overcome the problem of multicollinearity, viz., ridge regression (RR), principal component analysis (PCA), stepwise regression, partial least square regression (Yeniay and Goktas 2002). Among them, the most common technique is RR proposed by Hoerl and Kennard (1970). Gunst and Webster (1974) examined the sources of multicollinearity and discussed some of its harmful effects. Hoerl et al. (1975) also gave an algorithm for selecting the biasing parameter (K) in RR. More significant regression coefficients can be used as a measure of finding the best solution of ridge trace according to Vinod (1976). Malthouse (1999) showed how RR can be used to improve the performance of direct marketing scoring models. Whittaker et al. (2000) used the RR method for the selection of marker for quantitative traits in crosses between inbred lines. Akdeniz and Erol (2003) compared the almost unbiased generalized RR estimator in the mean square error (MSE) matrix sense in the presence of multicollinearity. Jaiswal and Khanna (2004) have used some RR estimator for prediction of lactation yield in Indian buffaloes and compared with OLS. Pasha and Shah (2004) concluded that the RR method was found to be better than OLS in case of multicollinearity in the explanatory variable. Adnan et al. (2006) provide a better solution to deal with the problem of multicollinearity in some simulated data sets as compared to RR, principal component regression and partial least square regression. Principal component regression is a combination of PCA and OLS regression. RR technique although biased provides estimates of the regression coefficients with greater precision than the simple OLS. Basarir et al. (2006) used RR technique to study productivity growth and technical changes in Turkish agriculture measured for 1961–2001 period using Cobb–Douglas production function. Sufian (2010) used 'q' value as a measure to detect multicollinearity in the data, while VIF and RR techniques have been used to detect interrelation among the internal components of the regression model.

Thus, in the presence of multicollinearity, least square estimates are unbiased, but their variances are large so they may be far from the true value. By adding a small degree of bias to the regression estimates, RR reduces the standard errors and the net effect will give estimates that are more reliable. Using RR, it is easier to find optimal values of ridge parameter, i.e., values for which the MSE of the ridge estimator is minimum. In addition, if the optimal values for biasing constants differ significantly from each other then this estimator has the potential to save a greater amount of MSE than the OLS estimator (Stephen and Christopher, 2001).

2 Materials and Methods

The study area stretched between longitude $74^{\circ}\,21'$–$75^{\circ}\,45'$ E and latitude $32^{\circ}\,22'$–$32^{\circ}\,55'$ N covering three districts, namely Jammu, Samba and Kathua. The landscape comprised of undulating topography, steep and irregular slopes, erodible and low water retentive soils and badly dissected terrain by numerous gullies. The vegetation in forest land use has been classified into various forest types as per the classification made by Champion and Seth (1968); northern dry mixed deciduous forest is major forest type of the study area with *Acacia catechu, Dalbergia sissoo, Grewia optiva, Dendrocalamus strictus, Acacia modesta, Mallotus philippensis*, *Bombax ceiba, Carissa spinarum, Dodonaea viscosa*, etc., being the main species. The fringe zone toward its higher reaches comprises Himalayan subtropical pine forest sub-type (*lower Shivalik Chir Pine Forest*). The general floristic composition in the study area of this sub-type includes *Pinus roxburghii*, *Acacia catechu, Dalbergia sissoo, Butea monosperma, Mallotus philippensis, Ziziphus jujuba, Syzygium cumini, Ficus glomerata,* etc.

2.1 Carbon Estimation in Trees

In order to estimate the total carbon sequestration in forest species, nondestructive method was used as it is more rapid and much larger area and the number of trees can be sampled, reducing the sampling error encountered with the destructive method (Hairiah et al. 2011). Carbon concentration in plants was calculated using combustion method as suggested by Negi et al. (2003). The oven-dried plant components (stem, branches, leaves, bark, etc.) were burnt in muffle furnace at 400 °C temperature, ash content (inorganic elements in the form of oxides) left after burning was weighed, and carbon was calculated by using the following equation (Negi et al. 2003).

Carbon % = 100—(ash weight + molecular weight of O_2 (53.3) in $C_6H_{12}O_6$).

Carbon stock in different plant components (stem, branches, leaves, bark, etc.) were obtained by multiplying the dry weight of different plant components by their average carbon concentration. Total carbon stock has been obtained by taking sum of the carbon stocks in different plant components. The carbon stock in roots was estimated using root–shoot ratios along with the methodology suggested by Mokany et al. (2006). Forest species *Acacia catechu* was taken for the study because of its high relative dominance, relative density, relative frequency and importance value index (IVI) in forest land-use system. In all, 15 sample plots of size 10 m x 10 m in forest land-use system were laid across the study area covering all three districts, viz., Jammu, Samba and Kathua. Total carbon (TC) sequestered (Mg/ha) has been taken as dependent and a set of regressors such as diameter at breast height (DBH (cm)), height (HT (cm)), stem biomass (SB (Mg/ha)), branch biomass (BB (Mg/ha)), leaf biomass (LB (Mg/ha)), below-ground biomass (BGM (Mg/ha)), total above-ground biomass (TB (Mg/ha). In order to study the effect of regressors on dependent variable, RR model was used.

2.2 Ridge Regression (RR) Method

A multiple linear regression model in terms of the observations may be written as:

$$\boldsymbol{Y} = \boldsymbol{X\beta} + \boldsymbol{\varepsilon}, \boldsymbol{\varepsilon} \sim \mathrm{N}(0, \sigma^2 \boldsymbol{I}_n)$$

where

$\boldsymbol{Y} = \begin{bmatrix} y_1 \\ y_2 \\ . \\ . \\ . \\ y_n \end{bmatrix}_{n\times 1}$ is the vector of the observations, $\boldsymbol{\beta} = \begin{bmatrix} \beta_1 \\ \beta_2 \\ . \\ . \\ . \\ \beta_p \end{bmatrix}_{p\times 1}$ is the vector of the regression coefficients, $\boldsymbol{X} = \begin{bmatrix} 1 & x_{11} & x_{12} & . & x_{1p} \\ 1 & x_{21} & x_{22} & . & x_{2p} \\ . & . & . & . & . \\ . & . & . & . & . \\ 1 & x_{n1} & x_{n2} & . & x_{np} \end{bmatrix}_{n\times p}$ is the matrix of the level of the regressor variables and $\boldsymbol{\varepsilon} = \begin{bmatrix} \varepsilon_1 \\ \varepsilon_2 \\ . \\ . \\ . \\ \varepsilon_n \end{bmatrix}_{n\times 1}$ is the vector of random errors.

When terms are correlated and the columns of the design matrix $\boldsymbol{X}$ have an appropriate linear dependence, the matrix $(\boldsymbol{X}'\boldsymbol{X})^{-1}$ becomes close to singular. As a result, the least square estimate $\hat{\boldsymbol{\beta}} = (\boldsymbol{X}'\boldsymbol{X})^{-1}\boldsymbol{X}'\boldsymbol{Y}$ becomes highly sensitive to random errors in the observed response $\boldsymbol{Y}$, producing a large variance. This gives rise to the problem of multicollinearity. RR addresses the problem by estimating regression coefficients using $\hat{\boldsymbol{\beta}} = (\boldsymbol{X}'\boldsymbol{X} + K\boldsymbol{I})^{-1}\boldsymbol{X}'\boldsymbol{Y}$, where K is the ridge parameter and $\boldsymbol{I}$ is the identity matrix. Small positive values of K improve the conditioning of the problem and reduce the variance of the estimates. While biased, the reduced variance of ridge estimates offers a smaller MSE when compared to least square estimates. When the predictor variables are highly correlated among themselves, the coefficients of the resulting least square fit may be very imprecise. By allowing a small amount of bias in the estimates, more reasonable coefficients may often be obtained. RR is one method to address these issues. Often, small amounts of bias lead to dramatic reductions in the variance of the estimated model coefficients. The optimum value of K can be determined through (i) plotting the standardized beta coefficients of the dependent variable with the values of K, (ii) plotting VIF values of Y against K, (iii) coefficient of determination R^2 and (iv) VIF values.

3 Results and Discussion

Minimum, maximum, mean, standard deviation and coefficient of variation (CV) of explanatory variables for TC have been listed in Table 1. The result showed that the variability of explanatory variables of TC ranges from 34.78 to 180.67%. Maximum CV was observed in SB followed by BGM, TB, LB, BB and SB, whereas minimum CV was observed in HT.

From Table 2, it has been observed that the variables SB, BB, LB, BGM and TB were significantly positively correlated with one another. Further, the correlation among these variables was found about 0.90 which is an indication of the presence of multicollinearity in the data. Further, maximum correlation was observed between TB and SB and it was 0.99.

Table 3 show that the VIF of SB (1690.32), BB (232.23) and TB (3494.69) were greater than 10 which indicates that multicollinearity is present in the data (Marquardt 1970). Also, it has been observed that the condition index value was coming to be more than 1000 which indicated the presence of severe multicollinearity problem in the data (Johnston 1984).

The ridge trace of standardized betas of explanatory variables has been plotted against the values of K; it was found that the standardized beta estimates started

Table 1 Descriptive statistics of dependent and explanatory variable

Variable	Minimum	Maximum	Mean	Standard deviation	CV
DBH	2.70	62.00	12.45	8.29	66.58
HT	165.00	820.00	405.04	140.85	34.78
SB	0.22	63.01	4.62	8.34	180.67
BB	0.20	20.48	2.47	2.85	115.54
LB	0.09	8.53	1.11	1.30	116.94
BGM	0.15	30.21	2.42	3.41	140.69
TB	0.50	89.40	8.16	12.22	149.78
TC	0.20	45.72	4.26	6.09	142.92

Table 2 Correlation matrix of the dependent and explanatory variable

	TC	DBH	HT	SB	BB	LB	BGM	TB
TC	1.00							
DBH	0.11	1.00						
HT	0.01	0.52*	1.000					
SB	0.98*	0.04	−0.050	1.00				
BB	0.95*	0.05	0.000	0.93*	1.00			
LB	0.89*	0.11	−0.03	0.87*	0.91*	1.00		
BGM	0.86*	0.05	−0.02	0.86*	0.91*	0.83*	1.00	
TB	0.98*	0.05	−0.04	0.99*	0.97*	0.91*	0.89*	1.00

*significant at 1% level of significance

Table 3 VIF and condition index of the explanatory variables

Explanatory variable	VIF	Condition index
DBH	1.46	1.00
HT	1.48	3.05
SB	1690.32	9.59
BB	232.23	27.06
LB	25.57	32.12
BGM	6.11	97.92
TB	3494.69	25,087.11

stabilizing in the interval of K between the ranges from 0.01 to 1.00 as shown in Fig. 1. Similar range has been observed during the plotting of K against the VIF values of all dependent variables as shown in Fig. 2.

The different values of K, R^2 and maximum VIF have been listed in Table 4. The values of K and R^2 were critically observed, and the values of R^2 were found to be increasing at faster rate from the values of $K = 1.00$ to $K = 0.02$ and then became somewhat constant or increasing at slower rate. Hence, the optimum value of K was taken as 0.02 for the problem under study lies in the interval obtained through ridge trace. Moreover, the value of maximum VIF has also been reduced to less than 10 after $K = 0.02$ which may be considered as optimum value of K.

The perusal of the data depicted in Table 5 revealed that RR model for total carbon content was statistically significant and adequate with respect to the explanatory variables. $R^2 = 0.973$ indicates that 97.3% of the total variation in total carbon content was explained by the explanatory variables taken under consideration. The functional analysis of total carbon content revealed that DBH, SB and TB were found to be positively significant. The values of regression coefficients for DBH, SB and TB were 0.043, 0.337 and 0.183, respectively, and are significant variables to increase the total carbon content sequestered by the species.

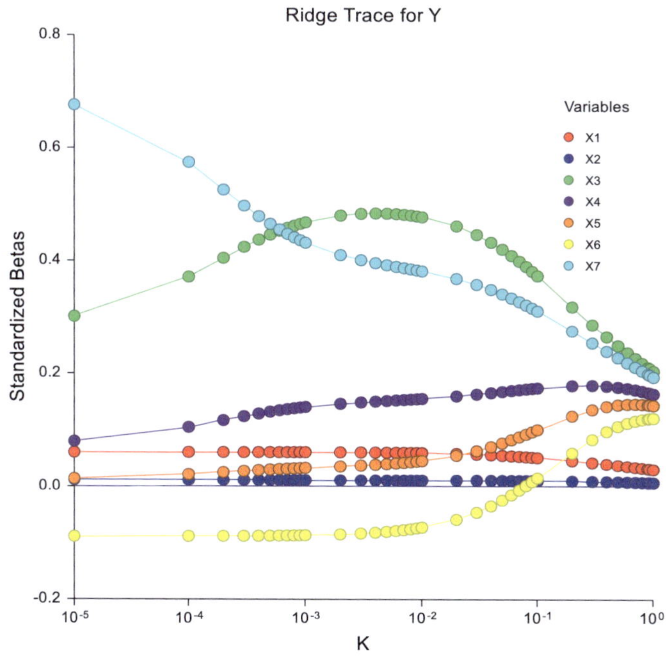

Fig. 1 Ridge trace of standardized beta coefficients versus *K*

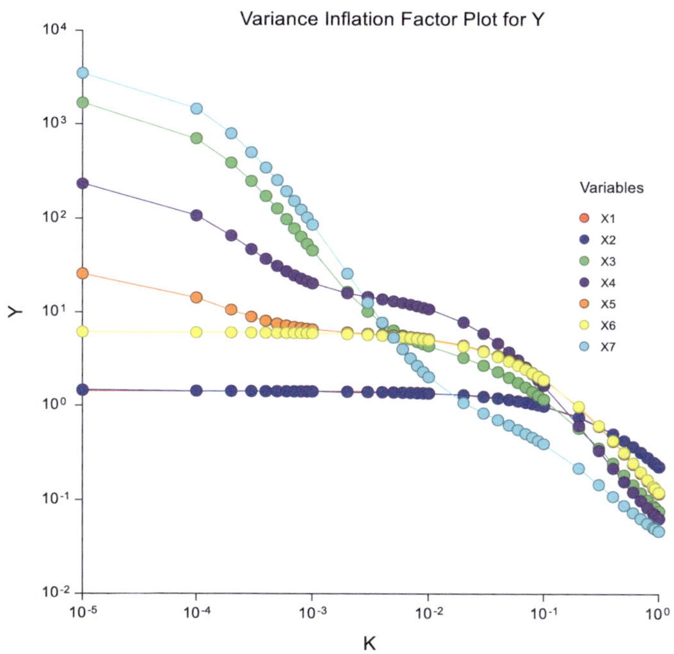

Fig. 2 Plot of VIFs for *Y* versus *K*

Table 4 Values of R^2 and maximum VIF at different values of K

K	R^2	Max VIF
0.0000	0.9819	3494.695
0.0001	0.9818	1471.286
0.0002	0.9818	806.2222
0.0003	0.9817	508.2334
0.0004	0.9817	349.5674
0.0005	0.9816	255.1876
0.0006	0.9816	194.5314
0.0007	0.9815	153.255
0.0008	0.9815	123.9025
0.0009	0.9815	102.2871
0.0010	0.9814	85.9104
0.0020	0.9810	25.9648
0.0030	0.9805	14.6949
0.0040	0.9801	13.8523
0.0050	0.9797	13.1995
0.0060	0.9793	12.6402
0.0070	0.9789	12.1387
0.0080	0.9784	11.6786
0.0090	0.978	11.2512
0.0100	0.9776	10.8511
0.0200	**0.9737**	**7.8517**
0.0200	0.9737	7.8517
0.0300	0.9699	5.9620
0.0400	0.9664	4.6862
0.0500	0.9630	3.7834
0.0600	0.9598	3.1211
0.0700	0.9566	2.6206
0.0800	0.9536	2.2952
0.0900	0.9507	2.1030
0.1000	0.9479	1.9353
0.2000	0.9227	1.0054
0.3000	0.9009	0.6283
0.4000	0.8811	0.5140
0.5000	0.8626	0.4343
0.6000	0.8452	0.3733
0.7000	0.8286	0.3253
0.8000	0.8128	0.2866

(continued)

Table 4 (continued)

K	R^2	Max VIF
0.9000	0.7977	0.2548
0.9200	0.7933	0.2465
1.0000	0.7832	0.2284

Table 5 Estimates of regression coefficients through RR model at $K = 0.02$

Explanatory variable	Regression coefficient	VIF	R^2	MSE
Intercept	−0.382		0.973	1.047
DBH	0.043*	1.328		
HT	0.001	1.315		
SB	0.337**	3.293		
BB	0.341	7.852		
LB	0.257	4.462		
BGM	−0.105	4.346		
TB	0.183**	1.094		

*significant at 5%, **significant at 1%

Table 6 Comparison of RR model with OLS model

Explanatory variable	Ridge coefficients (standard error)	L. S. coefficients (standard error)
Intercept	−0.382	−0.368
DBH	0.043* (0.014)	0.044** (0.012)
HT	0.001 (0.001)	0.001 (0.001)
SB	0.337** (0.022)	0.220 (0.413)
BB	0.341 (0.099)	0.170 (0.447)
LB	0.257 (0.164)	0.065 (0.326)
BGM	−0.105 (0.062)	−0.159** (0.061)
TB	0.183** (0.009)	0.337 (0.405)
R^2	0.973	0.982

*significant at 5%, **significant at 1% level

Table 6 shows the comparison of RR with OLS method. Both models were found to be significant. It is evident from Table 6 that in case of RR model, DBH, SB and TB had significant effect on TC whereas in case of least square model only DBH and BGM were found out to be significant. While comparing the two methods, it was found that variable SB was nonsignificant in case of least square turnout to be significant in case of RR model may be due to the presence of the multicollinearity. Hence, it can be clearly seen that explanatory variables which were found to be nonsignificant in OLS came out to be significant in RR as the effect of these variables was masked due to the presence of multicollinearity which gives enough proof to apply such models to overcome the problem of multicollinearity.

4 Conclusions

From this study, while estimating the total carbon sequestration from the forest species *Acacia catechu* through the different explanatory variables, it has been observed the multicollinearity has adverse and drastic effect on sign, size and significance of the regression coefficients of the model under study. Further, the ridge regression is a more efficient technique than its counterparts as supported by the results in the study. Thus, the proposed model for total carbon sequestration by *Acacia catechu* was found to be $TC = -0.382 + 0.043\ DBH^{*} + 0.001\ HT + 0.337\ SB^{**} + 0.341\ BB + 0.257\ LB - 0.105\ BGM + 0.183\ TB^{**}$ with positive and significant effect of DBH, SB and TB on total carbon sequestration for the species *Acacia catechu*.

Acknowledgements The authors are highly thankful to the reviewers for their valuable suggestions to improve the quality of the chapter.

References

Adnan, N., Ahmad, M. H., & Adnan, R. (2006). A comparative study on some methods for handling multicollinearity problems. *Matematika, 22*(2), 109–119.

Akdeniz, F., & Erol, H. (2003). Mean square error matrix comparison of some biased estimators in linear regression. *Communications in Statistics-Theory and Methods, 32*(12), 2389–2413.

Alin, A. (2010). Multicollinearity. *Wiley Interdisciplinary Reviews: Computational Statistics, 2*(3), 370–374.

Basarir, A., Karli, B., & Bilgic, A. (2006). An evaluation of Turkish agricultural production performance. *International Journal of Agriculture and Biology, 8*(4), 511–515.

Bowerman, B. L., & O' Connell, R. T. (1990). *Linear statistical models an applied approach* (2nd ed.). Boston: PWS-KENT Publishing Co.

Champion, H. G., & Seth, S. K. (1968). *A revised survey of forest types of India* (p. 404). New Delhi: Government of India Press.

Gunst, R. F., & Webster, J. T. (1974). Regression analysis and the problem of multicollinearity. *Communication in Statistics, 4*(3), 277–292.

Hairiah, K., Dewi, S., Agus, F., Velarde, S., Ekadinata, A., Rahayu, S. & van Noordwijk, M. (2011). Measuring carbon stocks across land use systems: A manual. Bogor, Indonesia. World Agroforestry Centre (ICRAF), SEA Regional Office, p. 154.

Hoerl, A. E., & Kennard, R. W. (1970). Ridge regression: Biased estimation to nonorthogonal problems. *Technometrics, 12,* 56–67.

Hoerl, A. E., Kennard, R. W., & Baldwin, K. F. (1975). Ridge regression: Some simulations. *Communication in Statistics, 4,* 105–123.

Jaiswal, U. C., & Khanna, A. S. (2004). Comparison of some biased estimators with ordinary least square in regression analysis of lactation records in buffaloes. *Indian Journal of Dairy Sciences, 57*(4), 255–258.

Johnston, J. (1984). *Econometric methods* (3rd ed.). New York: McGraw-Hill Publishing Company.

Malthouse, E. C. (1999). Ridge regression and direct marketing scoring models. *Journal of Interactive Marketing, 13*(4), 10–23.

Mokany, K., Raison, R. J., & Prokushkin, A. S. (2006). Critical analysis of root: Shoot ratios in terrestrial biomes. *Global Change Biology, 12,* 84–96.

Negi, J. D. S., Manhas, R. K., & Chauhan, P. S. (2003). Carbon allocation in different components of some tree species of India: A new approach for carbon estimation. *Current Science, 85*(11), 1528–1531.

Pasha, G. R., & Shah, M. A. A. (2004). Application of ridge regression to multicollinear data. *Journal of Research (science), 15*(1), 97–106.

Stephen, G. W., & Christopher, J. P. (2001). Generalized ridge regression and a generalization of the Cp statistic. *Journal of Applied Statistics, 28*(7), 911–922.

Sufian, A. J. M. (2010). An analysis of poverty-a ridge regression approach. In *Proceedings of 4th International multi-conference on Society, Cybernetics, and Informatics* (IMSCI-2010), June 29–July 2, 2010. Orlando, Florida, U.S.A.

Vinod, D. H. (1976). Application of new ridge regression methods to a study of bell system scale economies. *Journal of the American Statistical Association, 71*(356), 835–841.

Whittaker, J. C., Thompson, R., & Dentiam, M. C. (2000). Marker assisted selection using ridge regression. *Genetically-Research, 75*(2), 249–252.

Yeniay, O., & Goktas, A. (2002). A comparison of partial least squares regression with other prediction methods. *Hacettepe Journal of Mathematics and Statistics, 31,* 99–111.

Some Investigations on Designs for Mixture Experiments with Process Variable

Krishan Lal, Upendra Kumar Pradhan and V. K. Gupta

Abstract In mixture experiments, the response is assumed to depend on the relative proportions of the ingredients present in the mixture and not on total amount of the mixture. When these experiments are conducted with variables that do not form any portion of the mixture, the levels changed could affect the blending properties of the ingredients are called mixture experiments with process variables. Different models used in various situations for the analysis of mixture experiments with process variables have been discussed. Two methods of construction of mixture experiments with process variables have been developed. In the first method, an efficient response surface design with orthogonal blocks is taken. The mixture designs have been obtained by projecting a suitable unconstrained design (response surface design) onto the hyper-plane defined by the constraints (Prescott in Commun Stat Theory Methods, 29:2229–2253, 2000). The second method is developed by modifying the method given by Wu and Ding (J Stat Plan Infer 71:331–348, 1998) for response surface designs for qualitative and quantitative factors.

Keywords Box–Behnken design · Central composite design · *G*-efficiency · Orthogonal blocking · Process variable · Projection designs

1 Introduction

A mixture experiment involves the study of performance of various mixtures formed by mixing two or more components called ingredients. Let x_i represent the proportion of ith ingredient in a mixture. Evidently, $0 \leq x_i \leq 1, i = 1, 2, \ldots, q$ and $\sum_{i=1}^{q} x_i = 1$, where q is the number of components. In experiments with the mixture, the factors are ingredients of a mixture, and their levels are not independent. Scheffé (1958, 1963) was the first to introduce models and designs for mixture experiments. Sometimes, the mixture experiments are conducted with process variables. In mixture experiments,

K. Lal (✉) · U. K. Pradhan · V. K. Gupta
Indian Agricultural Statistics Research Institute, Library Avenue, Pusa, New Delhi 110012, India
e-mail: klkalra@gmail.com

G. Chandra et al. (eds.), *Statistical Methods and Applications in Forestry and Environmental Sciences*, Forum for Interdisciplinary Mathematics,
https://doi.org/10.1007/978-981-15-1476-0_12

the process variables are factors that do not form any portion of the mixture but whose levels when changed could affect the blending properties of the ingredients.

When the mixture experiments are conducted with process variables, the experimenter is not only interested in blending properties of the mixture components but also to see blending behaviour with the change of levels of the process variable. Mixture experiments with process variables have commonly been used in agricultural, forestry, veterinary and industrial experiments. For example, an experiment was conducted under the project Agricultural Field Experiments Information System (AFEIS) on the sugarcane crop (*Saccharum Officinarum* L.). The objective of the experiment was to study the effect of organic and inorganic sources of nitrogen on growth, quality and yield of sugarcane. For this, nitrogen (N) was applied by three sources [urea, farm yard manure (FYM) and sulphonation press mud cakes (SPMC)] in five different combinations (C_1, C_2, C_3, C_4, C_5), and there were two rates (levels) of quantity of nitrogen applied as A_1 (150 kg/ha) and A_2 (225 kg/ha). Details of the experiment are given in Table 1.

In this experiment, the interest of the experimenter is to fit the suitable model of the mixture ingredients along with how the two levels of nitrogen (process variable) are affecting the mixture ingredients. Such experiments in National Agricultural Research System (NARS) of India are generally conducted and analysed using a factorial randomized block/split plot designs. This provides the inference on the best treatment tried during the experiment and does not provide any information on the relationship of the proportions with the response variable. These questions can be answered by drawing an analogy of these experiments with mixtures.

Similarly, we consider an agricultural experiment in which a fixed quantity of nitrogen is applied in split doses to the crop at different growth stages. In it, different levels of irrigation (say at the depth of 10, 20 cm) are applied. This factor (irrigation)

Table 1 Proportion of nitrogen from different sources with two levels

Treatment combination	Source of nitrogen			Levels of nitrogen	Yield under replication		
	Urea (x_1)	FYM (x_2)	SPMC (x_3)		I	II	III
C_1	1	0	0	A_1	1.45	1.26	1.33
C_2	1/2	1/2	0	A_1	1.36	1.41	1.59
C_3	1/2	0	1/2	A_1	1.55	1.81	1.50
C_4	2/3	1/3	0	A_1	1.36	1.53	1.36
C_5	2/3	0	1/3	A_1	1.43	1.41	1.49
C_1	1	0	0	A_2	1.61	1.34	1.40
C_2	1/2	1/2	0	A_2	1.65	1.67	1.95
C_3	1/2	0	1/2	A_2	1.70	1.78	2.03
C_4	2/3	1/3	0	A_2	1.52	1.80	1.67
C_5	2/3	0	1/3	A_2	1.84	1.64	1.69

is not a part of the mixture (split doses) but affects the yield and is called the process variables, and the whole experiment is called *mixture experiment with process variable*.

Scheffé (1963) first introduced the experimentation of simplex-centroid × process variables with q mixture variables and n (>0) process variables. Daniel (1963) suggested the method by using fractional factorials at each vertex of a lattice in the mixture simplex. Cornell (1971) pointed out that Scheffé' assumption of homogeneous variance among the responses may not be valid. Another approach to this problem was shown by Cornell (1971) who introduced process variables in mixture problems where the components are categorized. Hare (1979) used a specialized geometric approach to obtain some designs in which the same set of blends is not repeated. Snee and Rayner (1982) and John (1984) have expressed concern about ill-conditioned matrices in mixture problems. Czitrom (1988, 1989) obtained experimental designs for mixture components with process variables for three blending components in two blocks that are orthogonal for quadratic (or linear) blending. Kumari and Singh (1988) developed mixture × process variables design with restricted simplex region. Draper et al. (1993), Prescott et al. (1993) and Lewis et al. (1994) obtained mixture designs in orthogonal blocks. Draper and Pukelsheim (1998) gave mixture models based on homogeneous functions. Dhekale (2001) fitted the quadratic polynomial of Scheffé (1958) with one process variable. Goldfarb et al. (2005), Chung et al. (2007) and Rodriquez et al. (2009) generated mixture experiment with process variable involving control and noise variables. Lal et al. (2010) discussed the methods of construction of mixture experiments, mixture experiments under constrained region and with process variables. Pradhan et al. (2017) constructed mixture experiments with one process variable in minimum number of runs. A detailed bibliography of mixture designs may be seen at http://iasri.res.in/design/mixture/mixture.htm. The methods of construction of mixture experiments with process variables given in the literature by using Latin squares or some other methods are complicated. These methods are, in general, not applicable to practical situation when there are three or more components with at least one process variable because such experiments require large number of runs.

This chapter, therefore, investigates two methods of construction of mixture experiments with process variables along with the analysis of these designs with an example have been illustrated. In Sect. 2, models of mixture experiments with process variables have been discussed. Construction of mixture designs with process variables has been given in Sect. 3. The analysis of these designs has been illustrated with an experimental data in Sect. 4. Section 5 gives the catalogue of the designs obtained by two methods with G-efficiency. The conclusion of the chapter is given in Sect. 6.

2 Models of Mixture Experiment with Process Variables

The model of mixture experiment with process variables is the combination of both the models for mixture experiment and process variables. The combined designs are used for collecting data to fit the combined model in the mixture components and the process variables. The combined quadratic model for q mixture components and m process variables with interaction in an easily interpretable form can be written as

$$E[y(\mathbf{x},\mathbf{z})] = \sum_{i=1}^{q}\gamma_i^0 x_i + \sum\sum_{i<j}^{q}\gamma_{ij}^0 x_i x_j + \sum_{l=1}^{m}\left[\sum_{i=1}^{q}\gamma_i^1 x_i + \sum\sum_{i<j}^{q}\gamma_{ij}^1 x_i x_j\right] z_l$$
$$+ \sum\sum_{l<p}^{m}\left[\sum_{i=1}^{q}\gamma_i^{lp} x_i + \sum\sum_{i<j}^{q}\gamma_{ij}^{lp} x_i x_j\right] z_l z_p \quad (2.1)$$

where x_i, $i = 1, \ldots, q$ are the q mixture components; z_l, $l = 1, \ldots, m$ are m process variables; γ_i^0 and γ_{ij}^0, $i, j = 1, \ldots, q$ are the parameters of mixture components for linear and quadratic terms, respectively; γ_i^1 and γ_{ij}^1 are the parameters of mixture experiments for linear and quadratic terms, respectively, associated with process variables z_l, $l = 1, \ldots, m$, and similarly γ_i^{lp} and γ_{ij}^{lp} are the parameters of mixture experiments associated with two factor interaction of process variables $z_l\, z_p$, $l, p = 1, \ldots, m$. This model for q components and two process variables becomes

$$E[y(\mathbf{x},\mathbf{z})] = \sum_{i=1}^{q}\gamma_i^0 x_i + \sum\sum_{i<j}^{q}\gamma_{ij}^0 x_i x_j + \sum_{l=1}^{2}\left[\sum_{i=1}^{q}\gamma_i^1 x_i + \sum\sum_{i<j}^{q}\gamma_{ij}^1 x_i x_j\right] z_l$$
$$+ \left[\sum_{i=1}^{q}\gamma_i^{lp} x_i + \sum\sum_{i<j}^{q}\gamma_{ij}^{lp} x_i x_j]\right] z_1 z_2 \quad (2.2)$$

For $q = 3$, the model (2.2) simplifies as

$$\begin{aligned} E[y(\mathbf{x},\mathbf{z})] = {} & \gamma_1^0 x_1 + \gamma_2^0 x_2 + \gamma_3^0 x_3 + \gamma_{12}^0 x_1 x_2 + \gamma_{13}^0 x_1 x_3 + \gamma_{23}^0 x_2 x_3 \\ & + \gamma_1^1 x_1 z_1 + \gamma_2^1 x_2 z_1 + \gamma_3^1 x_3 z_1 + \gamma_{12}^1 x_1 x_2 z_1 + \gamma_{13}^1 x_1 x_3 z_1 \\ & + \gamma_{23}^1 x_2 x_3 z_1 + \gamma_1^2 x_1 z_2 + \gamma_2^2 x_2 z_2 + \gamma_3^2 x_3 z_2 + \gamma_{12}^2 x_1 x_2 z_2 \\ & + \gamma_{13}^2 x_1 x_3 z_2 + \gamma_{23}^2 x_2 x_3 z_2 + \gamma_1^{12} x_1 z_1 z_2 + \gamma_2^{12} x_2 z_1 z_2 \\ & + \gamma_3^{12} x_3 z_1 z_2 + \gamma_{12}^{12} x_1 x_2 z_1 z_2 + \gamma_{13}^{12} x_1 x_3 z_1 z_2 + \gamma_{23}^{12} x_2 x_3 z_1 z_2 \end{aligned}$$

In the above model, first q and ${}^qC_2$ terms on the right side of equality side are the linear and quadratic blending portion of the model since these terms involve the component proportions only. The remaining terms represent the effects of changing the processing conditions (i.e. the levels or settings of the process variables) on the linear and quadratic blending properties of the mixture components.

When there is only one process variable, then the above model becomes

$$E[y(\mathbf{x},\mathbf{z})] = \sum_{i=1}^{q} \gamma_i^0 x_i + \sum_{i<j}^{q}\sum \gamma_{ij}^0 x_i x_j + \left[\sum_{i=1}^{q} \gamma_i^1 x_i + \sum_{i<j}^{q}\sum \gamma_{ij}^1 x_i x_j\right] z_1 \tag{2.3}$$

When only the linear blending is assumed to exist and quadratic blending is assumed to be negligible, the model simplifies as

$$E[y(\mathbf{x},\mathbf{z})] = \sum_{i=1}^{q} \gamma_i^0 x_i + \left[\sum_{i=1}^{q} \gamma_i^1 x_i\right] z_1 \tag{2.4}$$

Now, if the effect of z_1 is constant for all the blends, then the respective linear and quadratic models are simplifies as

$$E[y(\mathbf{x},\mathbf{z})] = \sum_{i=1}^{q} \gamma_i^0 x_i + \sum_{i=1}^{q} \gamma_i^1 x_i + z_1 \tag{2.5}$$

$$E[y(\mathbf{x},\mathbf{z})] = \sum_{i=1}^{q} \gamma_i^0 x_i + \sum_{i<j}^{q}\sum \gamma_{ij}^0 x_i x_j + z_1 \tag{2.6}$$

These models can be used in different situations as per the requirement of the experimenter. The designs for mixture experiments with process variables have been obtained by two methods and efficiency of these designs have been discussed.

3 Construction of Mixture Experiments with Process Variable

In mixture experiments with process variables, the standard type of design is a lattice/centroid mixture design at each point of a factorial arrangement (Cornell 2002). For the construction of mixture designs with process variables, one of the trivial methods is to take a lattice or centroid arrangement in the mixture component set-up at each point of a factorial arrangement. In this, the total number of runs is the number of points in mixture experiment multiplied by the total number of level combinations of the process variables in the experiment. In another approach, if a large run size is affordable, we start with an efficient response surface design (central composite design or Box–Behnken design) say ***d***. We repeat the design ***d*** for each level of process variable. Transform this repeated design into mixture design by projection method. This mixture design with level of process variable is the mixture design with process variable. Box and Hau (2001) and Prescott (2000) have given the method of construction of projection designs for situations when the design variables are subjected to linear constraints. The idea of the construction is to project an appropriate unconstrained design onto the constrained space.

Here, we give two methods of construction of efficient mixture designs with process variables in a desired number of runs.

3.1 Orthogonal Blocking

In this method, we take the response surface design with orthogonal blocking. The response surface design is either central composite designs or Box–Behnken designs. The design is of minimum resolution V plan so that main effects and two factor interactions can be estimated.

We start with the response surface designs in orthogonal blocks. Then by the projection method, we project this design onto the constrained space separately for each block separately. Now, taking the levels of factorial experiment and annexing in the different blocks as the level of process variables of the mixture experiment, we get the mixture experiment with process variable. We call this method as the method of orthogonal blocking. The advantage of projection design is that the projection designs have some useful properties, such as orthogonal blocking and rotatability are retained in the projected designs which make these designs suitable for mixture experiments. A catalogue of such designs for $q = 2$–5 has been constructed. The G-efficiency for the obtained designs has been worked out and is given along with the designs constructed.

3.2 Mixture Components as Quantitative and Process Variable as Qualitative Factor

There are many experimental situations where both the qualitative and quantitative variables are involved. For example, in fertilizer trials, the response or yield of a crop not only depends on the proportions of fertilizer from various sources but also on the method of application, viz. foliar application, behind the plough, broadcasting, etc. Here, the method of application is a qualitative variable while the proportions of fertilizer applied are quantitative variables.

The qualitative-cum-quantitative experiments differ from experiments involving only quantitative factors in the sense that we may often have dummy treatments. Here, the experimenter is mainly interested in the types or forms that are more responsive as well as in the interaction of different qualities with quantities. The important feature of these experiments which makes them different from ordinary factorials is that some of the level combinations, namely those where the quantitative factor is at zero level are indistinguishable.

Draper and John (1988) were the first to tackle the problem of obtaining response surface designs for qualitative-cum-quantitative factors. They discussed the relations between designs and models and gave designs for some specific situations.

Wu and Ding (1998) have given a systematic method of construction of such designs of economical size and discussed the underlying objectives and models. Aggarwal and Bansal (1998) further extended this method of construction for the situations where some of the quantitative factors are uncontrollable or noise factors.

The process variables in mixture experiments are similar to the qualitative factors in response to surface experiments. The qualitative factors will be taken as process variable, and the quantitative factors will be transformed to get the mixture components. Here, the experimenter is mainly interested in the types or forms that are more responsive as well as in the interaction of different process variables with mixture components. The method given by Wu and Ding (1998) has been modified for the construction of mixture design with process variable. We start with the design given by Wu and Ding (1998) based on central composite design. For one process variable with two levels, the initial design is given in Table 2.

For the construction of mixture design with process variable, we start with the efficient second-order response surface designs for qualitative and quantitative factors as discussed above. For the first t runs, the interaction having the maximum resolution will be given for the qualitative (z) variate. The response surface design will be transformed into mixture experiment by the projection method as discussed above. Now, the levels for z for the remaining $t+1$ to $t+2v+2$ runs will be allotted

Table 2 Qualitative-cum-quantitative central composite response surface design (Wu and Ding 1998)

Run	X_1	X_2	…	X_v	Z
1 2 . . . $t(=2^{v-m})$	$\pm$ 1 according to 2^{v-p} design				See Note 1
$t+1$ $t+2$	0 0	0 0	… …	0 0	1 or −1 −1 or 1
$t+3$ $t+4$ $t+5$ $t+6$ . . . $t+2v+1$ $t+2v+2$	B $-\beta$ 0 0 . . . 0 0	0 0 β $-\beta$ . . . 0 0	… … … …. . . . … …	0 0 0 0 . . . β $-\beta$	$z=-1$ or $z=1$

Note 1 $z=1$ or -1 by equating the z to the coefficients of interaction of highest order among the x_i's

by some criteria. In the present study, the criteria will be the minimization of trace of $(\mathbf{X}'\mathbf{X})^{-1}$ and determinant of $(\mathbf{X}'\mathbf{X})^{-1}$ and maximization of the G-efficiency. With this, we will be able to obtain the efficient mixture designs in minimum runs. The efficient mixtures for $q = 2$–5 have been constructed and these are catalogued.

G-efficiency is the efficiency based on per point so it can be computed independently for a mixture experiment. It is defined as

$$G\text{-efficiency} = \frac{p}{n \times d} 100$$

where n is the number of design points in the design, p is the number of parameter in the model and $d = \max\,\{v = x(\mathbf{x}'\mathbf{x})^{-1}\,x'\}$ over a specified set of design points x (the row vector) in $\mathbf{x}$, where $\mathbf{x}$ is the design matrix depending on model to be fitted. As a practical rule of thumb, Wheeler (1972) suggested that any design with a G-efficiency $\geq 50\%$ could be called *good* for practical purpose and shows that pursuit of higher efficiencies is not generally justified in practice.

4 Analysis of Mixture Experiments with Process Variables

The analysis of mixture experiments with process variables depends on the interests of the experimenter and the model one opts. In Sect. 2, different models have been discussed. The combined quadratic model in (2.1) for q mixture components and m process variables with interaction can be re-written in matrix notations as

$$\mathbf{y} = (\mathbf{X} \circ \mathbf{Z})\boldsymbol{\beta} + \boldsymbol{\varepsilon} \tag{4.1}$$

or

$$\mathbf{y} = \mathbf{U}\boldsymbol{\beta} + \boldsymbol{\varepsilon}; \quad \mathbf{U} = \mathbf{X} \circ \mathbf{Z}$$

where $\mathbf{y}$ is the $n \times 1$ vector of responses, $\mathbf{X} \circ \mathbf{Z}$ is component wise product of the columns of matrix $\boldsymbol{X}$ and matrix $\mathbf{Z}$, matrix $\boldsymbol{X}$ is $n \times (q + {}^qC_2)$ design matrix of mixture components and functions of the component proportions, and $\mathbf{Z}$ is partitioned as $\mathbf{Z} = [\mathbf{Z}_1\ \mathbf{Z}_2\ \mathbf{Z}_3]$. $\mathbf{Z}_1$ is $n \times 1$ column vector of $\mathbf{1}$'s, $\mathbf{Z}_2$ is $n \times m$ matrix of having m column vectors for the levels of process variables, and $\mathbf{Z}_3$ is $n \times {}^mC_2$ matrix of having ${}^mC_2$ column vectors for the two factor interaction of levels of process variables. Thus, $\mathbf{U}$ is the matrix of order $n \times [(q + {}^qC_2)(m + {}^mC_2 + 1)]$, $\boldsymbol{\beta}$ is the $[(q + {}^qC_2)(m + {}^mC_2 + 1)] \times 1$ vector of unknown parameters, and $\boldsymbol{\varepsilon}$ is the $(n \times 1)$ vector of errors distributed normally with zero mean and variance–covariance matrix $\sigma^2\mathbf{I}$, i.e. $\boldsymbol{\varepsilon} \sim N(\mathbf{0}, \sigma^2\mathbf{I})$. We see that the size of matrix $\mathbf{X}$ increases very rapidly, and thus, the number of parameters to be estimated also increases. If we take only one process variable, in model (2.1), then $\mathbf{U}$ matrix will be of order $n \times 2(q + {}^qC_2)$, and $\boldsymbol{\beta}$ is the vector of order $2(q + {}^qC_2) \times 1$.

Table 3 Analysis of variance for the mixture design with process variables

Source of variations	df	Sum of squares	Mean sum of squares	F value
Regression (fitted model)	$\left[(q+{}^{q}C_2)(m+{}^{m}C_2+1)\right]-1=(v-1)$ (say)	$SSR=\hat{\mathbf{b}}'\mathbf{U}'\mathbf{y}-(\mathbf{1}'\mathbf{y})^2/n=\sum_{u=1}^{n}(\hat{y}_u-\bar{y})^2$	MSR = SSR/(v − 1)	MSR/MSE
Error	$n-v$	$SSE=\mathbf{y}'\mathbf{y}-\hat{\mathbf{b}}'\mathbf{U}\mathbf{y}=\sum_{u=1}^{n}\left(y_u-\hat{y}_u\right)^2$	MSE = SSE/(n − v)	
Total	$n-1$	$SST=\mathbf{y}'\mathbf{y}-\left(\mathbf{1}'\mathbf{y}\right)^2/n=\sum_{u=1}^{n}\left(y_u-\hat{y}\right)^2$		

The normal equations for estimating $\boldsymbol{\beta}$ are given by

$$\mathbf{U}'\mathbf{U}\boldsymbol{\beta}-\mathbf{U}'\mathbf{y}=\mathbf{0} \tag{4.2}$$

Hence,

$$\hat{\mathbf{b}}=\left(\mathbf{U}'\mathbf{U}\right)^{-1}\mathbf{U}'\mathbf{y} \tag{4.3}$$

The various sums of squares, degrees of freedom and mean squares are given in Table 3.

This analysis will be illustrated with an experiment given in Table 1. The experiment shown in Table 1 has analogy with mixture experiment with one process variable. The three sources urea (x_1), FYM (x_2) and SPMC (x_3) are mixture components, factor A has two levels ($A_1=150$ kg/ha and $A_2=225$ kg/ha), and it is the process variable (z). This data can be arranged (Table 4).

The data given in Table 4 is in the form of mixture experiment with three components (x_1, x_2 and x_3) and one process variable (z). In the data, there are ten points. But for quadratic model given in (2.3) requires minimum 12 points data. So, linear mixture model (2.4) with one process variable was applied as

$$E[y(\mathbf{x},\mathbf{z})]=\gamma_1^0x_1+\gamma_2^0x_2+\gamma_3^0x_3+\gamma_1^1x_1z_1+\gamma_2^1x_2z_1+\gamma_3^1x_3z_1$$

The results after fitting above linear regression model (without intercept) are obtained as (Tables 5, 6 and 7).

These results reveal that the R^2 value for the fitted model is more than 99%. It means that it is the best-fitted model. Also, the estimates of regression for the three components of the mixture, i.e. urea, FYM and SPMC are highly (more than 1%) significantly affecting the yield. Further, process variable associated with FYM is highly significant while with SPMC it is significant at 5% level of significant. These answers can be obtained only when we analyse the data as mixture experiments with

Table 4 Data of sugarcane yield (q/ha)

x_1	1	0.500	0.500	0.667	0.667	1	0.500	0.500	0.667	0.667
x_2	0	0.500	0	0.333	0	0	0.500	0	0.333	0
x_3	0	0	0.500	0	0.333	0	0	0.500	0	0.333
Z	−1	−1	−1	−1	−1	1	1	1	1	1
Yield	1.347	1.453	1.620	1.417	1.443	1.450	1.757	1.837	1.663	1.723

Table 5 Analysis of variance for sugarcane data

Source of variations	df	Sum of squares	Mean square	F Value	$Pr > F$
Model	6	24.93367	4.15561	3388.48	<0.0001
Error	4	0.00491	0.00123		
Uncorrected Total	10	24.93857			

Table 6 Root MSE and values of R^2

Fit statistics root MSE	0.03502	R^2	0.9998
Dependent mean	1.57100	Adjusted R^2	0.9995
CV	2.22915		

Table 7 Estimates of the mixture components and their testing

Parameter estimates

Variable	df	Estimate	Standard error	t value	$Pr > \|t\|$	Type I SS	Type II SS
x_1	1	1.39232	0.02305	60.41	<0.0001	22.18151	4.47578
x_2	1	1.82325	0.05107	35.70	<0.0001	0.66448	1.56311
x_3	1	2.03442	0.05107	39.84	<0.0001	1.94645	1.94645
x_{1z}	1	0.06254	0.02305	2.71	0.0533	0.10676	0.00903
x_{2z}	1	0.24227	0.05107	4.74	0.0090	0.01609	0.02760
x_{3z}	1	0.19771	0.05107	3.87	0.0180	0.01838	0.01838

process variable. A catalogue of such designs for $q = 3$–5 is given in Sect. 5. The G-efficiency, trace of $(\mathbf{X}'\mathbf{X})^{-1}$ and determinant of $(\mathbf{X}'\mathbf{X})^{-1}$ for different designs have been worked out and are given along with the designs constructed.

5 Catalogue of the Designs for $q = 3$–5

Here, we give the catalogue of the designs obtained by the two methods discussed above with G-efficiency. These designs are for varying in number of runs and for $q = 3$–5 with one process variable. All the designs are having G-efficiency more than 50%, and G-efficiency is per point efficiency. So, these designs may be considered as good designs.

I. **Mixture designs with process variables (efficient response surface design with orthogonal blocking)**

Design 01 $q = 3, n = 27$; G-efficiency = 43.63%				Design 02 $q = 4, n = 27$; G-efficiency = 58.18%				
x_1	x_2	x_3	Z	x_1	x_2	x_3	x_4	z
0.222	0.222	0.556	−1	0.125	0.125	0.375	0.375	−1
0	0.667	0.333	−1	0	0.500	0.250	0.250	−1
0.667	0	0.333	−1	0.500	0	0.250	0.250	−1
0.444	0.444	0.111	−1	0.375	0.375	0.125	0.125	−1
0.222	0.222	0.556	−1	0.375	0.375	0.125	0.125	−1
0.222	0.222	0.556	−1	0.250	0.250	0	0.500	−1
0.444	0.444	0.111	−1	0.250	0.250	0.500	0	−1
0.444	0.444	0.111	−1	0.125	0.125	0.375	0.375	−1
0.333	0.333	0.333	−1	0.250	0.250	0.250	0.250	−1
0.222	0.556	0.222	0	0.125	0.375	0.125	0.375	0
0	0.333	0.667	0	0	0.250	0.500	0.250	0
0.667	0.333	0	0	0.500	0.250	0	0.250	0
0.444	0.111	0.444	0	0.375	0.125	0.375	0.125	0
0.222	0.556	0.222	0	0.375	0.125	0.375	0.125	0
0.222	0.556	0.222	0	0.250	0	0.250	0.500	0
0.444	0.111	0.444	0	0.250	0.500	0.250	0	0
0.444	0.111	0.444	0	0.125	0.375	0.125	0.375	0
0.333	0.333	0.333	0	0.250	0.250	0.250	0.250	0
0.556	0.222	0.222	1	0.375	0.125	0.125	0.375	1
0.333	0	0.667	1	0.250	0	0.500	0.250	1
0.333	0.667	0	1	0.250	0.500	0	0.250	1
0.111	0.444	0.444	1	0.125	0.375	0.375	0.125	1
0.556	0.222	0.222	1	0.125	0.375	0.375	0.125	1
0.556	0.222	0.222	1	0	0.250	0.250	0.500	1
0.111	0.444	0.444	1	0.500	0.250	0.250	0	1
0.111	0.444	0.444	1	0.375	0.125	0.125	0.375	1
0.333	0.333	0.333	1	0.250	0.250	0.250	0.250	1

Design 03 $q = 4, n = 27$; G-efficiency = 53.70%					Design 04 $q = 5, n = 60$; G-efficiency = 61.22%					
x_1	x_2	x_3	x_4	Z	x_1	x_2	x_3	x_4	x_5	z
0.250	0.250	0.250	0.250	−1	0.200	0.200	0.200	0.200	0.200	−1
0	0	0.500	0.500	−1	0.350	0.350	0.100	0.100	0.100	−1

(continued)

(continued)

Design 03 $q = 4, n = 27$; G-efficiency = 53.70%					Design 04 $q = 5, n = 60$; G-efficiency = 61.22%					
x_1	x_2	x_3	x_4	Z	x_1	x_2	x_3	x_4	x_5	z
0	0.500	0	0.500	−1	0.350	0.100	0.350	0.100	0.100	−1
0.500	0	0	0.500	−1	0.350	0.100	0.100	0.350	0.100	−1
0.500	0	0.500	0	−1	0.350	0.100	0.100	0.100	0.350	−1
0.500	0.500	0	0	−1	0.100	0.350	0.350	0.100	0.100	−1
0	0.500	0.500	0	−1	0.100	0.350	0.100	0.350	0.100	−1
0.250	0.250	0.250	0.250	−1	0.100	0.350	0.100	0.100	0.350	−1
0.250	0.250	0.250	0.250	−1	0.100	0.100	0.350	0.350	0.100	−1
0.167	0.167	0.167	0.500	0	0.100	0.100	0.350	0.100	0.350	−1
0.167	0.167	0.500	0.167	0	0.100	0.100	0.100	0.350	0.350	−1
0.167	0.500	0.167	0.167	0	0.250	0.250	0.250	0.250	0	−1
0.500	0.167	0.167	0.167	0	0.250	0.250	0.250	0	0.250	−1
0	0.333	0.333	0.333	0	0.250	0.250	0	0.250	0.250	−1
0.333	0	0.333	0.333	0	0.250	0	0.250	0.250	0.250	−1
0.333	0.333	0	0.333	0	0	0.250	0.250	0.250	0.250	−1
0.333	0.333	0.333	0	0	0.200	0.200	0.200	0.200	0.200	−1
0.250	0.250	0.250	0.250	0	0.200	0.200	0.200	0.200	0.200	−1
0.500	0.167	0.167	0.167	1	0.200	0.200	0.200	0.200	0.200	−1
0	0.333	0.333	0.333	1	0.200	0.200	0.200	0.200	0.200	−1
0.167	0.500	0.167	0.167	1	0.400	0.150	0.150	0.150	0.150	0
0.333	0	0.333	0.333	1	0.150	0.400	0.150	0.150	0.150	0
0.167	0.167	0.500	0.167	1	0.150	0.150	0.400	0.150	0.150	0
0.333	0.333	0	0.333	1	0.150	0.150	0.150	0.400	0.150	0
0.167	0.167	0.167	0.500	1	0.150	0.150	0.150	0.150	0.400	0
0.333	0.333	0.333	0	1	0.300	0.300	0.300	0.050	0.050	0
0.250	0.250	0.250	0.250	1	0.300	0.300	0.050	0.300	0.050	0
					0.300	0.300	0.050	0.050	0.300	0
					0.300	0.050	0.300	0.300	0.050	0
					0.300	0.050	0.300	0.050	0.300	0
					0.300	0.050	0.050	0.300	0.300	0
					0.050	0.300	0.300	0.300	0.050	0
					0.050	0.300	0.300	0.050	0.300	0
					0.050	0.300	0.050	0.300	0.300	0
					0.050	0.050	0.300	0.300	0.300	0
					0.200	0.200	0.200	0.200	0.200	0
					0.200	0.200	0.200	0.200	0.200	0

(continued)

(continued)

Design 03 $q = 4, n = 27$; G-efficiency = 53.70%					Design 04 $q = 5, n = 60$; G-efficiency = 61.22%					
x_1	x_2	x_3	x_4	Z	x_1	x_2	x_3	x_4	x_5	z
					0.200	0.200	0.200	0.200	0.200	0
					0.200	0.200	0.200	0.200	0.200	0
					0.200	0.200	0.200	0.200	0.200	0
					0.400	0.150	0.150	0.150	0.150	1
					0	0.250	0.250	0.250	0.250	1
					0.150	0.400	0.150	0.150	0.150	1
					0.250	0	0.250	0.250	0.250	1
					0.150	0.150	0.400	0.150	0.150	1
					0.250	0.250	0	0.250	0.250	1
					0.150	0.150	0.150	0.400	0.150	1
					0.250	0.250	0.250	0	0.250	1
					0.150	0.150	0.150	0.150	0.400	1
					0.250	0.250	0.250	0.250	0	1
					0.400	0.150	0.150	0.150	0.150	1
					0	0.250	0.250	0.250	0.250	1
					0.150	0.400	0.150	0.150	0.150	1
					0.250	0	0.250	0.250	0.250	1
					0.150	0.150	0.400	0.150	0.150	1
					0.250	0.250	0	0.250	0.250	1
					0.150	0.150	0.150	0.400	0.150	1
					0.250	0.250	0.250	0	0.250	1
					0.150	0.150	0.150	0.150	0.400	1
					0.250	0.250	0.250	0.250	0	1

Design 05 $q = 5, n = 60$; G-efficiency = 57.14%						
x_1	x_2	x_3	x_4	x_5	z_1	z_2
0.200	0.200	0.200	0.200	0.200	−1	−1
0.100	0.100	0.100	0.350	0.350	−1	−1
0.050	0.300	0.300	0.050	0.300	−1	−1
0.050	0.300	0.300	0.300	0.050	−1	−1
0.300	0.050	0.300	0.050	0.300	−1	−1
0.300	0.050	0.300	0.300	0.050	−1	−1
0.350	0.350	0.100	0.100	0.100	−1	−1
0.250	0.250	0	0.250	0.250	−1	−1
0.200	0.200	0.200	0.200	0.200	−1	−1

(continued)

(continued)

Design 05
$q = 5, n = 60$; G-efficiency = 57.14%

x_1	x_2	x_3	x_4	x_5	z_1	z_2
0.200	0.200	0.200	0.200	0.200	−1	−1
0.150	0.150	0.150	0.150	0.400	−1	1
0.150	0.150	0.150	0.400	0.150	−1	1
0.100	0.350	0.350	0.100	0.100	−1	1
0	0.250	0.250	0.250	0.250	−1	1
0.350	0.100	0.350	0.100	0.100	−1	1
0.250	0	0.250	0.250	0.250	−1	1
0.300	0.300	0.050	0.050	0.300	−1	1
0.300	0.300	0.050	0.300	0.050	−1	1
0.200	0.200	0.200	0.200	0.200	−1	1
0.200	0.200	0.200	0.200	0.200	−1	1
0.400	0.150	0.150	0.150	0.150	0	−1
0	0.250	0.250	0.250	0.250	0	−1
0.150	0.400	0.150	0.150	0.150	0	−1
0.250	0	0.250	0.250	0.250	0	−1
0.150	0.150	0.400	0.150	0.150	0	−1
0.250	0.250	0	0.250	0.250	0	−1
0.150	0.150	0.150	0.400	0.150	0	−1
0.250	0.250	0.250	0	0.250	0	−1
0.150	0.150	0.150	0.150	0.400	0	−1
0.250	0.250	0.250	0.250	0	0	−1
0.100	0.100	0.350	0.100	0.350	0	1
0.100	0.100	0.350	0.350	0.100	0	1
0.150	0.400	0.150	0.150	0.150	0	1
0.050	0.300	0.050	0.300	0.300	0	1
0.400	0.150	0.150	0.150	0.150	0	1
0.300	0.050	0.050	0.300	0.300	0	1
0.250	0.250	0.250	0	0.250	0	1
0.250	0.250	0.250	0.250	0	0	1
0.200	0.200	0.200	0.200	0.200	0	1
0.200	0.200	0.200	0.200	0.200	0	1
0.150	0.150	0.400	0.150	0.150	1	−1
0.050	0.050	0.300	0.300	0.300	1	−1
0.100	0.350	0.100	0.100	0.350	1	−1
0.100	0.350	0.100	0.350	0.100	1	−1

(continued)

(continued)

Design 05 $q = 5, n = 60$; G-efficiency = 57.14%						
x_1	x_2	x_3	x_4	x_5	z_1	z_2
0.350	0.100	0.100	0.100	0.350	1	−1
0.350	0.100	0.100	0.350	0.100	1	−1
0.300	0.300	0.300	0.050	0.050	1	−1
0.200	0.200	0.200	0.200	0.200	1	−1
0.200	0.200	0.200	0.200	0.200	1	−1
0.200	0.200	0.200	0.200	0.200	1	−1
0.400	0.150	0.150	0.150	0.150	1	1
0	0.250	0.250	0.250	0.250	1	1
0.150	0.400	0.150	0.150	0.150	1	1
0.250	0	0.250	0.250	0.250	1	1
0.150	0.150	0.400	0.150	0.150	1	1
0.250	0.250	0	0.250	0.250	1	1
0.150	0.150	0.150	0.400	0.150	1	1
0.250	0.250	0.250	0	0.250	1	1
0.150	0.150	0.150	0.150	0.400	1	1
0.250	0.250	0.250	0.250	0	1	1

II. Mixture designs with process variables (mixture components as quantitative and process variable as qualitative factor)

Design 06 $q = 2, n = 1$; G-efficiency = 57.14%			Design 07 $q = 3, n = 16$; G-efficiency 58.01%			
x_1	x_2	Z	x_1	x_2	x_3	z
0.500	0.500	−1	0.333	0.333	0.334	1
1.000	0	1	0.667	0.167	0.166	−1
0	1.000	1	0.167	0.667	0.166	−1
0.500	0.500	−1	0.500	0.500	0	1
0.500	0.500	−1	0.167	0.167	0.666	−1
0.500	0.500	1	0.500	0	0.500	1
0.854	0.146	−1	0	0.500	0.500	1
0.146	0.854	1	0.333	0.333	0.334	−1
0.146	0.854	−1	0.333	0.333	0.334	−1
0.854	0.146	1	0.333	0.333	0.334	1
			0.053	0.473	0.474	−1

(continued)

(continued)

Design 06 $q = 2, n = 1$; G-efficiency = 57.14%			Design 07 $q = 3, n = 16$; G-efficiency 58.01%			
x_1	x_2	Z	x_1	x_2	x_3	z
			0.614	0.193	0.193	1
			0.473	0.053	0.474	1
			0.193	0.614	0.193	−1
			0.473	0.474	0.053	−1
			0.193	0.193	0.614	1

Design 08 $q = 4, n = 26$; G-efficiency = 53.85%					Design 09 $q = 5, n = 44$; G-efficiency = 69.45%					
x_1	x_2	x_3	x_4	Z	x_1	x_2	x_3	x_4	x_5	z
0.250	0.250	0.250	0.250	1	0.200	0.200	0.200	0.200	0.200	1
0.500	0.167	0.167	0.167	−1	0.367	0.158	0.158	0.158	0.158	−1
0.167	0.500	0.167	0.167	−1	0.158	0.367	0.158	0.158	0.158	−1
0.417	0.417	0.083	0.083	1	0.326	0.326	0.116	0.116	0.116	1
0.167	0.167	0.500	0.167	−1	0.158	0.158	0.368	0.158	0.158	−1
0.417	0.083	0.417	0.083	1	0.326	0.116	0.326	0.116	0.116	1
0.083	0.417	0.417	0.083	1	0.116	0.326	0.326	0.116	0.116	1
0.333	0.333	0.333	0	−1	0.284	0.284	0.284	0.074	0.074	−1
0.167	0.167	0.167	0.500	−1	0.158	0.158	0.158	0.368	0.158	−1
0.417	0.083	0.083	0.417	1	0.326	0.116	0.116	0.326	0.116	1
0.083	0.417	0.083	0.417	1	0.116	0.326	0.116	0.326	0.116	1
0.333	0.333	0	0.333	−1	0.284	0.284	0.074	0.284	0.074	−1
0.083	0.083	0.417	0.417	1	0.116	0.116	0.326	0.326	0.116	1
0.333	0	0.333	0.333	−1	0.284	0.074	0.284	0.284	0.074	−1
0	0.333	0.333	0.333	−1	0.074	0.284	0.284	0.284	0.074	−1
0.250	0.250	0.250	0.250	1	0.200	0.200	0.200	0.200	0.200	1
0.250	0.250	0.250	0.250	−1	0.158	0.158	0.158	0.158	0.368	−1
0.250	0.250	0.250	0.250	1	0.326	0.116	0.116	0.116	0.326	1
0	0.333	0.333	0.334	−1	0.116	0.326	0.116	0.116	0.326	1
0.500	0.167	0.167	0.167	1	0.284	0.284	0.074	0.074	0.284	−1
0.333	0	0.333	0.333	1	0.116	0.116	0.326	0.116	0.326	1
0.167	0.500	0.167	0.167	−1	0.284	0.074	0.284	0.074	0.284	−1
0.333	0.333	0	0.333	−1	0.074	0.284	0.284	0.074	0.284	−1
0.167	0.167	0.500	0.167	1	0.242	0.242	0.242	0.032	0.242	1
0.333	0.333	0.333	0	1	0.116	0.116	0.116	0.326	0.326	1
0.167	0.167	0.167	0.500	−1	0.284	0.074	0.074	0.284	0.284	−1
					0.074	0.284	0.074	0.284	0.284	−1

(continued)

(continued)

Design 08 $q = 4, n = 26$; G-efficiency = 53.85%					Design 09 $q = 5, n = 44$; G-efficiency = 69.45%					
x_1	x_2	x_3	x_4	Z	x_1	x_2	x_3	x_4	x_5	z
					0.242	0.242	0.032	0.242	0.242	1
					0.074	0.074	0.284	0.284	0.284	−1
					0.242	0.032	0.242	0.242	0.242	1
					0.032	0.242	0.242	0.242	0.242	1
					0.200	0.200	0.200	0.200	0.200	−1
					0.200	0.200	0.200	0.200	0.200	−1
					0.200	0.200	0.200	0.200	0.200	1
					0	0.250	0.250	0.250	0.250	−1
					0.399	0.150	0.150	0.150	0.150	1
					0.250	0	0.250	0.250	0.250	1
					0.150	0.399	0.150	0.150	0.150	−1
					0.250	0.250	0	0.250	0.250	−1
					0.150	0.150	0.399	0.150	0.150	1
					0.250	0.250	0.250	0	0.250	1
					0.150	0.150	0.150	0.399	0.150	−1
					0.250	0.250	0.250	0.250	0	−1
					0.150	0.150	0.150	0.150	0.400	1

6 Conclusions

The experiments, in which the response is a function of the proportion of the ingredients in the mixture rather than the total amount, are called mixture experiments. When the mixture experiments are conducted with process variables, the experimenter is not only interested in blending properties of the mixture components but also in blending behaviour with the change of levels of the process variable. Two methods of construction of mixture experiments with process variables have been developed. Efficient mixture designs with the process are obtained by using these methods. Catalogues of the obtained designs with G-efficiency for varying in number of runs and for $q = 3$–5 factors with one process variable are also given. The analysis of mixture design with process variable is also illustrated with experimental data.

References

Aggarwal, M. L., & Bansal, A. (1998). Robust response surface designs for qualitative and quantitative factors. *Communication in Statistics: Theory and Methods, 27*(1), 89–106.

Box, G. E. P., & Hau, I. (2001). Experimental designs with one or more factor constraints. *Journal of Applied Statistics, 28,* 973–989.

Chung, P. J., Montgomery, D. C., & Goldfarb, H. B. (2007). Optimal designs for mixture-process experiments involving control and noise variables. *Journal of Quality Technology, 39,* 179–190.

Cornell, J. A. (1971). Process variables in the mixture problems categorized components. *Journal of the American Statistical Association, 66,* 42–88.

Cornell, J. A. (2002). *Experiments with mixtures* (3rd ed.). New York: Wiley.

Czitrom, V. (1988). Mixture experiments with process variables: D-optimal orthogonal experimental designs. *Communication in Statistics: Theory and Methods, 17,* 105–121.

Czitrom, V. (1989). Experimental design for four mixture components with process variables. *Communication in Statistics: Theory and Methods, 18,* 4561–4581.

Daniel, C. (1963). Discussion of Scheffé's article The simplex-centroid design for experiments with mixtures. *Journal of the Royal Statistical Society, B, 25,* 256–257.

Dhekale, J. S. (2001). Analytical techniques in intercropping experiments. Ph. D. thesis, IARI, New Delhi.

Draper, N. R., & John, J. A. (1988). Response surface designs for quantitative and qualitative variables. *Technometrics, 30,* 423–428.

Draper, N. R., Prescott, P., Lewis, S. M., Dean, A. M., John, P. W. M., & Tuck, M. G. (1993). Mixture designs for four components in orthogonal blocks. *Technometrics, 35,* 268–276.

Draper, N. R., & Pukelsheim, F. (1998). A mixture models based on homogeneous polynomials. *Journal of Statistical Planning and Inference, 71,* 303–311.

Goldfarb, H. B., Borror, C. M., Montgomery, D. C., & Anderson-Cook, C. M. (2005). Using genetic algorithms to generate mixture-process experimental designs involving control and noise variables. *Journal of Quality Technology, 37,* 60–74.

Hare, L. B. (1979). Design for mixture experiments involving process variables. *Technometrics, 21*(2), 159–173.

John, P. W. M. (1984). Experiments with mixtures involving process variables. Technical Report 8. Centre for Statistical Sciences, University of Texas, Austin, pp. 1–17.

Kumari, R., & Singh, S. (1988). Mixture × process variables design with restricted simplex. *Journal of the Indian Society of Agricultural Statistics, 40,* 15–22.

Lal, K., Sharma, V. K., Gupta, V. K., Batra, P. K., & Bhar, L. M. (2010). Designs for mixture experiments in agriculture. New Delhi: Project Report: I.A.S.R.I.

Lewis, S. M., Draper, N. R., Dean, A. N., & Prescott, P. (1994). Mixture designs for q components in orthogonal blocks. *Journal of the Royal Statistical Society, B, 56,* 457–467.

Pradhan, U. K., Lal, K., Dash, S., & Singh, K. N. (2017). Designs and analysis of mixture experiments with process variable in smaller number of runs. *Communication in Statistics: Theory and Methods, 46*(1), 259–270.

Prescott, P. (2000). Projection designs for mixture experiments in orthogonal blocks. *Communication in Statistics: Theory and Methods, 29,* 2229–2253.

Prescott, P., Draper, N. R., Dean, A. M., & Lewis, S. M. (1993). Mixture designs for five components in orthogonal blocks. *Journal of Applied Statistics, 20*(1), 105–117.

Rodriguez, M., Montgomery, D. C., & Borror, C. M. (2009). Generating experimental designs involving control and noise variables using genetic algorithms. *Quality and Reliability Engineering International, 25,* 1045–1065.

Scheffé, H. (1958). Experiments with mixtures. *Journal of the Royal Statistical Society, B, 20,* 344–360.

Scheffé, H. (1963). Simplex-centroid designs for experiments with mixtures. *Journal of the Royal Statistical Society, B, 25,* 235–263.

Snee, R. D., & Rayner, A. (1982). Assessing the accuracy of mixture and regression calculations. *Journal of Quality Technology, 14*(2), 67–79.
Wheeler, R. E. (1972). Efficient experimental design. Presented at the Annual Meeting of the American Statistical Association, Montreal, Canada.
Wu, C. F. J., & Ding, Y. (1998). Construction of response surface designs for qualitative and quantitative factors. *Journal of Statistical Planning and Inference, 71,* 331–348.

Development in Copula Applications in Forestry and Environmental Sciences

M. Ishaq Bhatti and Hung Quang Do

Abstract Recently, copula is being used in social, natural, and physical sciences due to its flexibilities in joint distributions and marginals and high computing power. They are the powerful instruments to model-dependent structures between various complex correlated variables in environmental sciences. This chapter studies the development of copula models and its applications in the area of environmental sciences. It reviews the literature on the types of copula models, including Gumbel, Clayton, and vine, and outlines the theoretical development of mixture bivariate and multivariate copula distributions. A brief review of literature is done using DCC, ARCH, and GARCH copulas.

Keywords ARCH · Copulas · Environmental sciences · GARCH

1 Introduction

Recently, copula models are being used in many research areas, including business, economics, insurance, geosciences, irrigation, energy, safety analysis, environmental sciences, and climate change. This chapter covers historical background of copula models, reviews basic copula theory and its applications in environmental and forestry sciences, and sees how efficient copulas are to forecast future environmental events. It summarizes various articles in the area of forestry and environmental sciences in relation to dynamic conditional correlation, ARCH, and GARCH econometric models in conjunction with copula theory.[1] The rest of this chapter is structured as follows. Section 2 considers the basic copula theory and explains various types of copulas and theoretical developments. Moreover, its application in various areas is also given. Section 3 provides a literature review of copula application in forestry and environmental sciences. Section 4 contains some concluding remarks.

M. I. Bhatti (✉) · H. Q. Do
Department of Economics and Finance, La Trobe University, Bundoora, VIC 3083, Australia
e-mail: i.bhatti@latrobe.edu.au

[1] Some of the material in this chapter is in a summarize form from a published review paper authored by Bhatti and Do (2019).

G. Chandra et al. (eds.), *Statistical Methods and Applications in Forestry and Environmental Sciences*, Forum for Interdisciplinary Mathematics,
https://doi.org/10.1007/978-981-15-1476-0_13

2 Copula Theory

Copula was first introduced by Sklar (1959) which is defined as a function $C(u_1, \ldots, u_n)$ of multivariate joint distribution with uniform marginals in the interval [0, 1]. For any joint distribution $H(x_1, \ldots, x_n)$ with marginals $F_1(x_1), \ldots, F_n(x_n)$, we can express H as a function of $C(.)$ as in Eq. (1),

$$H(x_1, \ldots, x_n) = C(F_1(x_1), \ldots, F_n(x_n)), \quad \forall x_i \in \bar{\mathbf{R}} \times \bar{\mathbf{R}}. \tag{1}$$

If $F_1, \ldots, F_n$ are continuous, then the copula C associated with H is a unique function which can be obtained by the expression (2).

$$C(u_1, \ldots, u_n) = H\left(F_1^{(-1)}(u_1), \ldots, F_n^{(-1)}(u_n)\right), \tag{2}$$

where $u_1 = F_1(x_1), \ldots, u_n = F_n(x_n)$.

The density function (1) can be obtained because $F_1, \ldots, F_n$ and C are n-differentiable. Therefore, for a bivariate case, the density function can be written as,

$$h(x_1, x_2) = c(F_1(x_1), F_2(x_2)) \prod_{i=1}^{2} f_i(x_i), \tag{3}$$

where h is the density function related to H, f_i is the density function for each marginal, and the copula density c is obtained by differentiating (1) and can be written as

$$c(F_1(x_1), \ldots, F_n(x_n)) = \frac{h\left(F_1^{(-1)}(u_1), F_2^{(-1)}(u_2)\right)}{\prod_{i=1}^{2} f_i\left(F_1^{(-1)}(u_i)\right)} \tag{4}$$

Copula functions when applied in various sciences can have elliptical format and are called Gaussian (normal) and Student-t copulas, due to an associated quadratic form of correlation between the marginals. Their dependence structure is the linear correlation coefficient and takes values in the interval $[-1, 1]$, and their distribution functions are symmetric.

Other copulas like Archimedean are depending on the functional form of the generating factor[2] associated with them. For example, they may have a dependence measure belonging to the most diverse ranges of variations like in Gumbel (1960) copula which allows only positive dependence structures (or upper tail dependence), for which the parameter belongs to the interval $[1, \infty)$. This diversity of copula

[2]Archimedean copulas need to meet some varying criteria, for example, like convexity which are detailed in Nelsen (2007), Ahsanullah and Bhatti (2010), Al Rahahleh et al. (2017), Al Rahahleh and Bhatti (2017), Bhatti and Nguyen (2012), Nguyen and Bhatti (2012) and Nguyen et al. (2016).

functions with specific dependence structures makes it impossible to compare different functional forms for copulas. To make this comparison possible, we focus on a dependence measure known as tail dependence,[3] which is defined below.

Definition 1 If $\lim_{\varepsilon\to 0}\Pr(U_1 \le \varepsilon|U_2 \le \varepsilon) = \lim_{\varepsilon\to 0}\Pr(U_2 \le \varepsilon|U_1 \le \varepsilon) = \lim_{\varepsilon\to 0} C(\varepsilon,\varepsilon)/\varepsilon = \tau^L$ exists, then copula C has a lower tail dependence if $\tau^L \in (0, 1]$. Otherwise, C has no lower tail dependence. If $\lim_{\delta\to 1}\Pr(U_1 > \delta|U_2 > \delta) = \lim_{\delta\to 1}\Pr(U_2 > \delta|U_1 > \delta) = \lim_{\delta\to 1}(1-2\delta + C(\delta,\delta))/(1-\delta) = \tau^U$ exists, then copula C has upper tail dependence if $\tau^U \in (0, 1]$. Otherwise, C has no upper tail dependence. Various functional forms can be used as copulas, see, for example, Nelsen (2007) and Ahsanullah and Bhatti (2010). In this chapter, we review of constructing the following *four types* of copula functions which are routinely being used in forestry and environmental sciences.

(i) ***Gumbel Copula***

Gumbel copulas are upper tail dependence and can be defined as,

$$C_G(u_1, u_2|\theta) = \exp\left\{-\left[(-\log u_1)^\theta + (-\log u_2)^\theta\right]^{\frac{1}{\theta}}\right\}, \quad \text{where } \theta \in [1, +\infty).$$

Rotated Gumbel copula has only lower tail dependence which is defined as,

$$C_{RG}(u_1, u_2|\theta) = u_1 + u_2 - 1 + C_G(1-u_1, 1-u_2|\theta), \quad \text{where } \theta \in [1, \infty).$$

(ii) ***Symmetrized Joe-Clayton (SJC) Copula***

SJC copula is defined as,

$$\begin{aligned} C_{\mathrm{SJC}}\left(u_1, u_2|\tau^U, \tau^L\right) = 0.5\big(&C_{\mathrm{JC}}\left(u_1, u_2|\tau^U, \tau^L\right) \\ &+ C_{\mathrm{JC}}\left(1-u_1, 1-u_2|\tau^U, \tau^L\right) + u_1 + u_2 - 1\big), \end{aligned}$$

where C_{JC} is the Joe-Clayton copula and is given by,

$$\begin{aligned} C_{\mathrm{JC}}\left(u_1, u_2|\tau^U, \tau^L\right) = 1 - \Big(1 - \Big\{&\left[1-(1-u_1)^\kappa\right]^{-\gamma} \\ &+ \left[1-(1-u_2)^\kappa\right]^{-\gamma} - 1\Big\} wo^{-1/\gamma}\Big)^{-1/\kappa}, \end{aligned}$$

with value of κ and γ is expressed as,

$$\kappa = 1/\log_2\left(2-\tau^U\right), \quad \text{and} \quad \gamma = -\frac{1}{\log_2\left(\tau^L\right)} \quad \text{and} \quad \tau^U, \tau^L \in (0, 1)$$

[3]Other measures of dependence have been used in the literature to compare copulas, such as the measures of concordance Kendall's τ, Spearman's ρ, and Gini's co-graduation index. For further details, see Cherubini et al. (2004), Patton (2006) and Nelsen (2007).

The SJC copulas are better than Gumbel because SJC got both upper and lower tail dependence parameter structures. Its own dependence parameters, τ^U and τ^L, are the measures of dependence of the upper and lower tail, respectively. Furthermore, τ^U and τ^L range freely and are not dependent on each other.

(iii) ***Normal or Gaussian Copula***

Gaussian copula has no tail dependence, and its dependence parameter is the linear correlation coefficient which is defined as,

$$C_N(u_1, u_2|\rho) = \int_{-\infty}^{\phi^{-1}(u_1)} \int_{-\infty}^{\phi^{-1}(u_2)} \frac{1}{2\pi\sqrt{(1-\rho^2)}} \times \exp\left\{\frac{-(r^2 - 2\rho rs + s^2)}{2(1-\rho^2)}\right\} dr ds, \quad \rho \in (-1, 1).$$

(iv) ***Student-t Copula***

Student-t copula has the linear correlation coefficient as measure of dependence, as in the case of Gaussian. However, unlike the normal copula, it shows some tail dependence. These are symmetric tail dependence and can be defined as:

$$C_t(u_1, u_2|\rho, v) = \int_{-\infty}^{t_v^{-1}(u_1)} \int_{-\infty}^{t_v^{-1}(u_2)} \frac{1}{2\pi\sqrt{(1-\rho^2)}} \times \left(1 + \frac{r^2 - 2\rho rs + s^2}{v(1-\rho^2)}\right)^{-\frac{v+2}{2}} dr ds.$$

3 Copula Applications in Forestry and Environmental Sciences

3.1 Copulas in Forestry Studies

Recently, copulas are being used in the research area of forestry to understand the relationships between tree diameters, height, and volume in harvesting and maintaining sustainable forest growth. For example, Kershaw et al. (2010) used normal copula in generating the desired spatial dependency. However, Kershaw et al. (2017) showed that multiple imputations and copula sampling can be used as a benchmark to evaluate statistical and process models for tree growth and yield projection. They observed that copulas can provide a mechanism to separate signal from noise and copulas are useful in a large regional sample size. To examine wood volume, Serinaldi et al. (2012) find that copula-based model with nonparametric marginals shows accurate point estimates but biased interval estimates. Similarly, Fortin et al. (2013)

illustrate that Spearman's correlation coefficients decrease as the distance between the trees increases, and copula model fit data better than traditional models. Pothier et al. (2013) suggest that to increase harvested volume and improve future stand, partial cutting should focus on small low-vigor (LV)-high-quality (HQ) trees rather than large LV-HQ trees. Using similar variables of tree vigor/quality, Delisle-Boulianne et al. (2014) find that generalized linear mixed-effect models and a copula approach could improve maximum log-likelihood by 12%. Meanwhile, Wang et al. (2010) find that log-logistic best describes diameter and volume distributions although both logit-logistic and Weibull marginal model provide same fitting for height distribution. Ogana et al. (2018) find the importance of copula in distribution modeling in quantitative forestry and simulate wood harvesting regime for the forest stands. Moreover, MacPhee et al. (2018) observed that copula model performs well in predicting tree height and shows the least loss of functionality when applied to species with sparse data.

In examining forest resources, Ene et al. (2013) find that copula approach is superior to the bootstrap resampling for all cases and the best results are obtained using the copula approach and *k-NN* imputations with $k = 1$. Saarela et al. (2017) used Monte Carlo (MC) simulation and find that the estimator is more stable than other traditional estimators.

Dong et al. (2017b) reveal that copula model for multivariate distribution and alpha-stable distribution is useful for real-valued polarimetric features, which avoid complex matrix operations, and this model is flexible in constructing joint statistical model because it takes advantages of both copula and the alpha-stable distribution. Meanwhile, Moradian et al. (2017) find that copula-graphic estimator improves the estimation of the survival function with dependent censoring better than Kaplan-Meier estimator. Kangas et al. (2016) give a warning that an overfitted kernel model could result in a serious underestimation of true variance of a difference estimator in a *C*-vine copula.

Recently, forestry research focuses on other subfields, such as Arya and Zhang (2017) who study water quality in different watersheds. They find copula-based Markov process which is an efficient method in assessing water quality and risks. Musafer and Thompson (2017) used spatial vine copula for nonlinear spatial dependence. Some authors examine bidding behavior in forestry, as Tatoutchoup (2017) proposes optimal contract for correlated harvesting cost and harvesting age. Others like Ali et al. (2018) and Sarmiento et al. (2018) observe that hybridized model is more efficient than standalone models in forecasting rainfall and suggest future work to use their multistage probabilistic learning model for water resources management in arid regions. Sarmiento et al. (2018) note flexibility of copula model in modeling nonlinear time series. Note that most of the literature reviews above use simple bivariate copulas rather than complicated multivariate due to lack of expertise.

3.2 *Copulas in Environmental Sciences*

3.2.1 Copula *Applications in Hydrometeorology*

Copula theory is applied in hydrometeorology by Yee et al. (2016) who used copula models to analyze extreme value of rainfall data in Malaysia. They find that Gumbel copula is the best fitted copula. Moreover, Klein et al. (2016) cascade bivariate copulas by pair-copula construction. They apply a mixture of probability distributions to estimate the marginal densities and distributions of daily flows of meteorological and hydrological situations. They also proposed multi-model ensemble involving two hydrological and one statistical flow models at two gauge stations in the Moselle River Basin. Their empirical evidences suggest copulas are best suited for hydrological multi-model predictions.

Durocher et al. (2016) apply spatial copulas to predict extreme flood quantiles at ungauged locations to overcome a bias compared to employing traditional interpolation methods. This study finds that the spatial copula framework is able to deal with the problem of bias. It is also robust to the presence of problematic stations and may improve the quality of quantile predictions while reducing the level of complexity of the models used in. Furthermore, Sarhadi et al. (2016) develop a Bayesian, dynamic conditional copula to model the time-varying dependence structure between mixed continuous and discrete multidimensional hydrometeorological phenomena. The empirical evidence reveals that the nature and the risk of extreme-climate multidimensional processes are changed over time under the impact of climate change.

Abdi et al. (2016) developed an optimization-based method (OBM) to examine probability distribution of drought characteristics in Clayton, Frank, and Gumbel copula models. They find that the OBM has better performance than the conventional methods such as MOM and IFM. In addition, among the three considered copulas, the Gumbel copula is found to be the most appropriate copula for modeling the drought characteristics for the selected study area. Ozga-Zielinski et al. (2016) apply copula-based 2D probability distributions to analyze snowmelt flood frequency in Poland. Results showed that the 2D model for snowmelt floods built using the Gumbel–Hougaard copula is much better than the model built using the Gaussian copula. Specifically, the Archimedean copula in the form of Gumbel–Hougaard, coupled with the possibility to choose marginal distributions, could address several important issues related to the probabilistic description of snowmelt floods. An other study done by Salvadori et al. (2016) provides valuable tools for assessing the probability of threatening of natural occurrences. This paper finds that the outlined hazard scenarios well cope with the concept of failure probability and a structural approach. Moreover, Dai et al. (2016) apply copula-AR model to introduce season into the radar rainfall uncertainty model and suggest analyzing the relationship between specific synoptic regime and radar rainfall uncertainty. Requena et al. (2016) combine a distributed hydrometeorological model and a copula model to examine the risk in hydraulic structures such as dams based on flood series. They find that this method decreases the computation time and can be used for improving flood risk assessment studies.

Ahn and Palmer (2016) investigate stationary and nonstationary bivariate characteristics of annual low flow in the Connecticut River Basin, USA, in nonstationary copulas, whereas Chang et al. (2016) investigate drought risk in terms of joint probability and return period by constructing a multivariate integrated drought index and find that it could help to characterize future drought tendencies and build an early warning system for drought mitigation. Daneshkhah et al. (2016) highlight the usefulness of the D-vine copula model and minimum information D-vine copula to model the joint distribution of flood event properties. Atique and Attoh-Okine (2016) analyze dependency between several variables of pipe condition using vine copula models; in particular, it predicts pipe leakage due to climate condition. Razmkhah et al. (2016) study the effect of spatial correlation of rainfall on Hydrologic Engineering Center continuous rainfall-runoff simulation uncertainty, using bivariate copula. The paper suggests that the method could be used to predict future conditions based on climate change, land-use change, and other purposes of modeling and apply for three variate copula correlated rainfall uncertainty propagation. Zhang et al. (2016) apply Student-t copula function to construct multivariate joint probability of water supply and demand. The paper finds that the trivariate joint probability distribution is more reasonable than the bivariate one to reflect the water shortage risk, and it can provide water shortage risk evaluation technique in the irrigation district.

Bernardino and Palacios-Rodriguez (2017) proposed an explicit expression of the aforementioned multivariate risk measure in environmental sciences using Archimedean copula setting. The authors apply extreme value theory techniques to estimate this measure and study the asymptotic normality of the proposed estimator. The model is applied on real hydrological dataset of flood peak, volume, and initial water level of the Ceppo Morelli Dam in Italy, and the findings show that the model is useful in obtaining associated flood hydrograph and calculating maximum level of dam. The paper also suggests future study to consider the interaction between the hydrological load and the structure in evaluating the safety of dam.

Peng et al. (2017) propose a copula Monte Carlo (CMC) method to improve flood risk for confluence flooding control downstream of Xiluodu-Xiangjiaba reservoirs in China. Copula function is used to model the dependence between the mainstream and tributary, and the MC method is used to estimate the flood risk and the simulated tributary flood from the copula function. The paper finds that CMC method is more robust than the MC method because it can consider both the flood spatial correlations and the inside flood domain stochastic characteristics, as well as flood risk planning and management.

Wang et al. (2017) examine drought recurrence interval and its relationship with agricultural drought disaster using run theory and copula function on drought duration and severity data in the northern Shaanxi, China, from 1960 to 2015. The study finds that Frank copula function fits data well and is reliable in analyzing drought characteristics and can be used for agricultural drought disaster assessment. Bracken et al. (2018) apply a Gaussian elliptical copula to model the joint distribution of multiple hydrologic variables (stream flow, snow level, and reservoir elevation) at

Taylor Park Dam in Colorado, USA. Results show that copula model better captures multivariate dependence compared to an independent model for the incorporation of climate information.

Gao et al. (2018) develop uncertainty-based water shortage risk assessment (UWSRAM) model to study the combined effect of multiple water resources and the shortage degree under uncertainty. The UWSRAM combines CMC stochastic simulation and the chance-constrained programming-stochastic multi-objective optimization model, using data of Lunan water-receiving area in China. The paper finds that UWSRAM is valuable for mastering the overall multi-water resource and water shortage degree, adapting to the uncertainty surrounding water resources, establishing effective water resource planning policies for managers, and achieving sustainable development.

Liu et al. (2018a) examine rainfall characteristics in Guangzhou, China, from 1961 to 2012 in a 3D copula-based multivariate frequency analysis (design rainfall depth, total rainfall depth, and peak rainfall depth). The empirical results show that this method can reflect urban rainstorm characteristics well and can serve a scientific reference for urban flood control and drainage planning. Liu et al. (2018c) develop a hydrological uncertainty processor (HUP) based on a copula function for hydrological forecasting in water resources management and decision-making processes in Three Gorges Reservoir, Yangtze River Basin, China. The paper finds that HUP is superior to deterministic forecasts in terms of continuous rank probability score, implying the effectiveness of copula-based HUP over meta-Gaussian HUP in probabilistic forecasts.

Liu et al. (2018b) study compound floods (precipitation and surface runoff) in Texas, USA, in 3D and 4D vine copulas using El Niño–Southern Oscillation and rising temperatures as underlying conditions that amply the compounding effects. Empirical results show that the models well represent the interrelationship between observed variables and display consistently pattern with observations. The paper also finds that the conditional framework of vine copula is capable in yielding predictive information of compound events.

Manning et al. (2018) analyze soil moisture drought on multiple time scales related to both meteorological drought and heat waves in wet, transitional, and dry climates in Europe during summer. The study applies a pair-copula model to data from FluxNet sites in Europe and finds at all sites that precipitation exerts the main control over soil moisture drought. Mortuza et al. (2019) apply copula model to evaluate bivariate drought characteristics (drought duration and severity) and forecast future drought trends in three homogenous drought regions in Bangladesh (west, middle, and east and south). The paper finds that the bivariate drought frequency analysis is more precise than the standard univariate frequency analysis. It also suggests future studies to develop severity-areal extent frequency curves of droughts for a better understanding of drought spatial coverage and extent, as well as discern data uncertainties in multivariate analysis.

Pappadà et al. (2018) investigate specific spatial sub-regions (clusters) flood risk behavior in Po River Basin in Italia using a copula-based agglomerative hierarchical clustering algorithm. The paper comprises both univariate and bivariate approaches on flood peak and flood volume variables. It finds that clusters detected by the model adequately capture the distinction between different meteorological forcing and hydrological flow contributions. The paper states that the proposed algorithm could provide promising and valuable investigation tool for hydrological risks management.

Qian et al. (2018)propose a new method of parameter estimation (maximum entropy estimation) for both Gumbel and Gumbel–Hougaard copula in situations when insufficient data are available in a bivariate hydrological extreme frequency analysis. Yin et al. (2018) examine a hybrid model for hydrological prediction by combining copula entropy (CE) with wavelet neural network (WNN), in which CE theory permits to calculate mutual information to select input variables, whereas wavelet analysis can provide a good fit with the hydrological data. The results showed that the hybrid model produced better results in estimating the hydrograph properties than CE or WNN model.

You et al. (2018) examine the relationship between the buried depth of the phreatic water and driving factors for the reasonable planning of surface water resources and the water table for Jinghui Irrigation District in China from 1977 to 2013 using kernel distribution estimation and 2D and 3D Frank copula function. The paper finds that copula applications dominate other methods such as linear regression and ARIMA in modeling complicated functions of the water table and its driving factors as well as calculating relevant probabilities. Bezak et al. (2016) use Frank copula function to construct an intensity duration frequency (IDF) relationship for several rainfall stations using high-resolution rainfall data with an average subsample length of 34 years. The paper finds that a combination of several rainfall thresholds with an appropriate high-density rainfall measurement network can be used as part of the early warning system of the initiation of landslides and debris flows.

Das et al. (2018) examine the relationship between precipitation and meteorological drought in copula frameworks for regional water resource management in Beijing, China. The study shows that the frameworks not only reveal the occurrence of meteorological drought in Beijing but also provide a quantitative way to forecast future drought probability under different precipitation conditions. Zhang et al. (2018b) examine the concurrence of high/low flows and the ecological in stream flow of the nine reservoirs in the Liao River Basin, China, using Archimedean copula family (Gumbel–Hougaard copula, Clayton copula, and Frank copula) and nonparametric copula. The empirical results reveal that the general extreme value distribution model performs well in describing the probabilistic behavior of high/low flows in the basin. Particularly, the Gumbel and Frank copula functions perform better than other functions.

Arns et al. (2017) apply copula to examine the sensitivity of shallow coastal areas to changing nonlinear interactions between tides, surges, waves, and relative sea-level rise for coastal design heights. Fan et al. (2017) calculate standardized precipitation evapotranspiration index based on monthly total precipitation and mean temperature data to analyze the relationship between precipitation and meteorological drought in China in a copula framework. The paper finds that Clayton copula best fit the data.

3.2.2 Applications in Other Environmental Sciences

There is a variety of studies in other fields of environmental sciences that applies copulas. For example, Chen et al. (2016) propose a waste management planning (CCWMP) method using an optimal copula among Gaussian, Student's *t*, Clayton, Frank, Gumbel, and Ali-Mikhail-Haq copulas. However, the paper finds that this method has difficulties in handling uncertainties in the objective coefficients and suggests the future study to improve it.

Wang et al. (2016) apply pair-copula function to capture the multiple stochastic correlations among wind speed, solar insolation, and load power, in which the sequential MC simulation is used to assess a microgrid reliability and economic evaluation. Meanwhile, Cao and Yan (2017) explore the impacts of high-dimensional dependences of wind speed among wind farms on probabilistic optimal power flow (POPF), in which probability distribution of wind speed is estimated using kernel density estimate method, and the joint probability distribution function of wind speed among wind farms is obtained by pair-copula method.

Yazdi (2017) develops a stochastic model to assign the optimal sites and number of check dams on a stream network to reduce floods in downstream reaches of rivers on the Kan Basin in Tehran. The study employs copula method and an artificial neural network to handle uncertainty of rainfall variables. The empirical results reveal that optimal strategies are more efficient than traditional approaches in reducing the average of peak flood discharges (50% in comparison with 21%) with significantly lower costs or number of check dams.

Dong et al. (2017a, b) state that copulas are applying extensively in ocean engineering design in the literature recently because of their precision in describing statistical relations of different probability margins. Therefore, to reduce the costs of wind power at Point 2 in Lianyungang Harbour of China, the authors examine the efficiency of trivariate maximum entropy distributions based on trivariate Gumbel–Hougaard copula, Frank copula, Clayton copula, and normal copula using data of annual maximum significant wave height and corresponding wind speed and current velocity. The paper finds that normal copula fit the data well and conditional probability can present a joint design of significant wave height and corresponding wind speed and current velocity.

Das et al. (2018) apply a probabilistic approach and copula theory to analyze the stability of vegetated slopes, under the combined effect of univariate vegetation induced suction and bivariate mechanical parameters (cohesion and frictional angle). The paper finds that due to a higher amount of evapotranspiration in treed soil, treed slopes are more stable than grassed and bare slopes. In addition, it suggests that an assumption of independence between cohesion and frictional angle might lead to less realistic evaluation of vegetated slopes. Torres et al. (2017) propose a directional multivariate extreme identification procedure and analyze environmental phenomena in copula models to better describe an environmental catastrophe. The paper applies the proposed model to analyze flood incoming to the Ceppo Morelli Dam in Italy and finds that it can reduce the ratio of false positives. The paper also studies sea storms considering five variables (wave height, storm duration, storm magnitude, storm direction, and inter-arrival time) and shows relevant differences with previous study in the literature in terms of computational feasibility in 5D setting.

Um et al. (2017) apply copula models to investigate the optimal marginal distribution for the relations between the wind speed and the precipitation of typhoons at the Jeju weather station in South Korea. The marginal distributions for the copula models included the generalized extreme value, generalized logistic, generalized Pareto, and Weibull distributions and three copula models: Clayton, Frank, and Gumbel copulas. The Frank copula model was found having the best performance. This study presents the key steps needed to identify an optimum copula for a bivariate distribution in atmospheric sciences applications.

Lazoglou and Anagnostopoulou (2019) apply copula method to analyze the temperature and precipitation dependence among stations in the Mediterranean. The paper uses Kendall's tau correlation index to reveal temperature dependency among stations before calculating their marginal distributions. Then it applies several copula families, both Archimedean and elliptical, to model the dependence of the main climate parameters (temperature and precipitation). The empirical results show that Frank copula was identified as the best family to describe the joint distribution of temperature. For precipitation, the best copula families are BB1 and survival Gumbel.

Zhang et al. (2018a) apply asymmetric copulas for the modeling of multivariate ocean data; particularly, they focus on capturing asymmetric dependencies among the environmental parameters, both nonlinear and asymmetrically dependent variates. The paper finds that asymmetric copula models dominate over the traditional symmetric copula models, and they are found to be more realistic in describing ocean data. Jiang et al. (2017) propose MC simulation and copula theory to analyze uncertainty of a simulation model when parameters are correlated. Using the Akaike information criterion and the Bayesian information criterion for model selection, the paper finds that t copula is the optimal function for matching the relevant structure of the parameters (in a comparison with Gaussian copula function). Pappadà et al. (2017) implement a simulation study to evaluate the effects of the randomization of multivariate observations on the estimation of the structural risk using coastal engineering data.

Huiqun and Yong (2018) apply Gumbel–Hougaard copula method to discern the inherent relationship between chlorophyll-a and environmental variables of Chaohu Lake, China. The paper finds that the applied method presents an effective tool to analyze the interaction of eutrophic variables in complex water environment system, as well as provide reference for integrated management and treatment of lakes and reservoirs. Nguyen-Huy et al. (2018) used vine copula to model climate-yield dependence structures to investigate their spatiotemporal influence on the variability of seasonal wheat yield in five major wheat-producing states across Australia using data for the period 1983–2013. Particularly, the paper develops D-vine quantile regression model to forecast wheat yield at given different confidence levels. The empirical results reveal a comprehensive analysis of the spatiotemporal impacts of different climate mode indices on Australian wheat crops.

Han and De Oliveira (2016) proposed a class of random field models for geostatistical count data on Gaussian copulas, which allows for direct modeling of marginal distributions and association structure of the count data. The paper uses dataset of Japanese beetle larvae count for illustration and finds that the proposed models are more flexible than hierarchical Poisson model in terms of feasible correlation, sensitivity, and modeling of isotropy. Kanyingi et al. (2017) apply a robust Pair Copula-Point Estimation Method to examine the impact of wind power complex dependencies and uncertainties on small-signal stability of power system. The empirical results reveal that the approach provides a flexible probability model in power system analysis. Kelmendi et al. (2016) model joint exceedance probability of rain attenuation based on Gaussian copula to examine earth-space dual-site diversity system. The paper finds that Gaussian copula better results than Archimedean copulas.

Yanovsky et al. (2016) propose a three-dimensional copula to examine the dependence between different polarimetric parameters, and vine copulas are used to decompose the multivariate densities to bivariate linking copulas. The paper uses data received by the PARSAX radar, including sounding signal, precipitation, wet snow, and rain thunderstorm for empirical investigation, and finds that the set of the new linked bivariate copulas fully represents the marginal distribution dependence structure. Vezzoli et al. (2017) investigate behavior and performances of linked climate-hydrology model using both parametric (homogeneity and copula-equality tests) and nonparametric (Kendall and Spearman tests) approaches. The empirical results show that the proposed approaches are appropriate in evaluating.

Several studies have examined model selection for different fields. For example, Kim (2016) tries to capture the sparse output correlation among the output variables in sparse conditional copula models. The paper finds that the model has capability of representing complex input/output relationship without overfitting and demonstrates the superiority of the sparse copula model in prediction performance via several synthetic and real-world multiple-output regression problems. However, Prenen et al. (2016) develop a new formulation for the likelihood of Archimedean copula models for survival data to allow for clusters of large and variable size. Dou et al. (2016) proposed expectation maximization (EM) algorithms that utilize a representation of the Bernstein copula model. The paper uses three real datasets and a three-dimensional simulated data set to illustrate and finds that the Bernstein copula is able to represent various distributions flexibly and the EM algorithms work well for such data. It also finds that the Bernstein copula can also be incorporated into a vine copula to model data with more complicated correlation structures.

Meanwhile, Bilgrau et al. (2016) present and discuss an improved implementation in R package along with various alternative optimization routines to the EM algorithm. Kovács and Szántai (2016) express the regular-vine copulas using a special type of hypergraphs, which encodes the conditional independences. Musgrove et al. (2016) propose hierarchical copula regression models for areal data and develop an efficient computational approach to frequentist inference. Their approach allows for unbiased estimation of marginal parameters and intuitive conditional and marginal interpretations. Upcoming work might extend the framework to other copula families, e.g., Student's t copula or Bardossy's v-transform of the Gaussian copula. Perrone et al. (2016) used discrimination design techniques to solve the issue of copula selection. The paper suggests future work to generalize other discrimination criteria such as T-optimality and KL-optimality to flexible copula models and extend to multistage design procedures. Zhang and Wilson (2016) examine the influence of dependence structures on system reliability and component importance in coherent multistate systems using Gaussian copula. The paper suggests potential studies to extend their work in other copula functions such as vine copula if there is some nature ordering of the components.

Xu et al. (2016) develop a joint probability function of peak ground acceleration (PGA) and cumulative absolute velocity (CAV) for the strong ground motion data from Taiwan and apply a copula to model the joint probability distribution of PGA and CAV. The paper finds that the Gaussian copula provides adequate characterization of the PGA–CAV joint distribution observed in Taiwan, which verifies the validity of the copula application in predicting earthquake.

Sun et al. (2016) develop a fuzzy copula model to capture probabilistic uncertainty in wind speed correlation. The advantage of this model is that the copula parameters can be interval numbers, triangular, or trapezoidal fuzzy numbers based on the wind speed data and subjective judgment of decision makers. The results reveal that the fuzzy model is capable of describing the probabilistic uncertainty and evaluating its effect on wind curtailment. This study suggests that the fuzzy copula model can be applied in other fields such as reliability evaluation, reserve capacity determination, or economic dispatch. Cisty et al. (2016) apply a bivariate copula method to perform a joint analysis of the severity and duration of the most demanding potential annual irrigation periods in southwestern Slovakia. The empirical results verify the benefit of the proposed method for analysis of irrigation needs in a comparison with the typical one-dimensional analysis of individual climatic variables.

Li et al. (2016) propose a channel-pond joint water supply mode (CPJM) based on copula approaches (Plackett copula and No. 16 copula) to analyze a risk assessment of CPJM and determination of the water supply strategy given the pond water supply frequency in the Zhanghe Irrigation District, China. The results show that CPJM model is more realistic and practicable, and irrigation water allocation strategy based on CPJM can be determined for different hydrological years and predicted frequencies. Kanyingi et al. (2017) apply a robust Pair Copula-Point Estimation method to analyze the impact of wind power complex dependencies and uncertainties in both wind power and loads toward a power system's small-signal stability. The paper finds that the proposed approach provides a flexible probability model for effectively accounting for these factors in power system analysis.

4 Conclusions

In the area of environment and forestry study, there are various applications of copula. For example, in measuring relationship between tree height, diameter, and volume, researchers use copula. Moreover, in environmental science, copula models are applied widely in hydrometeorology, irrigation, wind speed, prediction of rainfall, and earthquake among others. Most of the studies have been found applying bivariate copula, and some apply three- or four-dimensional copulas. In general, the studies found that copula models perform well in characterizing joint dependent among variables, especially when there are extreme values. This chapter reveals that most of the studies apply basic copula models such as Gumbel copula and vine copula. There is lack of a mixture between copula and other models such as GARCH, DCC, or BEKK model to reveal a simultaneous dependency among a system of equations.

References

Abdi, A., Hassanzadeh, Y., Talatahari, S., Fakheri-Fard, A., & Mirabbasi, R. (2016). Parameter estimation of copula functions using an optimization-based method. *Theoretical and Applied Climatology, 129*(1–2), 1–12.

Ahn, K. H., & Palmer, R. N. (2016). Use of a nonstationary copula to predict future bivariate low flow frequency in the Connecticut river basin. *Hydrological Processes, 30*(19), 3518–3532.

Ahsanullah, M., & Bhatti, M. I. (2010). On the dependence functions of Copulas of Gumbel's bivariate extreme value and exponential distributions. *Journal of Statistical Theory and Applications, 9,* 615–629.

Al Rahahleh, N., & Bhatti, M. I. (2017). Co-movement measure of information transmission on international equity markets. *Physica A: Statistical Mechanics and its Applications, 470,* 119–131.

Al Rahahleh, N., Bhatti, M. I., & Adeinat, I. (2017). Tail dependence and information flow: Evidence from international equity markets. *Physica A: Statistical Mechanics and its Applications, 474*(12), 319–329.

Ali, M., Deo, R. C., Downs, N. J., & Maraseni, T. (2018). Multi-stage hybridized online sequential extreme learning machine integrated with Markov Chain Monte Carlo copula-Bat algorithm for rainfall forecasting. *Atmospheric Research, 213*(November), 450–464.

Arns, A., Dangendorf, S., Jensen, J., Talke, S., Bender, J., & Pattiaratchi, C. (2017). Sea-level rise induced amplification of coastal protection design heights. *Scientific Reports, 7*(January), 40171.

Arya, F. K., & Zhang, L. (2017). Copula-based Markov process for forecasting and analyzing risk of water quality time series. *Journal of Hydrologic Engineering, 22*(6), 1–12.

Atique, F., & Attoh-Okine, N. (2016). Using copula method for pipe data analysis. *Construction and Building Materials, 106*(March), 140–148.

Bernardino, E. D., & Palacios-Rodríguez, F. (2017). Estimation of extreme component-wise excess design realization: A hydrological application. *Stochastic Environmental Research and Risk Assessment, 31*(10), 2675–2689.

Bezak, N., Šraj, M., & Mikoš, M. (2016). Copula-based IDF curves and empirical rainfall thresholds for flash floods and rainfall-induced landslides. *Journal of Hydrology, 514*(Part A-October), 272–284.

Bhatti, M. I., & Do, H. Q. (2019). Recent development in copula and its applications to the energy, forestry and environmental sciences. *International Journal of Hydrogen Energy, 44,* 19453–19473.

Bhatti, M. I., & Nguyen, C. C. (2012). Diversification evidence from international equity markets using extreme values and stochastic copulas. *Journal of International Financial Markets Institutions and Money, 22*(3), 622–646.

Bilgrau, A. E., Eriksen, P. S., Rasmussen, J. G., Johnsen, H. E., Dybkær, K., & Bøgsted, M. (2016). GMCM: Unsupervised clustering and meta-analysis using gaussian mixture copula models. *Journal of Statistical Software, 70*(2), 1–23.

Bracken, C., Holman, K. D., Rajagopalan, B., & Moradkhani, H. (2018). A Bayesian hierarchical approach to multivariate nonstationary hydrologic frequency analysis. *Water Resources Research, 54*(1), 243–255.

Cao, J., & Yan, Z. (2017). Probabilistic optimal power flow considering dependences of wind speed among wind farms by pair-copula method. *International Journal of Electrical Power & Energy Systems, 84*(1), 296–307.

Chang, J., Li, Y., Wang, Y., & Yuan, M. (2016). Copula-based drought risk assessment combined with an integrated index in the Wei River Basin, China. *Journal of Hydrology, 540*(9), 824–834.

Chen, F., Huang, G., Fan, Y., & Wang, S. (2016). A copula-based chance-constrained waste management planning method: An application to the city of Regina, Saskatchewan, Canada. *Journal of the Air and Waste Management Association, 66*(3), 307–328.

Cherubini, U., Luciano, E., & Vecchiato, W. (2004). *Copula methods in finance*. Chichester: Wiley.

Cisty, M., Becova, A., & Celar, L. (2016). Analysis of irrigation needs using an approach based on a bivariate copula methodology. *Water Resources Management, 30*(1), 167–182.

Dai, Q., Han, D., Zhuo, L., Zhang, J., Islam, T., & Srivastava, P. K. (2016). Seasonal ensemble generator for radar rainfall using copula and autoregressive model. *Stochastic Environmental Research and Risk Assessment, 30*(1), 27–38.

Daneshkhah, A., Remesan, R., Chatrabgoun, O., & Holman, I. P. (2016). Probabilistic modeling of flood characterizations with parametric and minimum information pair-copula model. *Journal of Hydrology, 540*(9), 469–487.

Das, G. K., Hazra, B., Garg, A., & Ng, C. W. W. (2018). Stochastic hydro-mechanical stability of vegetated slopes: An integrated copula based framework. *CATENA, 160*(January), 124–133.

Delisle-Boulianne, S., Fortin, M., Achim, A., & Pothier, D. (2014). Modelling stem selection in northern hardwood stands: Assessing the effects of tree vigour and spatial correlations using a copula approach. *Forestry: An International Journal of Forest Research, 87*(5), 607–617.

Dong, S., Chen, C., & Tao, S. (2017a). Joint probability design of marine environmental elements for wind turbines. *International Journal of Hydrogen Energy, 42*(29), 18595–18601.

Dong, H., Xu, X., Sui, H., Xu, F., & Liu, J. (2017b). Copula-Based joint statistical model for polarimetric features and its application in PolSAR image classification. *IEEE Transactions on Geoscience and Remote Sensing, 55*(10), 5777–5789.

Dou, X., Kuriki, S., Lin, G. D., & Richards, D. (2016). EM algorithms for estimating the Bernstein copula. *Computational Statistics & Data Analysis, 93*(1), 228–245.

Durocher, M., Chebana, F., & Ouarda, T. B. (2016). On the prediction of extreme flood quantiles at ungauged locations with spatial copula. *Journal of Hydrology, 533*(2), 523–532.

Ene, L. T., Næsset, E., & Gobakken, T. (2013). Model-based inference for *k*-nearest neighbours predictions using a canonical vine copula. *Scandinavian Journal of Forest Research, 28*(3), 266–281.

Fan, L., Wang, H., Wang, C., Lai, W., & Zhao, Y. (2017). Exploration of use of copulas in analysing the relationship between precipitation and meteorological drought in Beijing, China. *Advances in Meteorology, 2017,* 1–11.

Fortin, M., Delisle-Boulianne, S., & Pothier, D. (2013). Considering spatial correlations between binary response variables in forestry: An example applied to tree harvest modeling. *Forest Science, 59*(3), 253–260.

Gao, X., Liu, Y., & Sun, B. (2018). Water shortage risk assessment considering large-scale regional transfers: a copula-based uncertainty case study in Lunan, China. *Environmental Science and Pollution Research, 25*(23), 23328–23341.

Gumbel, E. J. (1960). Distributions des valeurs extrêmes en plusieurs dimensions. *Publications de l'Institut de statistique de l'Université de Paris, 9,* 171–173.

Han, Z., & De Oliveira, V. (2016). On the correlation structure of Gaussian copula models for geostatistical count data. *Australian and New Zealand Journal of Statistics, 58*(1), 47–69.

Huiqun, M., & Yong, W. (2018). Correlation between chlorophyll-a and related environmental factors based on copula in Chaohu Lake, China. *IOP Conference Series: Earth and Environmental Science, 108*(4), 042076.

Jiang, X., Na, J., Lu, W., & Zhang, Y. (2017). Coupled Monte Carlo simulation and copula theory for uncertainty analysis of multiphase flow simulation models. *Environmental Science and Pollution Research, 24*(31), 24284–24296.

Kangas, A., Myllymäki, M., Gobakken, T., & Næsset, E. (2016). Model-assisted forest inventory with parametric, semiparametric, and nonparametric models. *Canadian Journal of Forest Research, 46*(6), 855–868.

Kanyingi, P., Wang, K., Li, G., & Wu, W. (2017). A robust pair copula-point estimation method for probabilistic small signal stability analysis with large scale integration of wind power. *Journal of Clean Energy Technologies, 5*(2), 85–94.

Kelmendi, A., Kourogiorgas, C. I., Hrovat, A., Panagopoulos, A. D., Kandus, G., & Vilhar, A. (2016). Modeling of joint rain attenuation in earth-space diversity systems using Gaussian copula. In *Proceedings of the 2016 10th European Conference on Antennas and Propagation (EuCAP)* (pp. 1–5). Davos, https://doi.org/10.1109/eucap.2016.7481617.

Kershaw, J. A., Richards, E. W., McCarter, J. B., & Oborn, S. (2010). Spatially correlated forest stand structures: A simulation approach using copulas. *Computers and Electronics in Agriculture, 74*(1), 120–128.

Kershaw, J. A., Weiskittel, A. R., Lavigne, M. B., & McGarrigle, E. (2017). An imputation/copula-based stochastic individual tree growth model for mixed species Acadian forests: A case study using the Nova Scotia permanent sample plot network. *Forest Ecosystems, 4*(1), 15.

Kim, M. (2016). Sparse conditional copula models for structured output regression. *Pattern Recognition, 60*(12), 761–769.

Klein, B., Meissner, D., Kobialka, H.-U., & Reggiani, P. (2016). Predictive uncertainty estimation of hydrological multi-model ensembles using pair-copula construction. *Water, 8*(4), 125.

Kovács, E., Szántai, T. (2016). Hypergraphs in the characterization of regular vine copula structures. In *Proceeding of the 13th Conference on Mathematics and its Application* (pp. 335–344). University "Politehnica" of Timisoara, arXiv:1604.02652.

Lazoglou, G., & Anagnostopoulou, C. (2019). Joint distribution of temperature and precipitation in the Mediterranean, using the Copula method. *Theoretical and Applied Climatology, 135*(3–4), 1399–1411.

Li, H., Shao, D., Xu, B., Chen, S., Gu, W., & Tan, X. (2016). Failure analysis of a new irrigation water allocation mode based on copula approaches in the Zhanghe Irrigation District, China. *Water, 8*(6), 251.

Liu, Z., Cheng, L., Hao, Z., Li, J., Thorstensen, A., & Gao, H. (2018a). A Framework for exploring joint effects of conditional factors on compound floods. *Water Resources Research, 54*(4), 2681–2696.

Liu, Z., Guo, S., Xiong, L., & Xu, C.-Y. (2018b). Hydrological uncertainty processor based on a copula function. *Hydrological Sciences Journal, 63*(1), 74–86.

Liu, C., Zhou, Y., Sui, J., & Wu, C. (2018c). Multivariate frequency analysis of urban rainfall characteristics using three-dimensional copulas. *Water Science and Technology, 2017*(1), 206–218.

MacPhee, C., Kershaw, J. A., Weiskittel, A. R., Golding, J., & Lavigne, M. B. (2018). Comparison of approaches for estimating individual tree height-diameter relationships in the Acadian forest region. *Forestry: An International Journal of Forest Research, 91*(1), 132–146.

Manning, C., Widmann, M., Bevacqua, E., Loon, A. F. V., Maraun, D., & Vrac, M. (2018). Soil moisture drought in Europe: A compound event of precipitation and potential evapotranspiration on multiple time scales. *Journal of Hydrometeorology, 19*(8), 1255–1271.

Moradian, H., Larocque, D., & Bellavance, F. (2017). Survival forests for data with dependent censoring. *Statistical Methods in Medical Research, 28*(2), 445–461.

Mortuza, M. R., Moges, E., Demissie, Y., & Li, H.-Y. (2019). Historical and future drought in Bangladesh using copula-based bivariate regional frequency analysis. *Theoretical and Applied Climatology, 135*(3–4), 855–871.

Musafer, G. N., & Thompson, M. H. (2017). Non-linear optimal multivariate spatial design using spatial vine copulas. *Stochastic Environmental Research and Risk Assessment, 31*(2), 551–570.

Musgrove, D., Hughes, J., & Eberly, L. (2016). Hierarchical copula regression models for areal data. *Spatial Statistics, 17*(10), 38–49.
Nelsen, R. B. (2007). *An introduction to copulas*. New York: Springer Science and Business Media.
Nguyen, C. C., & Bhatti, M. I. (2012). Copula model dependency between oil prices and stock markets: Evidence from China and Vietnam. *Journal of International Financial Markets Institutions and Money, 22*(4), 758–773.
Nguyen, C. C., Bhatti, M. I., Komornikova, M., & Komornik, J. (2016). Gold price and stock markets nexus under mixed-copulas. *Economic Modelling, 58,* 283–292.
Nguyen-Huy, T., Deo, R. C., Mushtaq, S., An-Vo, D.-A., & Khan, S. (2018). Modeling the joint influence of multiple synoptic-scale, climate mode indices on Australian wheat yield using a vine copula-based approach. *European Journal of Agronomy, 98*(10), 65–81.
Ogana, F. N., Osho, J. S. A., & Gorgoso-Varela, J. J. (2018). An approach to modeling the joint distribution of tree diameter and height data. *Journal of Sustainable Forestry, 37*(5), 475–488.
Ozga-Zielinski, B., Ciupak, M., Adamowski, J., Khalil, B., & Malard, J. (2016). Snow-melt flood frequency analysis by means of copula based 2D probability distributions for the Narew River in Poland. *Journal of Hydrology: Regional Studies, 6*(6), 26–51.
Pappadà, R., Durante, F., & Salvadori, G. (2017). Quantification of the environmental structural risk with spoiling ties: Is randomization worthwhile? *Stochastic Environmental Research and Risk Assessment, 31*(10), 2483–2497.
Pappadà, R., Durante, F., Salvadori, G., & De Michele, C. (2018). Clustering of concurrent flood risks via Hazard Scenarios. *Spatial Statistics, 23*(3), 124–142.
Patton, A. J. (2006). Modelling asymmetric exchange rate dependence. *International Economic Review, 47*(2), 527–556.
Peng, Y., Chen, K., Yan, H., & Yu, X. (2017). Improving flood-risk analysis for confluence flooding control downstream using copula Monte Carlo Method. *Journal of Hydrologic Engineering, 22*(8), 04017018.
Perrone, E., Rappold, A., & Müller, W. G. (2016). Optimal discrimination design for copula models. arXiv:1601.07739.
Pothier, D., Fortin, M., Auty, D., Delisle-Boulianne, S., Gagné, L.-V., & Achim, A. (2013). Improving tree selection for partial cutting through joint probability modelling of tree vigor and quality. *Canadian Journal of Forest Research, 43*(3), 288–298.
Prenen, L., Braekers, R., & Duchateau, L. (2016). Extending the Archimedean copula methodology to model multivariate survival data grouped in clusters of variable size. *Journal of the Royal Statistical Society, B, 79*(2), 483–505.
Qian, L., Wang, H., Dang, S., Wang, C., Jiao, Z., & Zhao, Y. (2018). Modelling bivariate extreme precipitation distribution for data-scarce regions using Gumbel-Hougaard copula with maximum entropy estimation. *Hydrological Processes, 32*(2), 212–227.
Razmkhah, H., AkhoundAli, A. M., Radmanesh, F., & Saghafian, B. (2016). Evaluation of rainfall spatial correlation effect on rainfall-runoff modeling uncertainty, considering 2-copula. *Arabian Journal of Geosciences, 9*(4), 1–15.
Requena, A. I., Flores, I., Mediero, L., & Garrote, L. (2016). Extension of observed flood series by combining a distributed hydro-meteorological model and a copula-based model. *Stochastic Environmental Research and Risk Assessment, 30*(5), 1363–1378.
Saarela, S., Andersen, H.-E., Grafström, A., Schnell, S., Gobakken, T., Næsset, E., et al. (2017). A new prediction-based variance estimator for two-stage model-assisted surveys of forest resources. *Remote Sensing of Environment, 192*(4), 1–11.
Salvadori, G., Durante, F., De Michele, C., Bernardi, M., & Petrella, L. (2016). A multivariate copula based framework for dealing with hazard scenarios and failure probabilities. *Water Resources Research, 52*(5), 3701–3721.
Sarhadi, A., Burn, D. H., Concepción Ausín, M., & Wiper, M. P. (2016). Time varying nonstationary multivariate risk analysis using a dynamic Bayesian copula. *Water Resources Research, 52*(3), 2327–2349.

Sarmiento, C., Valencia, C., & Akhavan-Tabatabaei, R. (2018). Copula autoregressive methodology for the simulation of wind speed and direction time series. *Journal of Wind Engineering and Industrial Aerodynamics, 174*(3), 188–199.

Serinaldi, F., Grimaldi, S., Abdolhosseini, M., Corona, P., & Cimini, D. (2012). Testing copula regression against benchmark models for point and interval estimation of tree wood volume in beech stands. *European Journal of Forest Research, 131*(5), 1313–1326.

Sklar, A. (1959). Distribution functions of n dimensions and margins. *Publications of the Institute of Statistics of the University of Paris, 8,* 229–231.

Sun, C., Bie, Z., Xie, M., & Jiang, J. (2016). Fuzzy copula model for wind speed correlation and its application in wind curtailment evaluation. *Renewable Energy, 93*(10), 68–76.

Tatoutchoup, F. D. (2017). Forestry auctions with interdependent values: Evidence from timber auctions. *Forest Policy and Economics, 80*(7), 107–115.

Torres, R., De Michele, C., Laniado, H., & Lillo, R. E. (2017). Directional multivariate extremes in environmental phenomena. *Environmetrics, 28*(2), e2428.

Um, M. J., Joo, K., Nam, W., & Heo, J. H. (2017). A comparative study to determine the optimal copula model for the wind speed and precipitation of typhoons. *International Journal of Climatology, 37*(4), 2051–2062.

Vezzoli, R., Salvadori, G., & De Michele, C. (2017). A distributional multivariate approach for assessing performance of climate-hydrology models. *Scientific Reports, 7*(1), 12071.

Wang, M., Upadhyay, A., & Zhang, L. (2010). Trivariate distribution modeling of tree diameter, height, and volume. *Forest Science, 56*(3), 290–300.

Wang, X., Zhang, Y., Feng, X., Feng, Y., Xue, Y., & Pan, N. (2017). Analysis and application of drought characteristics based on run theory and copula function. *Transactions of the Chinese Society of Agricultural Engineering, 33*(10), 206–214.

Wang, S., Zhang, X., & Liu, L. (2016). Multiple stochastic correlations modeling for microgrid reliability and economic evaluation using pair-copula function. *International Journal of Electrical Power & Energy Systems, 76*(3), 44–52.

Xu, Y., Tang, X. S., Wang, J., & Kuo-Chen, H. (2016). Copula-based joint probability function for PGA and CAV: A case study from Taiwan. *Earthquake Engineering and Structural Dynamics, 45*(13), 2123–2136.

Yanovsky, F. J., Rudiakova, A. N., & Sinitsyn, R. B. (2016). Multivariate copula approach for polarimetric classification in weather radar applications. In *Proceedings of the 2016 17th International Radar Symposium (IRS)* (pp. 1–5). Krakow.

Yazdi, J. (2017). Check dam layout optimization on the stream network for flood mitigation: Surrogate modelling with uncertainty handling. *Hydrological Sciences Journal, 62*(10), 1669–1682.

Yee, K. C., Suhaila, J., Yusof, F., & Mean, F. H. (2016). Bivariate copula in Johor rainfall data. *AIP Conference Proceedings, 1750,* 1–6. (Malaysia).

Yin, W., JiGuang, Y., ShuGuang, L., & Li, W. (2018). Copula entropy coupled with wavelet neural network model for hydrological prediction. *IOP Conference Series: Earth and Environmental Science, 113*(1), 012160.

You, Q., Liu, Y., & Liu, Z. (2018). Probability analysis of the water table and driving factors using a multidimensional copula function. *Water, 10*(4), 472.

Zhang, Y., Kim, C.-W., Beer, M., Dai, H., & Soares, C. G. (2018a). Modeling multivariate ocean data using asymmetric copulas. *Coastal Engineering, 135*(5), 91–111.

Zhang, J., Lin, X., & Guo, B. (2016). Multivariate copula-based joint probability distribution of water supply and demand in irrigation district. *Water Resources Management, 30*(7), 2361–2375.

Zhang, X., & Wilson, A. (2016). System reliability and component importance under dependence: A copula approach. *Technometrics, 59*(2), 215–224.

Zhang, Z., Zhang, Q., Singh, V. P., & Peng, S. (2018b). Ecohydrological effects of water reservoirs with consideration of asynchronous and synchronous concurrences of high- and low-flow regimes. *Hydrological Sciences Journal, 63*(4), 615–629.

Forest Cover-Type Prediction Using Model Averaging

Anoop Chaturvedi and Ashutosh Kumar Dubey

Abstract The forest cover-type prediction provides resource managers strategic advantages in management and conservation of forest in natural disasters. This can be done using predictive modeling. This chapter provides the methodology to apply model averaging technique in multinomial logistic regression model. The methodology is applied to forest cover-type dataset. We have also transformed the wilderness areas and soil types variables into two variables. Such type of transformation provides improved predictive performance as well as easy application of model averaging technique to the data.

Keywords Model averaging · Multinomial logistic regression · US forest service

1 Introduction

Linear models which involve different sets of explanatory variables for any given dataset can be fitted; however, selection of a suitable model is central to all statistical work. Rapid advances in this aspect can be seen in the past few decades. The usual approach of selecting a model based on some model selection criterion leads to a single *best* model and then for making the inferences, the selected model is treated as the true model. Such a single-model selection approach ignores the uncertainty about the model itself and leads to the underestimation of uncertainty about the quantities of interest. However, the fact that an analysis might have turned out differently if preceded by data with small modifications might have led to a different modeling route, prompted the basis of *model averaging*. Model averaging involves averaging models based on all possible combinations of predictors with appropriately selected weights, while making inferences or forecasting. Synthesizing the concept of shrinkage estimation with model averaging leads to the concept of using weighted combinations of estimators with different tuning parameters with the objective of overall stability, improved standard errors and predictive performance (see Ullah and Wang 2013).

A. Chaturvedi (✉) · A. K. Dubey
Department of Statistics, University of Allahabad, Allahabad, Uttar Pradesh 211002, India
e-mail: anoopchaturv@gmail.com

G. Chandra et al. (eds.), *Statistical Methods and Applications in Forestry and Environmental Sciences*, Forum for Interdisciplinary Mathematics,
https://doi.org/10.1007/978-981-15-1476-0_14

Natural resource managers responsible for ecosystem management, for which strategies require baseline information on forestry parameters including inventory data for forested lands (diversion land, etc.) to support their decision-making processes. However, this type of data for in holdings or neighboring lands that are outside their immediate jurisdiction is generally unavailable. Predictive models are one of the methods to obtain it. Kumar et al. (2014) and Nahib and Suryanta (2017) used logistic regression model to predict the forest cover type for Bhanupratappur forest division of Kanker district in Chhattisgarh province of India and Indragiri Hulu Regency, Riau Province of Indonesia, respectively. For predicting forest cover type from cartographic variables (no remotely sensed data), Blackard and Dean (1999) applied artificial neural network and discriminant analysis and compared their performances. The study area included four wilderness areas found in the Roosevelt National Forest of northern Colorado, USA. A total of twelve cartographic measures were utilized as independent variables in the predictive models, while seven major forest cover types were used as dependent variables. Several subsets of these variables were examined to determine the best overall predictive model.

2 Dataset Description

For analysis purpose, the forest cover type for a given observation (30m $\times$ 30m cell) was taken from US Forest Service (USFS) Region 2 Resource Information System data. The independent variables were derived from data obtained from US Geological Survey and USFS data. Data is in raw form (not scaled) containing binary (0 or 1) columns of data for wilderness areas and soil types, which are qualitative independent variables. The unprocessed dataset can be downloaded from the University of California, Irvine (UCI) machine-learning repository "https://archive.ics.uci.edu/ml/machine-learning-databases/covtypeUH/." The data set has 55 columns over 13 types of measures. First ten are quantitative variables, next four are binary wilderness areas and 40 columns for binary soil type variables and last column is polychotomous variable for forest cover type having seven classes $\{1, 2, 3, 4, 5, 6, 7\}$. The detailed attribute information is given in Table 1. The order of this listing corresponds to the order of numerals along the rows of the database. Soil types based on the USFS ecological land type units (ELUs) are given in Table 2, and in supplement to this table, the details of first digit for climatic zone and second digit for geologic zone of USFS ELU code given in Table 3.

The study area, as taken by Blackard and Dean (1999), includes four wilderness areas, Rawah, Neota, Comanche Peak and Cache la Poudre, located in the Roosevelt National Forest of northern Colorado, USA. Neota has the highest mean elevation value among these areas, Rawah and Comanche Peak have a lower mean elevation value, while Cache la Poudre has the lowest mean elevation value. In terms of primary major tree species, Neota would have spruce/fir (type 1), Rawah and Comanche Peak would have lodge pole pine (type 2), followed by spruce/fir and aspen (type 5). Cache la Poudre would tend to have Ponderosa pine (type 3), Douglas fir (type 6) and

Table 1 Brief description of dataset

Attribute	Data type	Description
Elevation	Quantitative	In meters (m)
Aspect	Quantitative	In degrees azimuth
Slope	Quantitative	In degrees
Horizontal_distance_to_hydrology	Quantitative	Horz dist to nearest surface water features (in m)
Vertical_distance_to_hydrology	Quantitative	Vert dist to nearest surface water features (in m)
Horizontal_distance_to_roadways	Quantitative	Horz dist to nearest roadway (in m)
Hillshade_9am	Quantitative	Hill shade index at 9am, summer solstice (0–255 index)
Hillshade_noon	Quantitative	Hill shade index at noon, summer solstice (0–255 index)
Hillshade_3 pm	Quantitative	Hill shade index at 3 pm, summer solstice (0–255 index)
Horizontal_distance_to_fire_points	Quantitative	Horz dist to nearest wildfire ignition points (in m)
Wilderness_area (4 binary columns)	Qualitative	0 (absence), 1 (presence)
Soil_type (40 binary columns)	Qualitative	0 (absence), 1 (presence)
Cover type (7 types)	Integer	Forest cover-type designation (1–7)

cottonwood/willow (type 4), see Blackard and Dean (1999) for detailed description. All these areas represent forests with minimal interference from forest management system and ecological processes mainly result in existing forest cover types.

(**Note:** The original owner of database is Remote Sensing and GIS Program, Department of Forest Sciences, College of Natural Resources, Colorado State University, Fort Collins, CO80523. Reuse of this database is unlimited with retention of copyright notice for Jock A. Blackard and Colorado State University).

The third and fourth ELU digits are unique to the mapping unit and have no special meaning to the climatic or geologic zones.

3 Methodology

This section described the three methods, multinomial logistic regression (MLR), model averaging and ridge model averaging in MLR.

Table 2 Soil types based on the USFS ELUs

Soil type	USFS ELU code	Description
1	2702	Cathedral family—rock outcrop complex, extremely stony
2	2703	Vanet—ratake families complex, very stony
3	2704	Haploborolis—rock outcrop complex, rubbly
4	2705	Ratake family—rock outcrop complex, rubbly
5	2706	Vanet family—rock outcrop complex, rubbly
6	2717	Vanet—wet more families—rock outcrop complex, stony
7	3501	Gothic family
8	3502	Supervisor—limber families complex
9	4201	Troutville family, very stony
10	4703	Bullwark—catamount families—rock outcrop complex, rubbly
11	4704	Bullwark—catamount families—rock land complex, rubbly
12	4744	Legault family—rock land complex, stony
13	4758	Catamount family—rock land—bullwark family complex, rubbly
14	5101	Pachic argiborolis—aquolis complex
15	5151	Unspecified in the USFS soil and ELU survey
16	6101	Cryaquolis—cryoborolis complex
17	6102	Gate view family—cryaquolis complex
18	6731	Rogert family, very stony
19	7101	Typic cryaquolis—borohemists complex
20	7102	Typic cryaquepts—typic cryaquolls complex
21	7103	Typic cryaquolls—leighcan family, till substratum complex
22	7201	Leighcan family, till substratum, extremely boulder
23	7202	Leighcan family, till substratum—typic cryaquolls complex
24	7700	Leighcan family, extremely stony
25	7701	Leighcan family, warm, extremely stony
26	7702	Granile—catamount families complex, very stony
27	7709	Leighcan family, warm—rock outcrop complex, extremely stony
28	7710	Leighcan family—rock outcrop complex, extremely stony
29	7745	Como—legault families complex, extremely stony
30	7746	Como family—rock land—legault family complex, extremely stony
31	7755	Leighcan—catamount families complex, extremely stony
32	7756	Catamount family—rock outcrop—leighcan family complex, extremely stony

(continued)

Table 2 (continued)

Soil type	USFS ELU code	Description
33	7757	Leighcan—catamount families—rock outcrop complex, extremely stony
34	7790	Cryorthents—rock land complex, extremely stony
35	8703	Cryumbrepts—rock outcrop—cryaquepts complex
36	8707	Bross family—rock land—cryumbrepts complex, extremely stony
37	8708	Rock outcrop—cryumbrepts—cryorthents complex, extremely stony
38	8771	Leighcan—moran families—cryaquolls complex, extremely stony
39	8772	Moran family—cryorthents—leighcan family complex, extremely stony
40	8776	Moran family—cryorthents—rock land complex, extremely stony

Table 3 Details of first and second digits of USFS ELU code

First digit	Climatic zone	Second digit	Geologic zones
1.	Lower montane dry	1.	Alluvium
2.	Lower montane	2.	Glacial
3.	Montane dry	3.	Shale
4.	Montane	4.	Sandstone
5.	Montane dry and montane	5.	Mixed sedimentary
6.	Montane and subalpine	6.	Unspecified in the USFS ELU survey
7.	Subalpine	7.	Igneous and metamorphic
8.	Alpine	8.	Volcanic

3.1 Multinomial Logistic Regression (MLR)

MLR model considers n observations over the set of variables $\{X'_j, y_j\}$, $j \in \{1, 2, \ldots, n\}$, where, y_j is the response variable taking any value in total C classes $y_j \in \{1, 2, \ldots, C\}$ and X'_j is $k \times 1$ vector of observed regressors. The logistic multiclass regression can be defined as one of the following two alternative types:

Type I: with class probability p_{jc} defined as

$$p_{jc} = \frac{\exp(X'_j \boldsymbol{\beta}_c)}{1 + \sum_{c=1}^{C-1} \exp(X'_j \boldsymbol{\beta}_c)}; \ \forall c \in \{1, 2, \ldots, C-1\}, \quad \text{and} \quad p_{jC} = 1 - \sum_{c=1}^{C-1} p_{jc}$$

where, $\boldsymbol{\beta}_c$ is $k \times 1$ vector of parameters of the model.

Type II: with class probability p_{jc} defined as

$$p_{jc} = \frac{\exp(\boldsymbol{X}'_j \boldsymbol{\beta}_c)}{\sum_{c=1}^{C} \exp(\boldsymbol{X}'_j \boldsymbol{\beta}_c)}; \ \forall c \in \{1,\ 2,\ \ldots,\ C\}$$

The likelihood function for c-th class is given by:

$$\ell_c(\boldsymbol{\beta}_c | p_{jc}, y_{jc};\ j \in \{1,\ 2,\ \ldots,\ n\}) = \prod_{j=1}^{n} p_{jc}^{y_{jc}} (1 - p_{jc})^{(1-y_{jc})}$$

The log likelihood is

$$L_n = \ln \prod_{c=1}^{C} \prod_{j=1}^{n} p_{jc}^{y_{jc}} (1 - p_{jc})^{(1-y_{jc})} = \sum_{c=1}^{C} \sum_{j=1}^{n} y_{jc} \boldsymbol{X}'_j \boldsymbol{\beta}_c - \ln\{1 + \exp(\boldsymbol{X}'_j \boldsymbol{\beta}_c)\}$$

Now, the ridge penalized (Hoerl and Kennard 1970) likelihood is given by

$$L_n = \sum_{c=1}^{C} \sum_{j=1}^{n} y_{jc} \boldsymbol{X}'_j \boldsymbol{\beta}_c - \ln\{1 + \exp(\boldsymbol{X}'_j \boldsymbol{\beta}_c)\} - \frac{\lambda}{2} \boldsymbol{\beta}'_c \boldsymbol{\beta}_c$$

Differentiating L_n w.r.t. $\boldsymbol{\beta}_c$, we get

$$\nabla_{\boldsymbol{\beta}_c} L_n = \boldsymbol{X}'(\boldsymbol{y}_c - \boldsymbol{p}_c) - \lambda \boldsymbol{\beta}_c,$$

$$\nabla^2_{\boldsymbol{\beta}_c} L_n = -[\boldsymbol{X}' \boldsymbol{P}_c \boldsymbol{X} + \lambda \boldsymbol{I}_n].$$

where $\boldsymbol{p}_c$ is an $n \times 1$ vector of probabilities of n instances. $\boldsymbol{P}_c = \mathrm{diag}(\boldsymbol{p}'_c(1 - \boldsymbol{p}_c))$.
Using Newton's iterative method, $\boldsymbol{\beta}_c$ can be calculated as

$$\boldsymbol{\beta}_{c(s+1)} = \boldsymbol{\beta}_{cs} + \nu[\boldsymbol{X}' \boldsymbol{P}_{cs} \boldsymbol{X} + \lambda \boldsymbol{I}_p]^{-1}[\boldsymbol{X}'(\boldsymbol{y}_c - \boldsymbol{p}_{cs}) - \lambda \boldsymbol{\beta}_{cs}].$$

Here, $0 < \nu < 1$ is the learning rate.

3.2 *Model Averaging*

The dependence of a variable $\boldsymbol{y}$ on k variables $\boldsymbol{X}_1,\ \boldsymbol{X}_2,\ \ldots,\ \boldsymbol{X}_k$ each having n observations in form of linear model is given by

$$y = X_1\beta_1 + X_2\beta_2 + \cdots + X_k\beta_k + \varepsilon$$

where, β_i 's ($i \in \{1, 2, \ldots, k\}$) are termed as the coefficients of the model.

There exist several types of measures. Based on those measures the variables can be categorized as qualitative and quantitative and so the natural process. There are several methods to deal with such type of variables generated by natural processes. The selection of variables to include in model is also a challenging task. There are one or $\binom{k}{0}$ model with no explanatory variables, $\binom{k}{1}$ models with one explanatory variable, $\binom{k}{2}$ models with two explanatory variables, ..., $\binom{k}{k}$ model with k explanatory variables. In total, there are $m = \binom{k}{0} + \binom{k}{1} + \cdots + \binom{k}{k} = (1+1)^k = 2^k$ sub models given as

$$\begin{aligned}
&y = \beta_0 + \varepsilon_1;\\
&y = \beta_0 + X_1\beta_1 + \varepsilon_2;\\
&y = \beta_0 + X_2\beta_2 + \varepsilon_3;\\
&y = \beta_0 + X_1\beta_1 + X_2\beta_2 + \varepsilon_4;\\
&\vdots\\
&y = \beta_0 + X_1\beta_1 + \cdots + X_k\beta_k + \varepsilon_m;
\end{aligned}$$

In these sub models, only one fits best based on certain model selection criterion. The model selection criterion can be specified using model properties and predictive performances. Some of the model selection criteria are coefficient of determination (R^2) or adjusted R^2, Akaike information criteria (AIC) (Akaike 1998). Bayes information criteria (BIC) (Schwarz 1978), Takeuchi information criterion (TIC) (Takeuchi 1976) and t-tests, F-tests or pretesting. This approach discards all other models except *the one* (winning model). In model averaging, a suitable weight or probability is assigned to each model and then the weighted average of estimated parameters from different models is calculated. The basic methodology of frequentist model averaging assumes that ith model includes k_i number of regressors $X_1, X_2, \ldots, X_{k_i}$. The ith model is written as

$$y = X_1\beta_1 + X_2\beta_2 + \cdots + X_{k_i}\beta_{k_i} + \varepsilon_i = XS_iS_i'\beta + \varepsilon_i.$$

Here, S_i is the selection matrix of order $k \times k_i$ that includes k_i number of columns from the matrix having k columns of regressors $X_1, X_2, \ldots, X_k$ and ε is the $n \times 1$ vector of random errors/noise. By using a suitable criterion, let us have weight ω_i for the ith model from unit simplex

$$\Omega = \{\omega_i : 0 < \omega_i < 1 \quad \forall i \in \{1, 2, \ldots, m\}\}$$

$$\text{and } \sum_{i=1}^{m} \omega_i = 1.$$

Multiplying the ith model by ω_i and summing over all models we get

$$\begin{aligned} \boldsymbol{y} &= \boldsymbol{X}\sum_{i=1}^{m} \omega_i \boldsymbol{S}_i \boldsymbol{S}_i' \boldsymbol{\beta} + \sum_{i=1}^{m} \omega_i \boldsymbol{\varepsilon}_i \\ &= \boldsymbol{X}\sum_{i=1}^{m} \omega_i \boldsymbol{S}_i \boldsymbol{\beta}_i + \sum_{i=1}^{m} \omega_i \boldsymbol{\varepsilon}_i \\ &= \boldsymbol{X}\mathbb{B}\boldsymbol{\omega} + \boldsymbol{\varepsilon}\boldsymbol{\omega} \end{aligned}$$

Here, $\boldsymbol{\beta}_i = \boldsymbol{S}_i'\boldsymbol{\beta}$, ω is $m \times 1$ vector of model weights, $\mathbb{B}$ is $k \times m$ matrix of coefficients and $\boldsymbol{\varepsilon}$ is $n \times m$ matrix of errors of m models.

3.3 Ridge Model Averaging in MLR

The probability based on averaged model for cth class is given by

$$p_{\omega jc} = \frac{\exp(\boldsymbol{X}_j'\mathbb{B}_c\boldsymbol{\omega}_c)}{\sum_{c=1}^{C}\exp(\boldsymbol{X}_j'\mathbb{B}_c\boldsymbol{\omega}_c)}; \quad \forall c \in \{1, 2, \ldots, C\}.$$

Here, $\mathbb{B}_c$ is $k \times m$ matrix of coefficients of m models and $\boldsymbol{\omega}_c$ is vector of model weights.

Now, the ridge penalized likelihood is given by

$$L_n = \sum_{c=1}^{C}\sum_{j=1}^{n} y_{jc}\boldsymbol{X}_j'\mathbb{B}_c\boldsymbol{\omega}_c - \ln\{1 + \exp(\boldsymbol{X}_j'\mathbb{B}_c\boldsymbol{\omega}_c)\} - \frac{\lambda}{2}\boldsymbol{\omega}_c'\mathbb{B}_c'\mathbb{B}_c\boldsymbol{\omega}_c.$$

The algorithm for model averaging in MLR has following steps:

Step 1. Obtain the estimates for β_{ic}'s using Newton's iterative method as:

$$\boldsymbol{\beta}_{ic(s+1)} = \boldsymbol{\beta}_{ics} + \nu\left[\boldsymbol{S}_i'\boldsymbol{X}'\boldsymbol{P}_{ics}\boldsymbol{X}\boldsymbol{S}_i + \lambda\boldsymbol{I}_{k_i}\right]^{-1}\left[\boldsymbol{S}_i'\boldsymbol{X}'(\boldsymbol{y}_c - \boldsymbol{p}_{ics}) - \lambda\boldsymbol{\beta}_{ics}\right]$$

Step 2. Obtain weight vector ω_c for each class by convex optimization of

$$L_n = \sum_{c=1}^{C}\sum_{j=1}^{n} y_{jc}X'_j\hat{\mathbb{B}}_c\omega_c - \ln\left\{1 + \exp\left(X'_j\hat{\mathbb{B}}_c\omega_c\right)\right\} - \frac{\lambda}{2}\omega'_c\hat{\mathbb{B}}'_c\hat{\mathbb{B}}_c\omega_c$$

Step 3. Obtain model average estimator for each class using $\hat{\boldsymbol{\beta}}_{\omega_c} = \hat{\mathbb{B}}_c\hat{\boldsymbol{\omega}}_c$
Step 4. Calculate the probabilities of classes from log of probability odds ratio.
Step 5. Classify the observation in the class having maximum class probability.

The analysis of cover-type dataset using the method of MLR and ridge model averaging in MLR is given in Sect. 4.

4 Analysis and Results

We have processed data and used 12 measured attributes as 12 regressors by assigning the values 1, 2, 3, 4 to the four wilderness areas, Rawah, Neota, Comanche Peak and Cache la Poudre , respectively, and 1, 2, ..., 40 to the 40 soil types to reduce the dimension of the data. There are no two or more wilderness areas or soil types present simultaneously in a single instance. We have indexed them same as hill shade. Hill shade can take any indexed value between 0 and 255, i.e., it can be categorized as 256 variables. The number of instances for each class included in training dataset and validation dataset is given in Table 4. The model averaging over 2^{54} models is very time consuming and result of MLR with such type of transformation results in better prediction than previously reported by Kolasa and Raja (2016). In their report, several methods have been reported to the train set and validation set of the data. The correct classification reported by using logistic ridge regression with 70–30 and 85–15 training–validation set split is 55.70 and 59.55%, respectively.

We have applied the above-mentioned model averaging technique. The "glmnet" package by Friedman et al. (2009) for MLR gives 61.21% correct classification for

Table 4 Forest cover-type classes and no. of instances in data subsets

Cover type	Class code	Number of instances in			
		Training set	Validation set	Test set	Complete dataset
Spruce/fir	1	1620	540	2,09,680	2,11,840
Lodgepole pine	2	1620	540	2,81,141	2,83,301
Ponderosa pine	3	1620	540	33,594	35,754
Cottonwood/willow	4	1620	540	587	2747
Aspen	5	1620	540	7333	9493
Douglas fir	6	1620	540	15,207	17,367
Krummholz	7	1620	540	18,350	20,510
Total		11,340	3780	5,65,892	5,81,012

validation set with 100-fold cross validation. Our proposed method of ridge model averaging in MLR gives 67.72% correct classification with single or no fold cross validation over same data subset.

5 Conclusion

The ridge model averaging in MLR has been explored for predicting forest cover type and it has been observed that it provides improvement over the traditional full model-based MLR method. The single-fold cross validation applied in the paper can be further modified to multifold cross validation for improving predictive performance, though it will require much more computational resources.

References

Akaike, H. (1998). Information theory and an extension of the maximum likelihood principle. In E. Parzen, K. Tanabe, & G. Kitagawa (Eds.), *Selected papers of Hirotugu Akaike, 199–213, Springer Series in Statistics (Perspectives in Statistics)*. New York: Springer.

Blackard, J. A., & Dean, D. J. (1999). Comparative accuracies of artificial neural networks and discriminant analysis in predicting forest cover types from cartographic variables. *Computers and Electronics in Agriculture, 24*(3), 131–151.

Friedman, J., Hastie, T., & Tibshirani, R. (2009). Glmnet: Lasso and elastic-net regularized generalized linear models. R package version, *1*(4).

Hoerl, A. E., & Kennard, R. W. (1970). Ridge regression: Biased estimation for non orthogonal problems. *Technometrics, 12*(1), 55–67.

Kolasa, T., & Raja, A. K. (2016). Forest cover type classification study. Retrieved from: HUhttps://rpubs.com/aravindkr/160297UH.

Kumar, R., Nandy, S., Agarwal, R., & Kushwaha, S. P. S. (2014). Forest cover dynamics analysis and prediction modeling using logistic regression model. *Ecological Indicators, 45,* 444–455.

Nahib, I., & Suryanta, J. (2017). Forest cover dynamics analysis and prediction modelling using logistic regression model (case study: Forest cover at Indragiri Hulu Regency, Riau Province). In *IOP Conference Series: Earth and Environmental Science*, *54*(1), p. 012044, IOP Publishing.

Schwarz, G. (1978). Estimating the dimension of a model. *The Annals of Statistics, 6*(2), 461–464.

Takeuchi, K. (1976). Distribution of information statistics and validity criteria of models. *Mathematical Science, 153,* 12–18.

Ullah, A., & Wang, H. (2013). Parametric and nonparametric frequentist model selection and model averaging. *Econometrics, 1*(2), 157–179.

Small Area Estimation for Skewed Semicontinuous Spatially Structured Responses

Chiara Bocci, Emanuela Dreassi, Alessandra Petrucci and Emilia Rocco

Abstract When surveys are not originally designed to produce estimates for small geographical areas, some of these domains can be poorly represented in the sample. In such cases, model-based small area estimators can be used to improve the accuracy of the estimates by borrowing information from other sub-populations. Frequently, in surveys related to agriculture, forestry or the environment, we are interested in analyzing continuous variables which are characterized by a strong spatial structure, a skewed distribution and a point mass at zero. In such cases, standard methods for small area estimation, which are based on linear mixed models, can be inefficient. The aim of this chapter is to discuss small area estimation models suggested in literature to handle zero-inflated, skewed, spatially structured data and to present them under the unified approach of generalized two-part random effects models.

Keywords Generalized linear mixed model · Geoadditive models · Geographic information · Skewed distribution · Zero-inflated data

1 Introduction

In this chapter, we consider situations where the target response is continuous and shows a strong spatial structure, a skewed distribution and a point mass at zero. Many target variables in environmental, forestry or agricultural surveys follow such kind of distribution. Think, among others, to the forest environmental income, certain environmental pollutants, the yield and/or the surface allocated to specific crops or the potential distribution of ecological niches for which a continuous abundance index is recorded.

C. Bocci · E. Dreassi · A. Petrucci (✉) · E. Rocco
Department of Statistics, Computer Science, Applications "G. Parenti", University of Florence, Florence, Italy
e-mail: alessandra.petrucci@unifi.it

G. Chandra et al. (eds.), *Statistical Methods and Applications in Forestry and Environmental Sciences*, Forum for Interdisciplinary Mathematics,
https://doi.org/10.1007/978-981-15-1476-0_15

Our interest lies in how to obtain valid estimates of some parameters of interest, such as means or totals, for small areas for which only small samples or no sampled units are available. As in all small area estimation problems, the small sample sizes within the sampled areas and the existence of non-sampled areas require the use of model-based methods.

The most popular models for small area estimation (SAE) are linear mixed models which include independent random area effects to account for the variability between the areas exceeding that explained by the auxiliary variables. There are two main categories of SAE models, depending on whether the response variable is observed only at the small area level, or at respondent (unit) level. Fay and Herriot (1979) studied the area level model and proposed an empirical Bayes estimator. Battese et al. (1988) considered the unit level model and presented an empirical best linear unbiased predictor (EBLUP) for the small area means. Unfortunately, if the target variable's distribution is zero-inflated, highly skewed, or presents a spatial trend, neither models are directly applicable to produce reliable small area means estimates. To this end, over the last decade several approaches have been developed for producing model-based small area estimates for variables with one or more of these characteristics.

The aim of this chapter is to discuss some of these models by presenting them under a unified framework. In this, the basic assumptions that we maintain throughout our presentation are that the response variable is observed at unit level and that it is continuous on the positive real line, with a substantial point mass at zero. In addition, this approach allows for the possible skewness of the response's distribution and it exploits, when available, the information on the units' spatial location in order to produce more accurate estimates of a spatially related phenomenon.

The chapter is organized as follows. In Sect. 2, we discuss statistical strategies to handle zero-inflated, skewed and spatially structured data. Section 3 describes a unified statistical model to analyze data that presents these features. Finally, Sect. 4 reports our conclusions.

2 Handling Zero-Inflated, Skewed and Spatially Structured Data

When the data are characterized by a high occurrence of zeros, it is reasonable to assume that the data generation process follows a two-component mixture distribution, whose terms are a degenerate distribution (a point mass at 0) and some standard distribution on the positive real line. Thus, to analyze such data, two regression models are employed: a logit or probit model to estimate the mixing proportion, that is the probability of observing extra zeroes; and a conditional regression model, which depends on the nature of the data, for the mean of the standard distribution. Several models of this type have been introduced in literature, originally developed to account for the excess of zeros in count data. Usually, they are called zero-inflated models and include regression models for zero inflation relating to Poisson, Negative Binomial and Binomial distributions. In Lambert (1992), Hall (2000), Ridout et al. (2001) among others, such models are extensively investigated.

A large number of zeros can occur in continuous data as well. Such data are typically referred to as *semicontinuous*. Zero-inflated models for semicontinuous variables have been developed mainly for longitudinal data in biomedical applications (Olsen and Schafer 2001; Berk and Lachenbruch 2002; Tooze et al. 2002; Albert and Shen 2005; Gosh and Albert 2009) and are known as *two-part* models. Since in longitudinal data, the repeated measures produce a clustered structure; two-part models typically include a cluster-specific random effect in both the mixing and the conditional regression models. More in general, two-part random effects models may be used for any kind of semicontinuous clustered data. If the clusters consist of geographical small areas, then the two-part random effects model can be exploited to produce model-based estimates. Pfeffermann et al. (2008) firstly introduced such a model in the context of small area estimation by adopting a Bayesian approach. Chandra and Sud (2012) developed a small area estimator based on a two-part random effects model under a frequentist approach. The use of a two-part random effects models in order to produce small area estimates has been then further developed in different aspects by Bocci et al. (2012), Dreassi et al. (2014), Chandra and Chambers (2016) and investigated by Karlberg (2014) and Krieg et al. (2016).

The two-part SAE random effects model can therefore be considered as the general framework that manages all the basic assumptions we specified in Sect. 1. Moreover, by defining different functional forms for the conditional regression model, several distributions of the non-zero values can be represented. Pfeffermann et al. (2008) and Chandra and Sud (2012) considered a non-skewed shape for the non-zero responses, adopting a normal distribution to model their means. Instead, positive highly skewed variables have been handled in literature either by a logarithmic scale linear mixed model (Bocci et al. 2012; Karlberg 2014; Krieg et al. 2016; Chandra and Chambers 2016) or by a Gamma mixed model (Dreassi et al. 2014).

Another important characteristic of the data that should not be neglected, if available, is the spatial distribution of the phenomenon under study. Several methods have been proposed in literature to exploit the information on the geographical location of the population units. These models account for the spatial trend of the study variable and simultaneously allow for any additional covariate effects, both linear or nonlinear. They include, among others, the geoadditive model by Kammann and Wand (2003), the geographically weighted regression introduced by Fotheringham et al. (2002) and the generalized additive mixed models based on Markov Random Fields (Fahrmeir and Lang 2001). In particular, our work focus on geoadditive models since they can be formulated as a linear mixed model, following the linear mixed model representation of penalized spline regression illustrated by Wand (2003). Geadditive models are composed by an additive model (Hastie and Tibshirani 1990) and a kriging model (Cressie 1993); therefore, they jointly handle the influence of covariates and the spatial structure of the variable under study.

Throughout the unifying mixed model framework, it is therefore straightforward to combine a two-part SAE random effects model with a geoadditive model in order to exploit the available geographical information to get more accurate small area estimates for semicontinuous target variables that exhibit a spatial trend (Bocci et al. 2012; Dreassi et al. 2014). The connection between penalized spline regression and linear mixed models in a more general SAE context has been firstly considered by Opsomer et al. (2008), which included the spatial effects in a SAE linear mixed model. More recently, Bocci and Petrucci (2016) considers a geoadditive SAE log scaled linear mixed model for producing model-based direct estimates and Chandra et al. (2018) presents a nonlinear spatial SAE model at area level.

Finally, we note that the general structure of the two-part SAE random effects model, that contains area random effects in both its components, allows in theory to specify different assumptions on the correlation among the random effects in the two components. However, the hypothesis of a non-null correlation (considered by Cantoni et al. (2017) and Dreassi et al. (2014), among others) makes model fitting computationally intensive and, according to the empirical results of Pfeffermann et al. (2008), does not significantly improve small area estimation. On the other hand, the assumption of a negligible correlation implies that the resulting model is not appropriate if there is actual reason to believe that the distributions of the components are dependent.

3 Two-Part Geoadditive Small Area Model

Let U be a finite population of N units, partitioned in m subgroups (areas) of size N_i, with $\sum_{i=1}^{m} N_i = N$. A sample r of n units is selected from U according to a non-informative sampling design. r may be decomposed as $r = \bigcup_{i=1}^{m} r_i$ where r_i is the area-specific sample of size n_i. A response variable y is observed for each unit in the sample; y_{ij} denotes the value of y for the unit $j = 1, \ldots, N_i$ in small area $i = 1, \ldots, m$. We are interested is the estimation of the area means $\overline{y}_i = N_i^{-1} \sum_{j=1}^{N_i} y_{ij}$ (or some other area parameters). Such means may be decomposed as $\overline{y}_i = N_i^{-1}(\sum_{j\in r_i} y_{ij} + \sum_{j\in q_i} y_{ij})$ where q_i indicates the set of area non-sampled units, that is the complement of the area-specific sample r_i to the area population (of size $N_i - n_i$). In a SAE problem, it is assumed that the sample area sizes n_i are too small to calculate reliable direct estimates. Further, the values of some covariates are available at area and/or unit level for $j \in r_i$ and for $j \in q_i$. Therefore, generally speaking, indirect estimation could be considered.

In addition, we assume that the spatial location $\mathbf{s}_{ij}$ ($\mathbf{s} \in R^2$) of each unit is known and that y is a non-negative semicontinuous skewed spatially structured variable. To account for semicontinuity, the response variable can be recoded as the product $y_{ij} = \delta_{ij} y'_{ij}$ of two independent variables

$$\delta_{ij} = \begin{cases} 1 \text{ if } y_{ij} > 0 \\ 0 \text{ if } y_{ij} = 0 \end{cases} \quad \text{and} \quad y'_{ij} = \begin{cases} y_{ij} & \text{if } y_{ij} > 0 \\ \text{irrelevant} & \text{if } y_{ij} = 0. \end{cases}$$

The distribution function $F(\cdot)$ of y_{ij} can be written as

$$F(y_{ij}) = \pi_{ij}\, G(y'_{ij}) + (1 - \pi_{ij})\, I_0,$$

where $\pi_{ij} = P(\delta_{ij} = 1)$, $G(\cdot)$ is the distribution function of y'_{ij} and I_0 is the indicator function of $y_{ij} = 0$.

Let $z_{ij} = f(y'_{ij})$ be a generic transformation of the positive values y'_{ij} distributed as an exponential family with mean μ_{ij} and link function $g(\cdot)$. In order to complete the specification of the two-part model for y, we need to choose a model specification for its two components: the mixing proportion π_{ij} and $g(\mu_{ij})$. In the following, both parts are specified conditionally on two sets of covariates $\mathbf{t}_{ij}$ and $\mathbf{t}^*_{ij}$, on the geographical coordinates $\mathbf{s}_{ij}$ and on two sets of area random effects $\{u_1, \ldots, u_m\}$ and $\{u^*_1, \ldots, u^*_m\}$. It is worth noting that $\mathbf{t}_{ij}$ and $\mathbf{t}^*_{ij}$, the covariates for the first and the second part respectively, may coincide or may be partially or completely different.

The mixing proportion π_{ij} is modeled as a geoadditive logistic mixed model

$$\eta_{ij} = \log \frac{\pi_{ij}}{1 - \pi_{ij}} = \beta_{0t} + \mathbf{t}^T_{ij} \boldsymbol{\beta}_t + h(\mathbf{s}_{ij}) + u_i, \tag{1}$$

where $h(\cdot)$ is some bivariate smooth function depending on the geographical unit coordinates $\mathbf{s}_{ij}$. The nonlinear function $h(\cdot)$ is estimated with a penalized spline (Eilers and Marx 1996). In particular, we assume that $h(\cdot)$ can be approximated with a low-rank thin-plate spline (Ruppert et al. 2003) with K knots $(\boldsymbol{\kappa}_1, \ldots, \boldsymbol{\kappa}_K)$

$$h(\mathbf{s}_{ij}) = \beta_{0s} + \mathbf{s}^T_{ij}\, \boldsymbol{\beta}_s + \sum_{k=1}^{K} \gamma_k\, b(\mathbf{s}_{ij}, \boldsymbol{\kappa}_k)$$

where $(\beta_{0s}, \boldsymbol{\beta}_s)$ and $(\gamma_1, \ldots, \gamma_K)$ are the coefficients for the 'linear' and the 'spline' portions of the model, respectively, and $b(\mathbf{s}_{ij}, \boldsymbol{\kappa}_k)$ are the spline basis functions, defined as

$$\mathbf{B} = \left\{b(\mathbf{s}_{ij}, \boldsymbol{\kappa}_k)\right\} = \left\{ \left[C\left(\mathbf{s}_{ij} - \boldsymbol{\kappa}_k\right)\right]_{\substack{1 \le j \le N_i \\ 1 \le i \le m \\ 1 \le k \le K}} \left[C\left(\boldsymbol{\kappa}_h - \boldsymbol{\kappa}_k\right)\right]^{-1/2}_{\substack{1 \le h \le K \\ 1 \le k \le K}} \right\} \tag{2}$$

where $C(\mathbf{r}) = \|\mathbf{r}\|^2 \log \|\mathbf{r}\|$.

We follow Ruppert et al. (2003) for the choice of the number and of the location of knots: They recommend the use of a space-filling design to ensure coverage of the covariate space as well as parsimony in the number of knots. Therefore, we choose the number of knots with Ruppert's rule of thumb and we use the *clara* space-filling algorithm of Kaufman and Rousseeuw (1990) to select the knots location.

With this representation for $h(\cdot)$, model (1) can be written as a linear mixed model (Kammann and Wand 2003; Opsomer et al. 2008)

$$\boldsymbol{\eta} = \mathbf{X}\boldsymbol{\beta} + \mathbf{B}\boldsymbol{\gamma} + \mathbf{D}\mathbf{u} \tag{3}$$

where $\mathbf{X}$ is the fixed effects matrix with rows $[1, \mathbf{t}_{ij}^T, \mathbf{s}_{ij}^T]$, $\mathbf{B}$ is the $N \times K$ matrix of the thin-plate spline basis functions, $\mathbf{D}$ is the $N \times m$ area-specific random effects matrix with rows $\mathbf{d}_{ij}$ containing indicators taking value 1 if observation j is in area i and 0 otherwise, $\boldsymbol{\beta} = (\beta_{0t}, \beta_{0s}, \boldsymbol{\beta}_t^T, \boldsymbol{\beta}_s^T)^T$ is a vector of unknown coefficients, $\mathbf{u}$ is the vector of the m area-specific random effects and $\boldsymbol{\gamma} = (\gamma_1, \ldots, \gamma_K)$ is the vector of the K thin-plate spline coefficients treated as random effects.

For the second part of the model, that is for the conditional mean value of $z_{ij} = f(y'_{ij})$, we define a geoadditive generalized linear mixed model by using the following linear mixed predictor

$$g(\mu_{ij}) = \beta_{0t}^* + \mathbf{t}_{ij}^{*T}\boldsymbol{\beta}_t^* + h^*(\mathbf{s}_{ij}) + u_i^*. \tag{4}$$

Representing $h^*(\cdot)$ with a low-rank thin-plate spline with K knots, as we did for $h(\cdot)$, model (4) becomes

$$g(\boldsymbol{\mu}) = \mathbf{X}^*\boldsymbol{\beta}^* + \mathbf{B}^*\boldsymbol{\gamma}^* + \mathbf{D}^*\mathbf{u}^* \tag{5}$$

where, in analogy with model (3), $\mathbf{X}^*$ is the fixed effects matrix, $\mathbf{B}^*$ is the matrix of the thin-plate spline basis functions, $\mathbf{D}^*$ is the area-specific random effects matrix, $\boldsymbol{\beta}^*$ is a vector of unknown coefficients, $\mathbf{u}^*$ is the vector of the area-specific random effects and $\boldsymbol{\gamma}^*$ is the vector of the thin plate spline coefficients treated as random effects.

Depending on the functional form chosen for $f(\cdot)$ and $g(\cdot)$, the generic model (5) can assume various formulations in order to analyze data with different characteristics. For example, the simplest case of the geoadditive linear mixed model is obtained when both $f(\cdot)$ and $g(\cdot)$ are identity functions. In many real applications, the positive values of the response variable are skewed but the use of a linear mixed model is still possible after an appropriate transformation of y', the most common be the logarithmic transformation. This is the case of the geoadditive log scaled linear mixed model, which is obtained from model (5) when $f(y') = \log(y)$ and $g(\cdot)$ is the identity link function. That is, for units with positive response we assume:

$$\mathbf{z} = \log(\mathbf{y}') = \mathbf{X}^*\boldsymbol{\beta}^* + \mathbf{B}^*\boldsymbol{\gamma}^* + \mathbf{D}^*\mathbf{u}^* + \mathbf{e}, \qquad \mathbf{e} \sim N(0, \sigma_e^2). \tag{6}$$

Instead of applying a transformation on the response variable, a different approach that could be used to take into account the skewness of the data is to use a Gamma model (mean parameterized) on the original scale. That is, to hypothesize that the positive values y'_{ij} follows a $Gamma(\nu, \mu_{ij})$ distribution, with density function

$$\frac{(\nu/\mu_{ij})(\nu y'_{ij}/\mu_{ij})^{\nu-1}\exp(-\nu y'_{ij}/\mu_{ij})}{\Gamma(\nu)},$$

where $\nu > 0$ and $\mu_{ij} > 0$. Here, μ_{ij} is the mean and $1/\sqrt{\nu}$ is the coefficient of variation of y'_{ij}. Note that the coefficient of variation does not depend on i or j, that is, it is constant over units (see page 287 McCullagh and Nelder 1989). Once the Gamma is mean parameterized, $g(\mu_{ij})$ could be model by a linear mixed predictor. Therefore, the geoadditive Gamma model can be derived from (5) by setting an identity function for $f(\cdot)$ and a logarithmic link function $g(\mu) = \log(\mu)$:

$$\log(\boldsymbol{\mu}) = \mathbf{X}^*\boldsymbol{\beta}^* + \mathbf{B}^*\boldsymbol{\gamma}^* + \mathbf{D}^*\mathbf{u}^*. \tag{7}$$

Formulations (6) and (7) are the more common models used in literature; however, some other transformations $f(\cdot)$ (e.g., Box-Cox or power transformations) and some other link functions $g(\cdot)$ could be more appropriate to reach the normality of the data or to handle other characteristics of the response variable. Model (5) unifies all these possible cases in its generic formulation.

Lastly, hypotheses need to be made on the distributions of the random effects: $\boldsymbol{\gamma}$, $\boldsymbol{\gamma}^*$, $\mathbf{u}$, $\mathbf{u}^*$. We assume that the area and spline random effects included in each components (3) and (5) of the two-part model are mutually independent, with the area effects independently and identically jointly distributed as

$$\begin{bmatrix} u_i \\ u_i^* \end{bmatrix} \sim N\left(\begin{bmatrix} 0 \\ 0 \end{bmatrix}, \Sigma_u = \begin{bmatrix} \sigma_u^2 & \sigma_{uu^*} \\ \sigma_{uu^*} & \sigma_{u^*}^2 \end{bmatrix}\right)$$

and the spline random effects independently and identically jointly distributed as

$$\begin{bmatrix} \gamma_k \\ \gamma_k^* \end{bmatrix} \sim N\left(\begin{bmatrix} 0 \\ 0 \end{bmatrix}, \Sigma_\gamma = \begin{bmatrix} \sigma_\gamma^2 & \sigma_{\gamma\gamma^*} \\ \sigma_{\gamma\gamma^*} & \sigma_{\gamma^*}^2 \end{bmatrix}\right).$$

Different assumptions can be adopted on the correlation structures of (u_i, u_i^*) and (γ_k, γ_k^*): If they are assumed to be independent, then $\sigma_{uu^*} = 0$ and $\sigma_{\gamma\gamma^*} = 0$ and the two components of the model can be fitted separately; conversely, if we cannot assume independence, we obtain a 'full' two-part model which must be jointly estimated.

The estimation of the two-part SAE model has been addressed in literature by adopting either a frequentist approach (as in Chandra and Sud 2012; Bocci et al. 2012) or a Bayesian approach (as in Pfeffermann et al. 2008; Dreassi et al. 2014). Both approaches have pros and cons and the choice usually depends on the specific model assumptions required for the analysis (Krieg et al. 2016). The Bayesian approach can be more computationally intensive, and it requires the definition of prior distributions for each parameter of the whole model given by (3) and (5). However, if the full two-part model is considered, this approach can straightforwardly handle the joint estimation of the two components. Conversely, if the independence assumption holds, the frequentist approach can easily produce reliable estimations.

Finally, it should be noted that the model specified through expressions (3) and (5) may be easily extended so as to include other random effects (due, for example, to other nonlinear covariate effects or to a clustering process inside the areas). Moreover, even if the area random effects and the spline random effects are present in both (3) and (5), there may be situations where a random effect is relevant for one part of the model but not for the other, and therefore it is only included in one of the two parts.

3.1 Small Area Mean Predictors

We are interested in the estimation of the area means $\overline{y}_i = N_i^{-1} \sum_{j=1}^{N_i} y_{ij}$. Notice that the means are computed over all the individuals, including individuals with zero y values, and that they may be decomposed as:

$$\overline{y}_i = N_i^{-1}(\sum_{j \in r_i} y_{ij} + \sum_{j \in q_i} y_{ij})$$

where r_i is the area-specific sample and q_i is its complement to the area population. In the frequentist approach, once the model coefficients are estimated via REML or ML estimators, the area means predictors are:

$$\widehat{\overline{y}}_i = N_i^{-1}(\sum_{j \in r_i} y_{ij} + \sum_{j \in q_i} \hat{y}_{ij}) \tag{8}$$

where $\hat{y}_{ij}$ are the predicted values under the two-part model (3) and (5).

Specifically, when a log transformed linear mixed model is assumed for the strictly positive y' values, a possible form for $\hat{y}_{ij}$, that includes the second-order bias correction for the log-back transformation suggested by Chandra and Chambers (2011), is obtained as:

$$\hat{y}_{ij} = \frac{\exp(\mathbf{x}_{ij}^T \hat{\boldsymbol{\beta}} + \mathbf{b}_{ij}^T \hat{\boldsymbol{\gamma}} + \hat{u}_i)}{1 + \exp(\mathbf{x}_{ij}^T \hat{\boldsymbol{\beta}} + \mathbf{b}_{ij}^T \hat{\boldsymbol{\gamma}} + \hat{u}_i)} \times \hat{\lambda}_{ij}^{-1} \exp\left(\mathbf{x}_{ij}^{*T} \hat{\boldsymbol{\beta}}^* + \frac{\hat{\nu}_{ij}}{2}\right) \tag{9}$$

where $\hat{\nu}_{ij} = \hat{\sigma}_e^2 + \hat{\sigma}_{u^*}^2 + \mathbf{b}_{ij}^{*T} \hat{\sigma}_{\gamma^*}^2 \mathbf{b}_{ij}^*$ is the estimated variance of $\log(y'_{ij})$ and $\mathbf{b}_{ij}$ and $\mathbf{b}_{ij}^*$ represent, respectively, the ij-row of the matrix $\mathbf{B}$ and $\mathbf{B}^*$ of the thin-plate spline basis. $\hat{\lambda}_{ij}$ is the bias adjustment factor for the log-back transformation and its expression is: $\hat{\lambda}_{ij} = 1 + 0.5[\hat{a}_{ij} + 0.25\hat{V}\left(\hat{\nu}_{ij}\right)]$ where $\hat{a}_{ij} = \mathbf{x}_{ij}^{*T} \hat{V}(\hat{\boldsymbol{\beta}}^*)\mathbf{x}_{ij}^*$, $\hat{V}(\hat{\boldsymbol{\beta}}^*)$ is the usual estimator of the variance of $\hat{\boldsymbol{\beta}}^*$ and $\hat{V}\left(\hat{\nu}_{ij}\right)$ is the estimated asymptotic variance of $\hat{\nu}_{ij}$.

When a Gamma generalized linear mixed model is assumed for the strictly positive y' values, the $\hat{y}_{ij}$ are obtained as:

$$\hat{y}_{ij} = \frac{\exp(\mathbf{x}_{ij}^T \hat{\boldsymbol{\beta}} + \mathbf{b}_{ij}^T \hat{\boldsymbol{\gamma}} + \hat{u}_i)}{1 + \exp(\mathbf{x}_{ij}^T \hat{\boldsymbol{\beta}} + \mathbf{b}_{ij}^T \hat{\boldsymbol{\gamma}} + \hat{u}_i)} \times \exp\left(\mathbf{x}_{ij}^{*T}\, \hat{\boldsymbol{\beta}} + \mathbf{b}_{ij}^{*T}\, \hat{\boldsymbol{\gamma}} + \hat{u}_i^*\right). \tag{10}$$

More in general, depending on how the specific functions $f(\cdot)$ and/or $g(\cdot)$ are defined, the predicted values $\hat{y}_{ij}$ can be obtained by considering the corresponding back-transformations of the functions $f(\cdot)$ and/or $g(\cdot)$.

In a Bayesian framework, once the posterior distribution of parameters has been obtained via Markov Chain Monte Carlo (MCMC) method, prediction is formally straightforward. In getting estimates of all parameters via MCMC (using the sample data for $j \in r_i$), all posterior distributions of parameters can be extended to those $j \in q_i$ so as to obtain the empirical predictive distribution for y_{ij} with $j \in q_i$. More precisely, for each MCMC algorithm iteration $l = 1, \ldots, L$ the empirical predictive distribution is $\hat{y}_{ij}^{(l)} = \hat{\pi}_{ij}^{(l)} \hat{y'}_{ij}^{(l)}$, for $i = 1, \ldots, m$ and $j \in q_i$. Then, the empirical predictive distribution for the mean of small area i can be written as

$$\hat{\bar{y}}_i^{(l)} = N_i^{-1} \left(\sum_{j \in r_i} y_{ij} + \sum_{j \in q_i} \hat{y}_{ij}^{(l)} \right).$$

Finally, a synthetic value of the predictive distribution, such as the mean, is taken as an estimate of the mean for each small area.

An important advantage of the Bayesian MCMC approach is that the draws can be used both for computing point estimates and for measures of accuracy. Usually, as a measure of precision, each estimate can be associated to their corresponding credibility interval, that is, the interval between the $\alpha/2$ and $(1 - \alpha/2)$ quantiles of the empirical predictive distribution (see, e.g., Pfeffermann et al. 2008; Dreassi et al. 2014). Alternatively, the posterior variance can be reported as a measure of uncertainty (see Datta and Ghosh 1991).

In the frequentist approach, analytic estimators of the MSE of nonlinear small area estimators are technically complex to derive and typically involve a considerable degree of approximation. As a consequence, a number of numerically intensive, but computationally tractable, methods for MSE estimation have been proposed, e.g., the jackknife method of Jiang et al. (2002) and the bootstrap methods described in Hall and Maiti (2006), Manteiga et al. (2008) and Manteiga et al. (2007) and references therein. Specifically, for two-part SAE mixed models the parametric bootstrap has been illustrated in Bocci et al. (2012), Chandra and Chambers (2016) and Chandra and Sud (2012), to which we refer.

4 Conclusions

In this chapter, we have discussed how SAE can be rightly addressed when dealing with zero-inflated, skewed, spatially structured data. This type of data is frequently found in environmental and agricultural surveys, as well as when studying socioeconomical phenomena. Various contributions have recently been presented in the literature on SAE, suggesting methods to address one or more of these characteristics. To illustrate their proposals, most of these papers present applications to real or pseudo-real datasets. For example, and without being exhaustive: Pfeffermann et al. (2008) estimates district-level average literacy scores of the adult population in Cambodia; Chandra and Sud (2012) predicts the district-level average value of amount of loan outstanding per household for the rural areas of the state of Uttar Pradesh in India; Karlberg (2014) estimates the regional-level average number of beef cattle on hand at the end of the financial year for the Australian farms; Bocci et al. (2012) and Dreassi et al. (2014) estimate the per-farm average grapevine production in Tuscany (Italy) at agrarian region level. We want to look back at these proposals and gather them together in the unified framework of generalized two-part random effects models, with the aim to present a general and useful tool for the analysis of zero-inflated phenomenon of different nature.

Taking into account the presence of zeros in the data is crucial. In fact, those approaches that ignore the accumulation of zeros usually lead to poor inference (with the relevance of this problem depending on the percentage of zeros in the data). Conversely, approaches that completely remove the zero values and take into account only the non-zero data are inefficient as well. In this way, the estimates are obtained neglecting the zero values and important information about some units is typically lost. Two-part random effects models allow to manage the excess of zeros in continuous clustered data by jointly modeling the excess of zeros and the positive response values.

The general structure of the model allows to analyze semicontinuous clustered data of different nature, and especially non-negative skewed data. This can be achieved by using a transformation of the data to obtain symmetry or by working on asymmetric data using some skewed distribution. The use of transformations for small area predictions under skewed data has been widely dealt with in literature even separately from zero inflation. Chandra and Chambers (2011) and Berg and Chandra (2012) investigated small area estimation methods for skewed variables, focusing on the case where a linear mixed model is appropriate after a logarithmic transformation. Nevertheless, other transformations could be proposed to reach the normality of the data. An alternative solution to the log-transformation of the skewed data is to model the response variable directly through an asymmetric distribution. Typically, a Gamma distribution is employed, but other asymmetric distribution could be used as well. For example, Moura et al. (2017) proposed the use of skew Normal and skew t models in order to produce small area estimates for skewed business survey data.

Moreover, by including a geoadditive structure in the generalized two-part random effects models, we can exploit, when available, the information on population units' spatial location in order to produce more accurate estimates of spatially related phenomenon. When the units' location is available for the whole population, as frequently occurs in many areas of observational sciences including environmental and forestry sciences, the inclusion of the geographical coordinates as an explicative variable of the model used for the analysis can substantially improve the understanding of the studied variable. Even if we have only a partial knowledge of the units' spatial location, that is when only the sampled units are exactly located, we can still apply geoadditive methodologies by adopting some opportune methods of imputation for the geographical coordinates of the non-sampled units, as suggested by Bocci and Rocco (2014).

Finally, some discussion on the model estimation approaches available in literature, Bayesian and frequentist, is given. Some findings (Krieg et al. 2016) indicate that both approaches seem to give equivalent results, with both having pros and cons. The choice between one or the other can be considered an additional element of flexibility of the general framework, and it is mostly related to the specific model assumptions required for the analysis. If the independence assumption holds, the frequentist approach easily produces reliable small area estimates, usually with ready-to-use statistical packages. However, since no analytic estimators of the MSE are available in literature, a measure of the estimates variability can only be given using bootstrap estimation. The Bayesian approach, on the other hand, requires the definition of prior distributions for each parameter of the model and the use of a M-CMC strategy, which is more computationally intensive but directly produces both the predictions and their measure of accuracy. Moreover, if the full two-part model is considered, this approach can straightforwardly handle the joint estimation of the two components, thanks to its flexibility on modeling complicated structure on the data.

References

Albert, P., & Shen, J. (2005). Modelling longitudinal semicontinuous emesis volume data with serial correlation in an acupuncture clinical trial. *Journal of the Royal Statistical Society Series C*, *54*, 707–720.

Battese, G., Harter, R., & Fuller, W. (1988). An error component model for prediction of county crop areas using survey and satellite data. *Journal of the American Statistical Association*, *83*, 28–36.

Berg, E. and Chandra, H. (2012). Small area prediction for a unit level lognormal model. In *Proceedings of the 2012 Federal Committee on Statistical Methodology Research Conference*. Washington: DC, USA.

Berk, K. N., & Lachenbruch, P. A. (2002). Repeated measures with zeros. *Statistical Methods in Medical Research*, *11*, 303–316.

Bocci, C., & Petrucci, A. (2016). Spatial information and geoadditive small area models. In M. Pratesi (Ed.), *Analysis of poverty data by small area estimation* (pp. 245–259). UK: Wiley.

Bocci, C., Petrucci, A., & Rocco, E. (2012). Small area methods for agricultural data, a two-part geoadditive model to estimate the agrarian region level means of the grapevines production in tuscany. *Journal of the Indian Society of Agricultural Statistics*, *66*, 135–144.

Bocci, C., & Rocco, E. (2014). Estimates for geographical domains through geoadditive models in presence of incomplete geographical information. *Statistical Methods and Applications*, *23*, 283–305.

Cantoni, E., Flemming, J. M., & Welsh, A. H. (2017). A random-effects hurdle model for predicting bycatch of endangered marine species. *The Annals of Applied Statistics*, *11*, 2178–2199.

Chandra, H., & Chambers, R. (2011). Small area estimation under transformation to linearity. *Survey Methodology*, *37*, 39–51.

Chandra, H., & Chambers, R. (2016). Small area estimation for semicontinuous data. *Biometrical Journal*, *58*, 303–319.

Chandra, H., Salvati, N., & Chambers, R. (2018). Small area estimation under a spatially non-linear model. *Computational Statistics and Data Analysis*, *126*, 19–38.
Chandra, H., & Sud, U. C. (2012). Small area estimation for zero-inflated data. *Communications in Statistics - Simulation and Computation*, *41*, 632–643.
Cressie, N. (1993). *Statistics for Spatial Data* (revised ed.). New York: Wiley.
Datta, G. S., & Ghosh, M. (1991). Bayesian prediction in linear models: applications to small area estimation. *The Annals of Statistics*, *19*, 1748–1770.
Dreassi, E., Petrucci, A., & Rocco, E. (2014). Small area estimation for semicontinuous skewed spatial data: An application to the grape wine production in Tuscany. *Biometrical Journal*, *56*, 141–156.
Eilers, P. H. C., & Marx, B. D. (1996). Flexible smoothing with B-splines and penalties. *Statistical Science*, *11*, 89–121.
Fahrmeir, L., & Lang, S. (2001). Bayesian inference for generalized additive mixed models based on markov random field priors. *Journal of the Royal Statistical Society Series C*, *50*, 201–220.
Fay, R. E., & Herriot, R. A. (1979). Estimation of income from small places: An application of james-stein procedures to census data. *Journal of the American Statistical Association*, *74*, 269–277.
Fotheringham, A. S., Brunsdon, C., & Charlton, M. E. (2002). *Geographically weighted regression: The analysis of spatially varying Relationships*. Chichester: Wiley.
Gosh, P., & Albert, P. S. (2009). A bayesian analysis for longitudinal semicontinuous data with an application to an acupuncture clinical trial. *Computational Statistics and Data Analysis*, *53*, 699–706.
Hall, D. B. (2000). Zero-inflated poisson and binomial regression with random effects: A case study. *Biometrics*, *56*, 1030–1039.
Hall, P., & Maiti, T. (2006). On parametric bootstrap methods for small area prediction. *Journal Royal Statistical Society Series B*, *68*, 221–238.
Hastie, T., & Tibshirani, R. (1990). *Generalized additive models*. London: Chapman & Hall.
Jiang, J., Lahiri, P., & Wan, S. (2002). A unified Jackknife theory for empirical best prediction with M-estimation. *Annals of Statistics*, *30*, 1782–1810.
Kammann, E. E., & Wand, M. P. (2003). Geoadditive models. *Applied Statistics*, *52*, 1–18.
Karlberg, F. (2014). Small area estimation for skewed data in the presence of zeros. *Statistics in Transition*, *16*, 541–562.
Kaufman, L., & Rousseeuw, P. J. (1990). *Finding groups in data: An introduction to cluster analysis*. New York: Wiley.
Krieg, S., Boonstra, H. J., & Smeets, M. (2016). Small-area estimation with zero-inflated data - a simulation study. *Journal of Official Statistics*, *32*, 963–986.
Lambert, D. (1992). Zero-inflated Poisson regression with an application to defects in manufacturing. *Technometrics*, *34*, 1–14.
Manteiga, G. W., Lombardìa, M. J., Molina, I., Morales, D., & Santamarìa, L. (2007). Estimation of the mean squared error of predictors of small area linear parameters under a logistic mixed model. *Computational Statistics and Data Analysis*, *51*, 2720–2733.
Manteiga, G. W., Lombardìa, M. J., Molina, I., Morales, D., & Santamarìa, L. (2008). Bootstrap mean squared error of a small-area EBLUP. *Journal of Statistical Computation and Simulation*, *78*, 443–462.
McCullagh, P., & Nelder, J. A. (1989). *Generalized linear models* (2nd ed.). London: Chapman and Hall.
Moura, F. A. S., Neves, A. F., & Silva, D. B. N. (2017). Small area models for skewed Brazilian business survey data. *Journal of the Royal Statistical Society Series A*, *180*, 1039–1055.
Olsen, M. K., & Schafer, J. L. (2001). A two-part random-effects model for semicontinuous longitudinal data. *Journal of the American Statistical Association*, *96*, 730–745.
Opsomer, J. D., Claeskens, G., Ranalli, M. G., Kauermann, G., & Breidt, F. J. (2008). Non-parametric small area estimation using penalized spline regression. *Journal of the Royal Statistical Society Series B*, *70*, 265–286.

Pfeffermann, D., Terryn, B., & Moura, F. A. S. (2008). Small area estimation under a two-part random effects model with application to estimation of literacy in developing countries. *Survey Methodology*, *34*, 235–249.

Ridout, M., Hinde, J., & Demetrio, C. G. B. (2001). A score test for testing a zero-inflated Poisson regression model against zero-inflated negative binomial alternative. *Biometrics*, *57*, 219–223.

Ruppert, D., Wand, M. P., & Carroll, R. J. (2003). *Semiparametric regression*. Cambridge: Cambridge University Press.

Tooze, J. A., Grunwald, G. K., & Jones, R. H. (2002). Analysis of repeated measures data with clumping at zero. *Statistical Methods in Medical Research*, *11*, 341–355.

Wand, M. P. (2003). Smoothing and mixed models. *Computational Statistics*, *18*, 223–249.

Small Area Estimation for Total Basal Cover in the State of Maharashtra in India

Hukum Chandra and Girish Chandra

Abstract This chapter describes small area estimation (SAE) approach to produce the small area estimates of the total basal cover (m^2/ha) for trees, shrubs and herbs for the state of Maharashtra in India. All seven forest types are defined as small areas. The analysis uses the data of survey conducted by Tropical Forest Research Institute, Jabalpur, India during the Indian Council of Forestry Research and Education's revisiting of forestry types of India in the year 2011–12. The nested quadrats of 10 m × 10 m, 3 m × 3 m and 1 m × 1 m size for tree, shrub and herb layers respectively are the sampling units. The auxiliary data, percentage of forest cover at small area level is available from India's State of Forest Report 2009 (FSI 2009). The results show that forest type-wise estimates of total basal cover for trees, shrubs and herbs generated by SAE approach are reliable as compared to direct survey estimates. Such disaggregate level estimates are invaluable policy information for state forest department and local resource managers.

Keywords Auxiliary variable · Basal cover · Direct estimates · Mean squared error · Precision

1 Introduction

Sample surveys are generally planned to produce reliable estimates for population characteristics of interest mainly at higher geographic (e.g. national, state) or larger domain levels. The sample size is fixed in such a way that direct estimators, which are calculated using only domain-specific sample data, of parameters for these larger domains provide reliable estimates. In many practical situations, the aim is to estimate parameters for domains that contain only a small number of sample observations or sometime no observation from the larger scale survey. Direct (or traditional) estimates

H. Chandra (✉)
Indian Agricultural Statistics Research Institute, New Delhi, India
e-mail: hchandra12@gmail.com

G. Chandra
Indian Council of Forestry Research and Education, Dehradun, Uttarakhand, India

G. Chandra et al. (eds.), *Statistical Methods and Applications in Forestry and Environmental Sciences*, Forum for Interdisciplinary Mathematics,
https://doi.org/10.1007/978-981-15-1476-0_16

use sample data specifically from the domain/area of interest and they lead unstable estimator when sample size is small. A domain or area is referred as small if domain or area specific sample size is not sufficient to support the reliable direct estimates.

Forestry parameters such as frequency, density and basal covers are important inventory attributes for forest resources management, particularly in India. Forest survey of India (FSI 2017) provides estimates of such forest parameters by the means of multistage sample plot techniques. The existing approach produces the statistics at higher aggregation level, for example at national or state level and provides unstable estimates at small area level due to very small sample sizes. This small sample size problem can be easily resolved provided auxiliary information is available to strengthen the limited sample data for the small area. The underlying theory is referred to as the small area estimation (SAE). The SAE techniques aim at producing reliable estimates for such small area(s) with small (or even zero) sample sizes by borrowing strength from data of other area(s). Generally, small areas in forestry can be taken as the categorized forest types/subtypes, forest divisions, ranges, compartments etc. The Forest types of India were classified in the year 1936 by Sir H. G. Champion which was later revised in 16 major forest types and more than 200 sub-types by the extensive forest survey (Champion and Seth 1968).

In forest sector, recently, the demand for small area statistics is increasing day by day due to formulating policies and programs and allocation of funds at national and regional levels for the sustainable forest management. Demand from private forestry and environmental organizations are also increasing because business decisions, particularly for small business holders, rely heavily on the local socio-economic, environmental and other conditions. Goerndt et al. (2010) assessed the strength of light detection and ranging (LiDAR) as auxiliary information for estimating selected forest attributes (number of trees/ha, basal area/ha, volume/ha, quadratic mean diameter) using area-level LiDAR metrics and single tree crown segmentation. The traditional composite estimator outperformed empirical best linear unbiased prediction in terms of bias but not precision. McRoberts (2012) used a combination of forest inventory observations and Landsat Thematic Mapper imagery for estimation of mean forest stem volume per unit area for small areas. Other important forestry applications of SAE techniques can be seen in Ohmann and Gregory (2002), Katila and Tomppo (2005), Tomppo (2006), Breidenback et al. (2010), Chandra and Chandra (2015).

The SAE techniques are generally based on model-based methods, see for example, Pfeffermann (2002) and Rao (2003). The idea is to use statistical models to link the variable of interest with auxiliary information, e.g. census and administrative data, for the small areas to define model-based estimators for these areas. Such small area models are generally classified into two broad types:

(i) Unit level random effect models, proposed originally by Battese et al. (1988). These models relate the unit values of a study variable to unit-specific covariates. When unit level data is not available, the SAE is carried out under the area level small area models.

(ii) Area level random effect models, which are used when auxiliary information is available only at area level. They relate small area direct estimates to area-specific covariates (Fay and Herriot 1979). This model is widely used area level model in SAE.

In this analysis, an application of SAE techniques has been explored to generate model-based estimates for the total basal cover of trees, shrubs and herbs at small area levels for the state of Maharashtra, India by linking survey data and percentage of forest cover. The survey was conducted by Tropical Forest Research Institute, Jabalpur during the Indian Council of Forestry Research and Education's (ICFRE) revisiting of forestry types of India in the year 2011-12. The auxiliary data, percentage of forest cover at small area (forest type-wise) level was from India's State of Forest Report (FSI 2009). For the present study, seven forest types of Maharashtra (Champion and Seth 1968) viz. Tropical Semi-Evergreen Forest, Tropical Moist Deciduous Forest, Littoral and Swamp Forest, Tropical Dry Deciduous Forest, Tropical Thorn Forest, Subtropical Broadleaved Hill Forest and Plantation/TOF were taken for disaggregation or small areas. The rest of the chapter is organized as follows.

Section 2 introduces the data used for the analysis and Sect. 3 describes the methodology applied for the analysis. In Sect. 4, the diagnostic procedures for examining the model assumptions and validating the small area estimates are presented and discuss the results. Section 5 finally sets out the main conclusions.

2 Data Description

Data from survey conducted by Tropical Forest Research Institute, Jabalpur during the ICFRE's revisiting of forest types of India in the year 2011–12 is used for the purpose. The nested quadrats of sizes 10 m × 10 m, 3 m × 3 m and 1 m × 1 m for tree, shrub and herb layers respectively are the sampling units. Here, the nested quadrat means outermost edges 10 m × 10 m, with 3 m × 3 m and 1 m × 1 m nested within (Fig. 1). The Central point was the random point taken by the global positioning system in the topographic sheet of Maharashtra.

The target is to produce forest type-wise reliable estimates of total basal cover for trees, shrubs and herbs for the state of Maharashtra using SAE approach. Total basal cover is calculated by πr^2, where, $\pi = 3.14$ and $r =$ radius of the species (average of two diameters at right angle). The total forest cover (km^2) corresponding to the 7 small areas (forest types) are shown in Fig. 2.

There is 16.93% of the total geographic area of Maharashtra state is under forest cover. The state has 8736, 20,652 and 21,294 km^2 area under very dense, moderately dense and open forest respectively (FSI 2017).

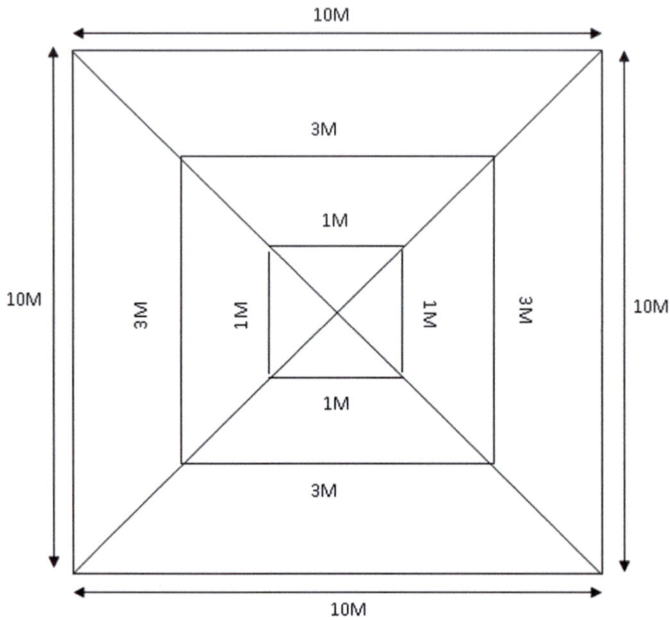

Fig. 1 Design of nested quadrat for tree, shrub and herbs

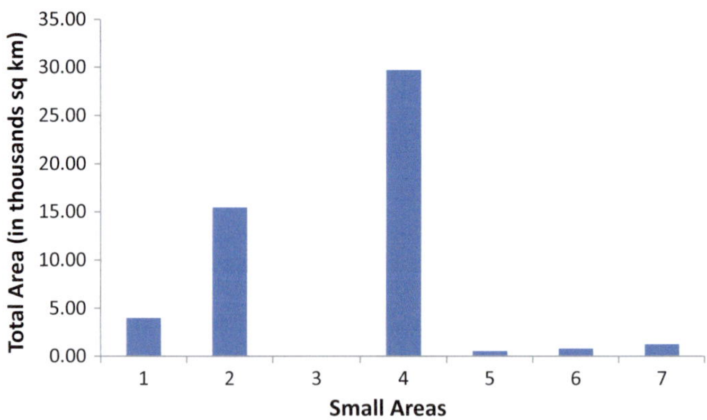

Fig. 2 Total area (in thousands km^2) corresponding to the 7 small areas

Brief descriptions about the important species found in each forest type are detailed as under:

1. *Tropical Semi-Evergreen Forest* in the state is found on the hill slopes of Western Ghats (450–1050 msl). The vegetation elements comprising of *Terminaliaelliptica, Terminaliapaniculata, Careyaarborea, Acacia catechu, Brideliaretusa, Terminaliachebula, Artocarpusintegrifolia, Gmelinaarborea, Oleadioica* and *Dalbergialanceolaria.*
2. *Tropical Moist Deciduous Forest* is of immense conservation value due to their biodiversity. These forests are known to support majority of the tiger population in Maharashtra. These forest are further classified into two main subtypes:

    a. *Moist teak bearing forest*: *Tectonagrandis* is the major species with *Terminaliatomentosa, Dalbergialatifolia, Adina Cordifolia, Madhucaindica, Pterocarpusmarsupium, Mitragynaparviflora, Bombaxmalabarica* and *Dendrocalamus* as associates.
    b. *Moist mixed deciduous forest*: *Tectonagrandis* is one of the associate species; the vegetation is primarily dominated by the evergreen species like *Terminaliatomentosa, Terminaliabellerica, Terminaliachebula, Lagerstroemia parviflora, Syzygiumcumini, Macarangadenticulata, Mallotusalbus, Tremaorientalis* etc.

3. *Littoral and Swamp Forest* is found along the creeks in Sindhudurg and Thane district. The forest cover is limited, however, these forest are rich in biodiversity. The forest plays an important role in protection of seacoast and marine biodiversity. The flora comprises of mangrove species like *Avicennia sp.* and *Rhizophora sp.*
4. *Tropical Dry Deciduous Forest* is the most common in several parts of Maharashtra. *Tectonagrandis*is dominant species with *Lagerstroemia parviflora, Careyaarborea, Terminaliaelliptica, Miliusatomentosa, Lanneagrandis Anogeissuslatifolia, Semecarpusanacardium, Diospyrosmelanoxylon* and *Emblicaofficinalis.*
5. *Tropical Thorn Forest* is a heavily degraded forest due to scanty rainfall and degraded soil. This forest is found in the areas of Marathwada, Vidharbha and Western Maharashtra. The main species in this area are *Acacia arabica, Acacia lecopholea, Balanitesroxburghii, Buteamonosperma and Ziziphus jujube.*
6. *Subtropical Broadleaved Hill Forest* has the vegetation elements like *Syzygiumcumini, Actinodaphnehookeri, Memecylonumbellatum, Randia dumetorum, Flacourtialatifolia, Terminaliachebula, Oleadioica, Glochidionhohenackeri, Xantolistomentosa.* These evergreen forests are found in the Western Ghats of Maharashtra.
7. *Plantation/TOF* comprises mainly the species *Tectonagrandis, Lagerstroemia parviflora, Careyaarborea, Terminaliaelliptica* and *Semecarpusanacardium.*

Serial numbers of the forest types mentioned above are used in each table and figures of this chapter. Area-wise distribution of sample sizes is presented in Fig. 3. The total sample size at the state level was 34. The forest type (or area) specific sample size is ranging from minimum of 1 to maximum of 12 with average of 5. The forest type wise sample size is small so the traditional sample survey estimation approaches lead to unstable estimate. Therefore, it is difficult to generate reliable

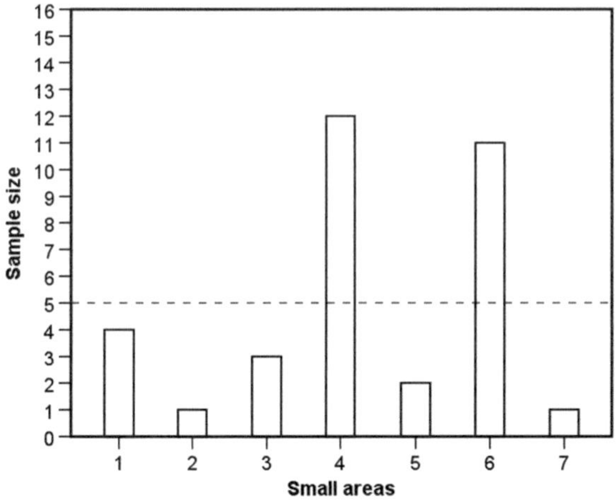

Fig. 3 Forest type-wise distribution of sample sizes from survey data

small area level direct survey estimates with associated standard errors from this survey data alone.

3 Small Area Estimation Methodology

This Section summarizes the underlining theory of SAE applied in the analysis presented in the chapter. There is a wide array of SAE techniques, but they primarily depend on the types of auxiliary data that are available. If there is relevant auxiliary data available for each unit in the population, then the small area models are formulated at the unit level. However, such unit level information is not usually available due to confidentiality or other reasons. In such circumstances, aggregate (or area) level small area models can be formulated that relate the direct survey estimates to the locally aggregated area-specific auxiliary information (Fay and Herriot 1979). The SAE under this model is one of the most popular methods used by private and public agencies because of its flexibility in combining different sources of information and explaining different sources of errors. We employed the area level small area model because auxiliary variable (covariate) for our study is available only at the area level. To start with, we first fix our notation. Throughout, we use a subscript d to index the quantities belonging to small area or forest type $d (d = 1, \ldots, D)$, where, D the number of small areas (or forest types) is in the population. Let $\hat{\theta}_d$ denotes the direct survey estimate of unobservable population value θ_d for area or forest type d $(d = 1, \ldots, D)$. Let $\mathbf{x}_d$ be the p-vector of known auxiliary variable, often obtained from various administrative and census records, related to the population mean θ_d. The simple area specific two stage model suggested by Fay and Herriot (1979) has the form

$$\hat{\theta}_d = \theta_d + e_d \text{ and } \theta_d = \mathbf{x}_d^T \boldsymbol{\beta} + u_d, d = 1, \ldots, D. \quad (3.1)$$

We can express model (1) as an area level linear mixed model given by

$$\hat{\theta}_d = \mathbf{x}_d^T \boldsymbol{\beta} + u_d + e_d;\ d = 1, \ldots, D. \quad (3.2)$$

Here $\boldsymbol{\beta}$ is a p-vector of unknown fixed effect parameters, u_d's are independent and identically distributed normal random errors with $E(u_d) = 0$ and $Var(u_d) = \sigma_u^2$, and e_d's are independent sampling errors normally distributed with $E(e_d|\theta_d) = 0$, $Var(e_d|\theta_d) = \sigma_d^2$. The two errors are independent of each other within and across areas. Usually, σ_d^2 is known while σ_u^2 is unknown and it has to be estimated from the data. Methods of estimating σ_u^2 include maximum likelihood (ML) and restricted maximum likelihood (REML) under normality, the method of fitting constants without normality assumption, See Rao (2003, Chapter 5) and Chandra (2013). Let $\hat{\sigma}_u^2$ denotes estimate of σ_u^2. Then under model (2), the Empirical Best Linear Unbiased Predictor (EBLUP) of θ_d is given by

$$\hat{\theta}_d^{EBLUP} = \mathbf{x}_d^T \hat{\boldsymbol{\beta}} + \hat{\gamma}_q(\hat{\theta}_d - \mathbf{x}_d^T \hat{\boldsymbol{\beta}}) = \hat{\gamma}_d \hat{\theta}_d + (1 - \hat{\gamma}_d)\mathbf{x}_d^T \hat{\boldsymbol{\beta}} \quad (3.3)$$

where, $\hat{\gamma}_d = \hat{\sigma}_u^2/(\sigma_d^2 + \hat{\sigma}_u^2)$ and $\hat{\boldsymbol{\beta}}$ is the generalized least square estimate of $\boldsymbol{\beta}$. It may be noted that $\hat{\theta}_d^{EBLUP}$ is a linear combination of direct estimate $\hat{\theta}_d$ and the model based regression synthetic estimate $\mathbf{x}_d^T \hat{\boldsymbol{\beta}}$, with weight $\hat{\gamma}_d$. Here $\hat{\gamma}_d$ is called "shrinkage factor" since it 'shrinks' the direct estimator, $\hat{\theta}_d$ towards the synthetic estimator, $\mathbf{x}_d^T \hat{\boldsymbol{\beta}}$. Prasad and Rao (1990) proposed an approximately model unbiased (i.e. with bias of order o(1/D)) estimate of mean squared error (MSE) of the EBLUP (3.3) given by

$$mse(\hat{\theta}_d^{EBLUP}) = g_{1d}(\hat{\sigma}_u^2) + g_{2d}(\hat{\sigma}_u^2) + 2g_{3d}(\hat{\sigma}_u^2)v(\hat{\sigma}_u^2), \quad (3.4)$$

where,

$g_{1d}(\hat{\sigma}_u^2) = \hat{\gamma}_d \sigma_d^2$,
$g_{2d}(\hat{\sigma}_u^2) = (1 - \hat{\gamma}_d)^2 \mathbf{x}_d^T v(\hat{\boldsymbol{\beta}}) \mathbf{x}_d$, and
$g_{3d}(\hat{\sigma}_u^2) = \{\sigma_d^4/(\sigma_d^2 + \hat{\sigma}_u^2)^3\} v(\hat{\sigma}_u^2)$

with $v(\hat{\sigma}_u^2) \approx 2D^{-2} \sum_{d=1}^{D} (\sigma_d^2 + \hat{\sigma}_u^2)^2$ when estimating $\hat{\sigma}_u^2$ by method of fitting constants. See Rao (2003, Chap. 5) for details about various theoretical developments.

4 Empirical Results

In this Section, some diagnostics to examine the reliability of small area estimates was carried out. Such diagnostics are suggested in Ambler et al. (2001), Chandra et al. (2011, 2018) and Das et al. (2019). The model-based small area estimates should be consistent with the unbiased direct estimates, be more precise than the direct survey estimates, and provide reasonable results to users. The values for the model-based small area estimates derived from the fitted model should be consistent with the unbiased direct estimates, wherever these are available, i.e. they should provide an approximation to the direct estimates that is consistent with these values being "close" to the expected values of the direct estimates. The model-based small area estimates should have mean squared errors significantly lower than the variances of corresponding direct estimates. For this purpose, the commonly used measures, a bias diagnostic and a percent coefficient of variation (% CV) diagnostic was considered. The objective of bias diagnostic is to examine if the model-based small area estimates are less extreme on the basis of comparison to the direct estimates, if available. If direct estimates are unbiased, their regression on the true values should be linear and correspond to the identity line. Further, if model-based small area estimates are close to the true values, the regression of the direct estimates on these model-based estimates should be similar. Therefore, direct estimates on the y-axis and corresponding model-based small area estimates on x-axis was plotted and we look for divergence of the fitted least squares regression line from the $y = x$ and test for intercept = 0 and slope = 1. Here, the straight line found by regressing the direct estimate against the model-based estimate provides an adequate fit of the small area estimates. The bias scatter plot of the district level direct estimates against the corresponding model-based small area estimates of total basal cover (m^2/ha) for tree (left plot), shrub (centre plot) and herbs (right plot) is given in Fig. 4, with fitted least squares regression line (dotted line) and line of equality (solid line) superimposed. The bias diagnostic plot in Fig. 4 indicates that the forest type-wise model-based estimates generated by SAE method given in Eq. (3.3) are less extreme when compared to the direct estimates, demonstrating the typical SAE outcome of shrinking more extreme values towards the average. The estimates of total basal cover (m^2/ha) for tree, shrub and herbs generated by SAE method given by expression (3.3) lies along the line $y = x$ for most of the small areas or forest types which indicates that they are approximately design unbiased. This is expected, since the SAE estimates are random variables and so the regression of the direct estimates on the SAE estimates is unbiased for a test of common expected values.

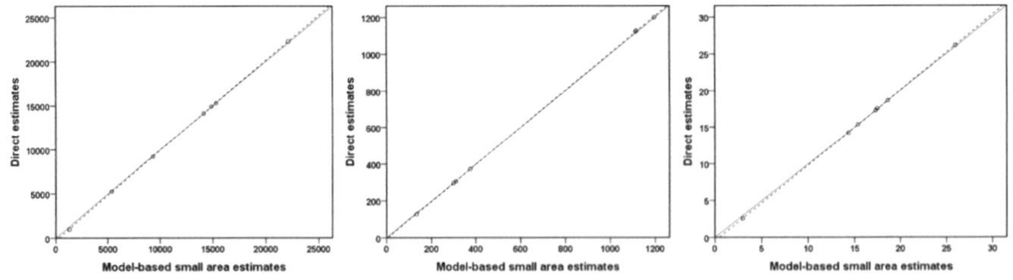

Fig. 4 Bias diagnostics plots with $y = x$ line (solid line) and regression line (dotted line) model-based small area estimate of total basal cover (m^2/ha) for tree (left), shrub (centre) and herbs (right)

We compute the % CV to assess the comparative precision of model-based small area estimates and direct estimates. The CVs show the sampling variability as a percentage of the estimate where % CV = (SE/Estimate) × 100. Estimates with large CVs are considered unreliable (i.e. smaller is better). In general, there are no internationally accepted tables available that allow us to judge what is "too large". Different organization used different cut off for CV to release their estimate for the public use. The % CV of direct and model-based small area estimates of total basal cover (m^2/ha) are given in Tables 1, 2 and 3 for trees, shrubs and herbs respectively. In Tables 1, 2 and 3, it can be observed that the CVs of SAE is smaller than those

Table 1 Forest type-wise estimates and % CV of total basal cover (m^2/ha) for trees generated from direct and model-based SAE methods

Trees		Direct method		SAE method	
Forest type	Sample size	Estimates	CV (%)	Estimate	CV (%)
1	4	22,300	25.8	22,090	4.2
2	1	14,941	–	14,863	12.5
3	3	5,275	44.7	5,398	20.0
4	12	15,286	26.7	15,299	3.5
5	1	1,946	–	2,542	73.0
6	11	14,130	13.1	14,104	4.0
7	1	9,233	–	9,322	19.9
Minimum		1,946	13.1	2,542	3.5
Maximum		22,300	44.7	22,090	73.0
Average		11,873	27.6	11,946	19.6

Table 2 Forest type-wise estimates and % CV of total basal cover (m^2/ha) for shrubs generated from direct and model-based SAE methods

Shrubs		Direct method		SAE method	
Forest type	Sample size	Estimates	CV (%)	Estimate	CV (%)
1	4	306	64.7	308	12.4
2	1	1,128	–	1,117	6.8
3	3	1,200	32.5	1,196	3.7
4	12	297	18.5	297	7.4
5	2	128	4.8	135	4.1
6	11	374	31.5	375	6.1
7	1	1,123	–	1,113	6.8
	Minimum	128	4.8	135	3.7
	Maximum	1,200	64.7	1,196	40.1
	Average	651	30.4	649	11.9

Table 3 Forest type-wise estimates and % CV of total basal cover (m^2/ha) for herbs generated from direct and model-based SAE methods

Herbs		Direct method		SAE method	
Forest type	Sample size	Estimates	CV (%)	Estimate	CV (%)
1	4	17.3	11.1	17.3	5.5
2	1	17.6		17.5	10.9
3	3	26.2	12.5	26.0	4.2
4	12	15.3	23.3	15.3	3.6
5	2	2.6	100.0	3.0	45.4
6	11	18.6	20.1	18.6	3.1
7	1	14.2		14.3	13.2
	Minimum	2.6	11.1	3.0	3.1
	Maximum	26.2	100.0	26.0	45.4
	Average	16.0	33.4	16.0	12.3

computed for direct estimates. Hence the estimates generated using SAE are more reliable than the direct estimate.

Basal cover in Tropical Semi-Evergreen Forest is found maximum (22,090 m^2/ha) for trees whereas that for shrubs is comparatively very less. This is due to that the shrubs are not growing well under the dense and evergreen forest and poor regeneration of trees. However, the herbs have reasonable growth (17.3 m^2/ha) in the forest type. For Littoral and Swamp Forest, the basal cover of the shrubs (1196 m^2/ha) and herbs (26 m^2/ha) was found maximum, whereas that for trees is very less (5398 m^2/ha). The reason behind it may be that the forest is mainly dominated by the mangrove trees. The minimum basal cover of trees is obtained in the Tropical Thorn Forest, viz. (2542 m^2/ha). In this forest type, the basal cover of herbs is also minimum (only 3 m^2/ha) and interestingly, that of shrubs is also limited. The main reason for this is the degraded forest due to scanty rainfall and degraded soil. If we see in the Tropical Moist Deciduous Forest, the estimates of basal cover showed that there is almost average basal cover for each of the category i.e. trees, shrubs and herbs. This indicates the rich biodiversity of plant species. Therefore, the maximum tiger population of Maharashtra is found in these areas.

From these observations, except the forest types Tropical Thorn and Tropical Moist Deciduous, one may see that the basal cover of trees is inversely proportional to that of shrubs and herbs.

5 Conclusions

Remarkable theoretical research has taken place in SAE in comparison with real applications. This is now high time for developed SAE methods to be applied and implemented in solving real life problems. In particular, to the best of our knowledge there is not much published research work in application of SAE in Indian forestry data. This chapter illustrates that the SAE technique can be satisfactorily applied to produce reliable small area (forest type) level estimates of forestry related parameter namely total basal cover using existing survey data. The results clearly show that forest type-wise estimates of total basal cover of trees, shrubs and herbs generated by using SAE approach are reliable as compared to direct estimates. State level estimates often mask variations at the disaggregate level and render little information for focused level planning and allocation of resources. SAE techniques can reveal striking differences and point to specific geographical areas where policy intervention should be strengthened. The results show the advantage of using SAE technique to cope up the small sample size problem in producing the reliable forest type-wise estimates and confidence intervals. The small area estimates are invaluable policy information for state forest department and local resource managers.

Acknowledgements The authors would like to acknowledge the valuable comments and suggestions of the reviewers. These led to a considerable improvement in the chapter. The work of Hukum Chandra was carried out under an ICAR-National Fellow Project at ICAR-IASRI, New Delhi, India.

References

Ambler, R., Caplan, D., Chambers, R., Kovacevic, M., & Wang, S. (2001). Combining unemployment benefits data and LFS data to estimate ILO unemployment for small areas: An application of a modified Fay-Herriot Method. In: *Proceedings of the International Association of Survey Statistician*. Meeting of the ISI: Seoul, August 2001.

Battese, G. E., Harter, R. M., & Fuller, W. A. (1988). An error component model for prediction of county crop areas using survey and satellite data. *Journal of the American Statistical Association, 83,* 28–36.

Breidenback, J., Northdurft, A., & Kändler, G. (2010). Comparison of nearest neighbor approaches for small area estimation of tree species-specific forest inventory attributes in central Europe using airborne laser scanner data. *European Journal of Forest Research, 129,* 833–846.

Champion, H. G., & Seth, S. K. (1968). *A revised survey of forest types of India*. Delhi: Govt. of India Press.

Chandra, H., Aditya, A., & Sud, U. C. (2018). Localised estimates and spatial mapping of poverty incidence in the state of Bihar in India—an application of small area estimation techniques. *PLoS ONE, 13*(6), e0198502.

Chandra, H. (2013). Exploring spatial dependence in area level random effect model for disaggregate level crop yield estimation. *Journal of Applied Statistics, 40,* 823–842.

Chandra, H., & Chandra, G. (2015). An overview of small area estimation techniques. In G. Chandra, R. Nautiyal, H. Chandra, N. Roychoudury, & N. Mohammad (Eds.), *Statistics in Forestry: Methods and Applications* (pp. 45–54). Coimbatoor: Bonfring Publication.

Chandra, H., Salvati, N., & Sud, U. C. (2011). Disaggregate-level estimates of indebtedness in the state of Uttar Pradesh in India-an application of small area estimation technique. *Journal of Applied Statistics, 38*(11), 2413–2432.

Das, S., Chandra, H., & Saha, U. R. (2019). District level prevalence of diarrhea disease among under-five children in Bangladesh: An application of small area estimation approach. *PLoS ONE, 14*(2), e0211062.

Fay, R. E., & Herriot, R. A. (1979). Estimation of income from small places: an application of James-Stein procedures to census data. *Journal of the American Statistical Association, 74,* 269–277.

FSI. (2009). *India's state of forest report 2009*, Forest Survey of India, Government of India, Dehradun, India.

FSI. (2017). *India's state of forest report 2017*, Forest Survey of India, Government of India, Dehradun, India.

Goerndt, M. E., Monleon, V., & Temesgen, H. (2010). Relating forest attributes with area-based and tree-based LiDAR metrics for western Oregon. *Western Journal of Applied Forestry, 25,* 105–111.

Katila, M., & Tomppo, E. (2005). Empirical errors of small area estimates from the multisource national forest inventory in eastern Finland. *Silva Fennica, 40*(729), 742.

McRoberts, R. E. (2012). Estimating forest attribute parameters for small areas using nearest neighbors techniques. *Forest Ecology and Management, 272,* 3–12.

Ohmann, J. L., & Gregory, M. J. (2002). Predictive mapping of forest composition and structure with direct gradient analysis and nearest neighbor imputation in coastal Oregon. U.S.A. *Canadian Journal of Forest Research, 32,* 725–741.

Pfeffermann, D. (2002). Small area estimation: new developments and directions. *International Statistical Review, 70,* 125–143.

Prasad, N. G. N., & Rao, J. N. K. (1990). The estimation of the mean squared error of the small area estimators. *Journal of the American Statistical Association, 85,* 163–171.

Rao, J. N. K. (2003). *Small area estimation*. New York: Wiley.

Tomppo, E. (2006). The Finnish national forest inventory. In A. Kangas & M. Maltamo (Eds.), *Forest Inventory: Methodology and Applications*. Dordrecht, the Netherlands: Springer.

Estimation of Abundance of Asiatic Elephants in Elephant Reserves of Kerala State, India

M. Sivaram, K. K. Ramachandran, E. A. Jayson and P. V. Nair

Abstract In this chapter, we presented the sampling methods employed for assessing the abundance of elephants in the four Elephant Reserves of Kerala State, India. Three different sampling methods, sampling block count and direct sighting and dung survey using line transect sampling were used. Dung survey appeared to be better method than the others. The total estimated elephant population using dung survey for all the reserves was 7490 with 95% confidence limit of 6280–9053. However, more studies to obtain defecation and decay rates in varying climatic conditions will help improving the accuracy of the estimates of the elephant population.

Keywords Animal density · Direct sighting · Dung survey · Line transect sampling · Wildlife population

1 Introduction

Elephant (*Elephas maximus*) is the largest land mammal and though occur in low numbers, contribute considerably to the total biomass of the system due to its large size. The population of wild elephants in India is about 30,000 (MoEF&CC 2018). Due to habitat fragmentation, elephants move out to agricultural land parcels leading to man–elephant conflict (Sukumar 2006; MoEF&CC 2017). Project Elephant is a conservation scheme of the Government of India launched to plan at landscape-scale for long-term viability of the elephant habitats and populations (MoEF&CC 2018). There are about 30 Elephant Reserves in India. Monitoring of elephant populations and their demographic characteristics such as age structure and sex ratio across landscapes is an important task for sustainable conservation and management of elephants. Commonly used methods to assess the abundance of wild elephants

M. Sivaram (✉)
Southern Regional Station, ICAR-National Dairy Research Institute, Bangalore, India
e-mail: sivaram.ndri@gmail.com

K. K. Ramachandran · E. A. Jayson · P. V. Nair
Wildlife Division, Kerala Forest Research Institute, Thrissur District, Kerala, India

G. Chandra et al. (eds.), *Statistical Methods and Applications in Forestry and Environmental Sciences*, Forum for Interdisciplinary Mathematics,
https://doi.org/10.1007/978-981-15-1476-0_17

include Total count, Sample block count, Waterhole count, Line transect sampling and Dung survey.

Based on the recommendation of the State Wildlife Advisory Board, Government of Kerala took a policy decision in 1993, to estimate the wildlife population periodically for the purpose of monitoring. The estimation results of wildlife population in Kerala State have been published from time to time covering major mammals of Kerala Forests (KFRI 1993; Easa and Jayaraman 1998; Easa et al. 2002; Sivaram et al. 2013), exclusively elephants as part of All India Synchronized Population Estimation of Elephants (Sivaram et al. 2006, 2007, 2010) and tigers, co-predators, prey and their habitat (Jhala et al. 2008, 2011). The present chapter elaborates the sampling methods employed for estimating elephant population in different Elephant Reserves of Kerala in 2011.

2 Materials and Methods

2.1 Sample Block Count Method—Direct Sighting

The forest area of each Protected Area/Territorial Forest Division in each Elephant Reserve (Fig. 1) was divided into number of small blocks based on identifiable physical features in forests such as stream, hill, etc., utilizing the toposheets of the Survey of India. In the present study, all the blocks were demarcated on geographic information system (GIS) platform and area of the individual blocks was measured. A systematic random sample of blocks was chosen in each Protected Area/Territorial Forest Division for enumeration. The total number of blocks sampled across different Elephant Reserves of Kerala is presented in Table 1.

In the present population estimation, in each of the selected sample blocks, search was made for the presence of selected wild animals including elephants by perambulation from 06.00 to 18.00 h. The search team for each blocks consisted of a trained volunteer, a forest staff and a tribal tracker. The animals sighted while traversing the area were counted in each block. Other details recorded were habitat type(s) of the sample blocks and the age–sex distribution of the animals sighted. The estimated elephant population in ith Elephant Reserve $(\widehat{N_i})$ is given by Lahiri-Choudhury (1991) as:

$$\widehat{N_i} = A_i \widehat{D_i}, \quad i = 1, 2, ..p$$

The sum of the estimates across Elephant Reserves provided state-level estimate. Standard error (SE) of $\widehat{N_i}$ is

$$\text{SE}(\widehat{N_i}) = \sqrt{\frac{A_i^2}{n_i(n_i - 1)} \sum_{j=1}^{n_i} \frac{\left(y_{ij} - y_i\right)^2}{\left(x_{ij} - x_i\right)^2}}$$

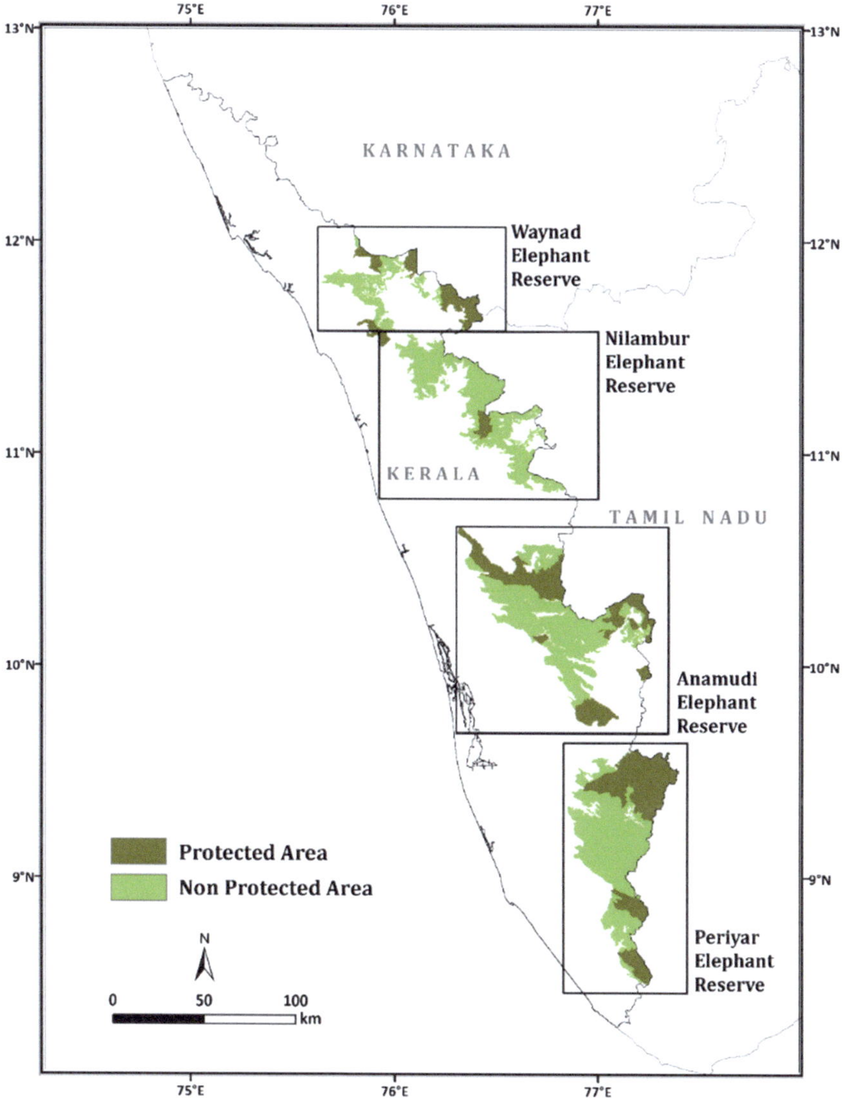

Fig. 1 Map showing Elephant Reserves of Kerala

95% confidence limit for $\widehat{N_i} = \widehat{N_i} \pm 1.96\,\mathrm{SE}(\widehat{N_i})$
where

$\widehat{D_i}$ = Estimated density of animals (No. of animals/km^2) in the ith Elephant Reserve
A_i = Total area under effective elephant habitat of the ith Elephant Reserve (after accounting for area devoid of elephants)
n_i = Number of blocks sampled in the ith Elephant Reserve

Table 1 Total number of sampled blocks and their area

Elephant Reserve	Number of blocks sampled	Total area sampled (km^2)
Wayanad	95	834.32
Nilambur	103	957.12
Anamudi	199	1859.00
Periyar	225	1864.49
Total	622	5514.94

y_{ij} = Number of elephants sighted in the *j*th block of the *i*th Elephant Reserve
y_i = Total number of elephants in sample blocks of the *i*th Elephant Reserve
x_{ij} = Area of *j*th sample block of the *i*th Elephant Reserve
x_i = Total area of the sample blocks in the *i*th Elephant Reserve
p = Number of Elephant Reserves.

2.2 *Line Transect Sampling (Direct Sighting)*

Line transect sampling technique has been in use for estimating the size of the biological populations such as deer (White et al. 1989), ungulates (Forcard et al. 2002), primates (Buckland et al. 2010a, b, 2015) and elephants (Easa et al. 2002). It is preferred over the other methods as it has a scientific basis to develop estimates of animal density based on detection probabilities of animals even without encountering all the animals in the study area (Burnham et al. 1980; Buckland et al. 2001). In line transect sampling, the observer(s) perform a standardized survey along a series of lines with total line length of L, searching for animals (clusters/herds/troop/groups) of interest (Fig. 2). The radial distance (r) from the observer to the geometric center of the animal herds along with the angle of sighting (θ) are recorded and perpendicular distances (x) are worked out; $x = r\ \sin(\theta)$. Animal density is estimated by using the formula $\widehat{D} = \frac{n\hat{f}(0)}{2L}$ (Buckland et al. 2015).
where

$$\hat{f}(0) = \frac{1}{\int_0^w g(x)}$$

$g(x)$ = the probability density function of detecting an animal (herd) in the survey area
x = perpendicular distance (m)
L = transect length (km)
w = perpendicular distance to animals.

The approximate variance of $\widehat{D}$ is

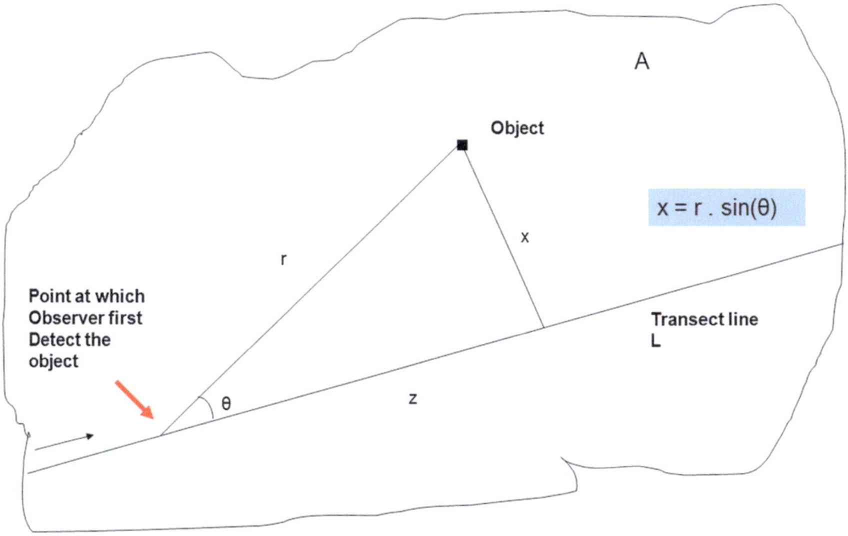

Fig. 2 Method of line transect sampling—direct sighting

$$\widehat{\mathrm{Var}}\left(\widehat{D}\right) = \widehat{D}^2 \left\{ \frac{\widehat{\mathrm{var}}(n)}{n^2} + \frac{\widehat{\mathrm{var}}\left[\widehat{E}(s)\right]}{\left[\widehat{E}(s)\right]^2} + \frac{\widehat{\mathrm{var}}\left[\hat{f}(0)\right]}{\left[\hat{f}(0)\right]^2} \right\}$$

where

n = number of herd/troop/groups encountered
s = expected herd/troop/group size.

An approximate $100(1 - 2\alpha)\%$ confidence interval is given by

$$\widehat{D} \pm Z\alpha\sqrt{\widehat{V}\left(\widehat{D}\right)}$$

($z = z_{0.025} = 1.96$ for a 95% confidence level).

Some of the functional forms of $g(x)$ considered include uniform, half normal, negative exponential and hazard rate. In order to improve the fit of the model, especially, the tail part of the detection curves, series expansion such as simple polynomial or higher-order polynomial is added with the key detection function. One among the functional forms is chosen based on fit statistics/model selection criterion such as Chi-square test, Akaike information criterion (AIC) and Bayesian information criterion (BIC) (Buckland et al. 2001, 2004).

In the present study, the line transect sampling was adopted in the selected sample blocks. In each block, a transect of about 2 km length was laid. These transects were covered on foot recording the sighting distance (r) and the sighting angle (θ)

Table 2 Number of line transects, total length and number of detections in direct sightings

Elephant Reserve	No. of transects	Total length (km)	No. of detections
Wayanad	95	189.05	27
Nilambur	103	205.25	17
Anamudi	199	394.24	80
Periyar	225	449.60	54
Total	622	1238.14	178

to the elephant or geometric center of the elephant herds. Ocular estimation of the sighting distance was made. The sighting angle was measured with a compass. These measurements formed the input data for the estimation of elephant density. The details of line transects employed for the population estimation are presented in Table 2.

The radial distances (r) to the elephant herds and angle of sighting (θ) formed the input data for the estimation of elephant density.

The density estimates were obtained by using the software DISTANCE 6.0 developed by Thomas et al. (2010). The univariate half-normal distribution with the series expansion of simple polynomial was used for estimating the elephant density. A 5% truncation of the largest distance values was adopted to improve the precision of the density estimates. The density estimates were developed after adjusting for herd size bias. The herd size estimation was done by regressing distance function $g(x)$ on logarithm of herd size. Whenever the regression approach was not possible, mean/median herd size was used.

2.3 Dung Survey Using Line Transect Sampling

Population estimation methods based on direct sightings of animals especially elephants suffer due to scattered occurrence, group behavior and its vast home range (Jachmann 1991). The indirect evidences of animals such as droppings (pellets and dung) present in the area survive for a considerable time period which can be used for the estimation of animal density (Marques et al. 2001). Standing crop and clearance plot methods are the possible indirect methods for estimating the elephant population. The clearance plot method involves clearing dung from marked plots at regular intervals, counting the dung piles, and correcting the counts by the defecation rate (Staines and Ratcliffe 1987). The standing crop method is the most commonly used method. One of the assumptions of this method is that there is a stable relationship between the amount of dung present and the number of elephants (Barnes and Barnes 1992). The elephant density using the method involves one-time survey of dung for estimating dung density which is corrected by defecation and decay rate. Estimated density of elephants, $\widehat{D_e}$ is given by

$$\widehat{D_e} = \frac{\text{Dung density}}{\text{Defecation rate}} \times \text{Dung decay rate}$$

Dung density (number of dung per unit area) is usually estimated through dung surveys using quadrat sampling, strip transect sampling or line transect sampling. Defecation rate (number of dung piles defecated per day per animal) can be estimated by monitoring captive elephants or by placing a known number of elephants in an enclosure previously cleared off dung and estimating the number of dung produced over a fixed time period. The dung decay rate is defined as the number of dung decayed per day and is expressed as the reciprocal of the estimated mean time to decay (Barnes and Barnes 1992). The literature on dung decay experiments are available (Barnes and Jenson 1987; Barnes and Barnes 1992; Laing et al. 2003; Sivaram et al. 2016).

In this study, the technique of line transect sampling was adopted in the sampled blocks for estimating the dung density of elephant. In each block, the transects were covered on foot recording the perpendicular distance to the geometric center of the dung piles of elephant. The perpendicular distance was measured using a tape. The number of line transects and total transect length used for the estimation are presented in Table 3.

The perpendicular distances to dung piles formed the input data for the estimation of dung density of elephants. Univariate half-normal distribution with the series expansion of simple polynomial was used as detection function for estimating the dung density. A 5% truncation of the largest perpendicular distance values was adopted to improve the precision of the density estimates. The defecation rate of 16.33 per day, as obtained from wild elephants in Mudumalai by Watve (1992), was used in the above formula. As far as dung decay rate is concerned, the rate 0.0102 per day, obtained from dung decay experiments conducted in Wayanad Elephant Reserve in the year 2005 was used. The elephant population in each Elephant Reserve was estimated by multiplying density estimates with their respective extent of elephant habitat.

Table 3 Number of line transects for the estimation of dung density

Elephant Reserves	Total no. of transects	Total length of transects (km)	No. of dung piles recorded
Wayanad	95	189.05	1941
Nilambur	103	205.25	1215
Anamudi	199	394.24	2571
Periyar	225	449.60	2003
Total	622	1238.14	7732

3 Results

3.1 Sample Block Count

The estimated elephant population based on sample block count method for various Elephant Reserves is given in Table 4 (also depicted in Fig. 3). The total number of elephant sightings using sample block count method was 1357 in 2011 and 1911 in 2010 while using total count method it was 2296 in 2002. The extrapolated number of elephants for the State was 1958 in 2011 against 3520 in 2010. The population estimation in 2010 was exclusively for elephants. In the year 2011, the focus was for all the major mammals including elephants and thus lesser number of detections and decrease in the estimate of total elephant population. The area sampled also varied from 4474 km^2 in 2010 to 5515 km^2 in 2011. In 2011, all the block maps were digitized and exact area statistics was calculated using GIS, to work out area sampled. Due to the use of digitized maps, there was an upside revision in the area sampled and thus affecting the estimation of elephant density and elephant population. Besides above, the variations in weather and field conditions might have also contributed to the differences in number of detections.

The highest elephant density is usually found in Wayanad Elephant Reserve. On the contrary, in 2011, the highest was in Anamudi Elephant Reserve followed by Wayanad, Periyar and Nilambur Elephant Reserve (Table 4). In terms of total number of elephants, Anamudi Elephant Reserve ranked first followed by Periyar, Wayanad and Nilambur Elephant Reserve. The population characteristics of the elephants sighted in different Elephant Reserves of Kerala are presented in Table 5. Lower confidence limit (LCL) and upper confidence limit (UCL) were worked out at 5% level of significance for all the methods.

3.2 Line Transect Sampling—Direct Sighting

The herd and elephant density are presented in Table 6. The highest elephant density was in Anamudi followed by Wayanad, Periyar and Nilambur.

Table 4 Estimated elephant population using sample block count method

Elephant Reserve	Effective elephant habitat (km^2)	Number of elephants sighted	Elephant density (No./km^2)	Extrapolated number of elephants	SE	LCL	UCL
Wayanad	934.160	225	0.2697	252	25.99	20.00	302.88
Nilambur	1142.30	167	0.1745	199	19.74	160.60	238.03
Anamudi	2817.45	555	0.2985	841	59.78	724.00	958.31
Periyar	3026.41	410	0.2199	666	44.47	578.30	752.66
Total	7920.32	1357		1958		1664.00	2251.88

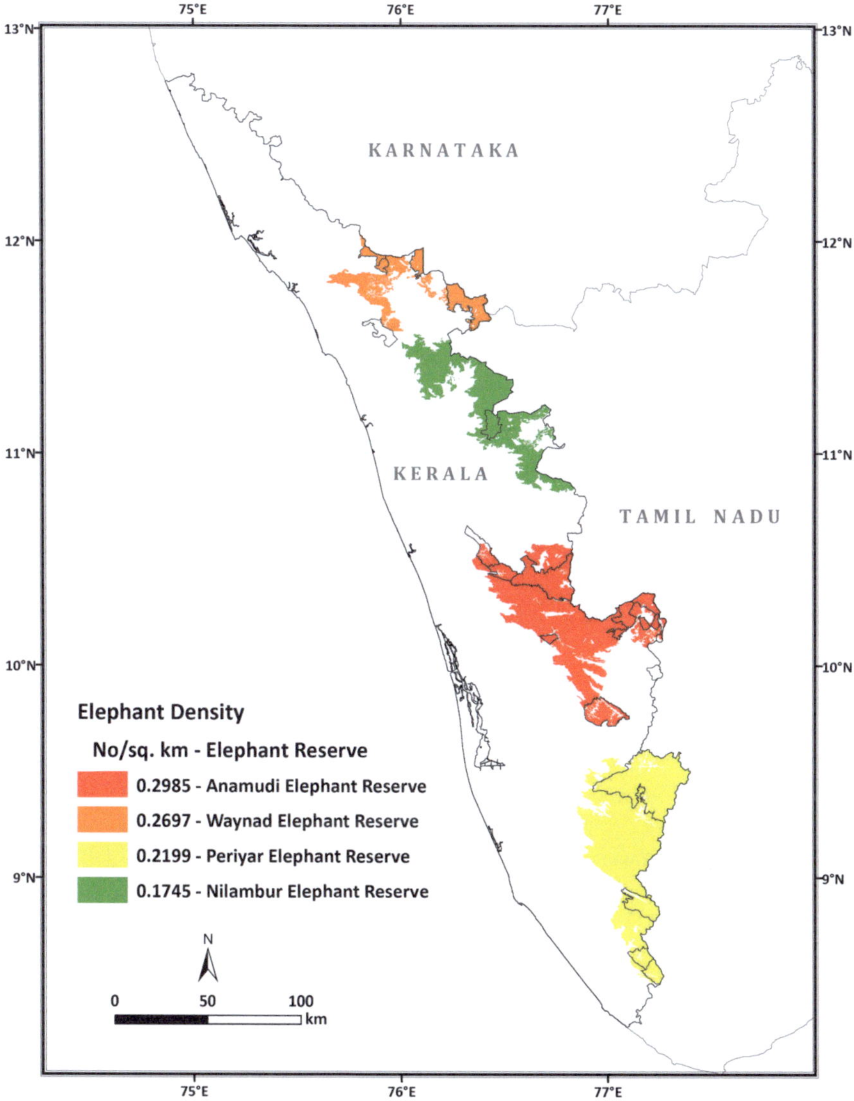

Fig. 3 Density of elephants using sample block count method

3.3 *Line Transect Sampling—Dung Survey*

The dung density estimates for various Elephant Reserves are presented in Table 7. The estimated density and population of elephants in various Elephant Reserves are presented in Table 8 and Fig. 4. The highest elephant density was in Wayanad followed by Anamudi, Periyar and Nilambur. Anamudi had the highest elephant population followed by Periyar, Wayanad and Nilambur.

Table 5 Population characteristics of elephants using sample block count method

Population characteristics	Wayanad	Nilambur	Anamudi	Periyar	State level
Adult male: adult female (bull: cow)	01:01.7	01:01.5	01:03.3	01:01.7	01:02.1
Sub-adult male: sub-adult female	01:01.3	01:01.4	01:02.2	01:01.0	01:01.5
% Tusker in adult and sub-adult population	33.86	37.59	21.46	34.50	29.38
Adult cow: calf	4.19:1	2.35:1	4.14:1	3.19:1	3.55:1
Tusker: makhna in adults	25:1	0	71:1	43:1	49.4:1

Adult male: >240 cm of height; Adult female: >210 cm of height; Sub-adult male: 151–240 cm of height; Sub-adult female: 151–210 cm of height; Calf: <120 cm of height

Table 6 Estimates of elephant density using line transect sampling—direct sightings

Elephant Reserve	Herd density herds/km^2	LCL	UCL	% CV	[a]Elephant density (No./km^2)	LCL	UCL
Wayanad	0.8782	0.5390	1.4310	25.10	2.6346	1.6169	4.2930
Nilambur	0.5076	0.2869	0.8980	29.34	1.5228	0.8608	2.6939
Anamudi	0.9424	0.6868	1.2930	16.20	2.8271	2.0604	3.8790
Periyar	0.5843	0.4138	0.8249	17.67	1.7528	1.2414	2.4747
State level	0.7281	0.5857	0.9051	11.11	2.1843	1.7571	2.7154

[a]Elephant density was worked out by multiplying the herd density by median herd size, three elephants per herd

Table 7 Estimated dung density using line transect—dung survey

Elephant Reserves	Dung density (No./km^2)	% CV	SE	LCL	UCL
Wayanad	2829.50	11.00	311.20	2276.70	3516.50
Nilambur	1336.90	10.90	145.90	1077.90	1658.10
Anamudi	1863.50	9.13	170.11	1557.30	2230.00
Periyar	849.44	8.39	71.26	720.32	1001.70

Table 8 Density and population of elephant based on dung density estimates

Elephant Reserves	Actual extent of elephant habitat (km^2)	Elephant density (No./km^2)	No. of elephants	LCL	UCL
Wayanad	934.16	1.7674	1651	1328	2052
Nilambur	1142.30	0.8351	954	769	1183
Anamudi	2817.45	1.1640	3279	2741	3924
Periyar	3026.41	0.5306	1606	1362	1894
Total	7920.32		7490	6200	9053

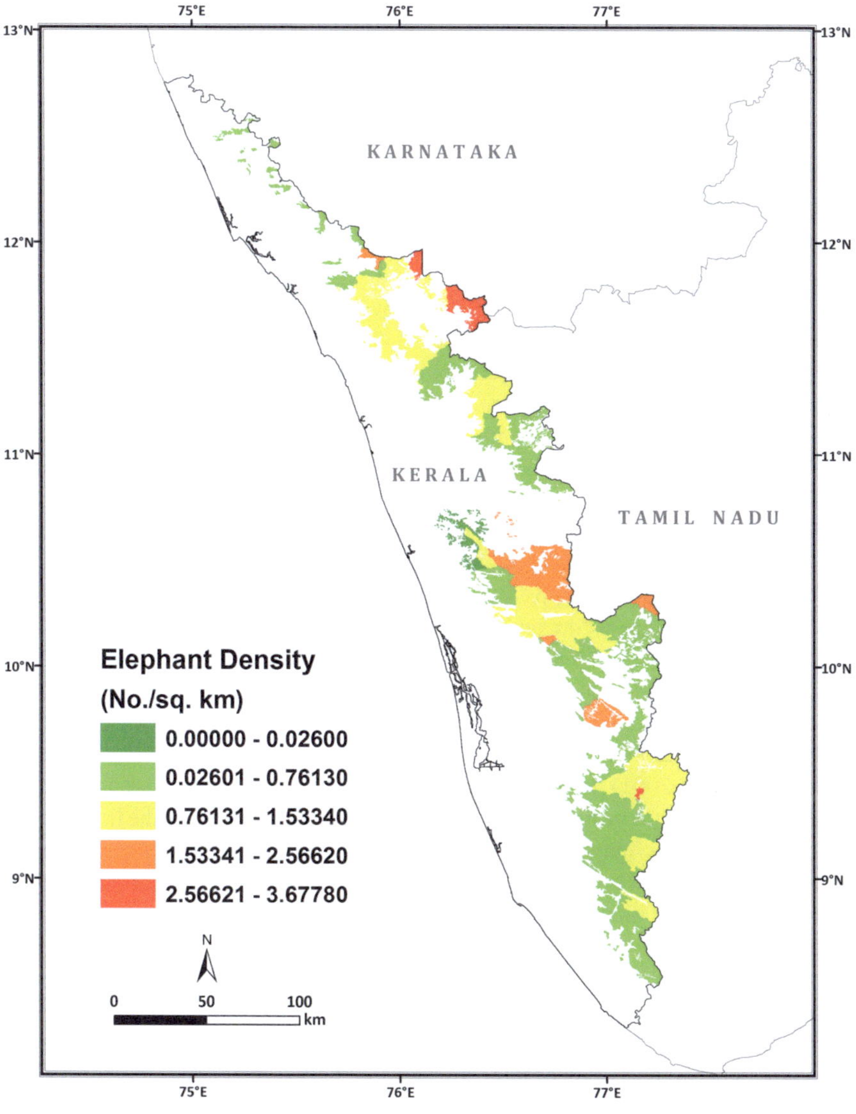

Fig. 4 Density of elephants using line transect sampling—dung survey

4 Discussion and Conclusions

The population estimation in 2011 covered major mammal species including elephants, while the estimation in 2010 was exclusively for elephants. The possible reason for the decrease in elephant population in 2011 is given in Sect. 3.1. We have not used line transect sampling—direct sighting for estimating the elephant population as it provided inflated population figures. The estimated elephant population using dung survey was 7490.

The sample block count method which is based on elephants sighted while traversing the sample block area may suffer from under counting and repetitive counting of elephants. A sampling frame of blocks was formed based on identifiable physical features in forests such as stream, hill, etc. Some of the sample blocks chosen were large in size and have difficult terrains and thus lesser geographical coverage while traversing. This might lead to under estimation of the population. The solitary elephants behind the bushes could not have been counted. The elephants have large home range and move across the blocks. This might lead to repetitive counting across the chosen blocks. The sample block count method may not be acceptable for the population estimation as it is not based on detection probability of the animals. Nonetheless, the sample block count method provides information on the characteristics of the population such as sex ratio among adults and sub adults, % tuskers and number of makhnas.

The line transect sampling—direct sighting has greater advantages than the sample block count method (Sivaram et al. 2016, 2018). However, there are many practical difficulties. Line transect may pass through different local forest conditions and thus may provide imprecise estimates. Detection probability of elephants varies across forest types, therefore, in thick forests (say, evergreen forests), the visibility of elephants at longer distance may lead to under estimation. The key assumptions of line transect sampling are that the detection probability $g(x)$ is the same for all the animals and that 100% of the animals are detected along the transect line, i.e., $g(0) = 1$ (Buckland et al. 2001). In the case of elephants, these assumptions are likely to be violated as there could be instances in which sighting of elephants on the transect line and closer to the line may danger to the enumerator. Such difficulties may produce imprecise estimates. Though line transects are cut before the estimation exercise, the enumerators are likely to deviate from walking on the line when elephants are sighted for personal security reasons. The elephant density using line transect sampling—direct sighting was overestimated when compared with the other methods and thus affecting the estimate of the total elephant population.

The estimation of elephant population using dung survey involves three estimates viz., dung density, decay rate of dung piles and defection rate. Estimation of each of these involves their own assumptions and practical difficulties (Barnes and Barnes 1992; Laing et al. 2003). The presence of dung piles and detection probability depend on the level of rain fall preceding the date of dung survey (Sivaram et al. 2016; Barnes et al. 1997). Dung decay rate varies across sites (Varman et al. 1995; Sivaram et al. 2018). The dung decay rates are mostly borrowed from other studies for the population estimation. For example, the present exercise used the dung decay rate estimated in parts of Wayanad Elephant Reserve during the year 2005 (Sivaram et al. 2006). It is advisable that the dung decay experiments be conducted during every population estimation across Elephant Reserves for obtaining reliable estimates. Similarly, the defecation rate used in the formula of elephant density was from Watve (1992) based on his observations in Mudumalai Sanctuary.

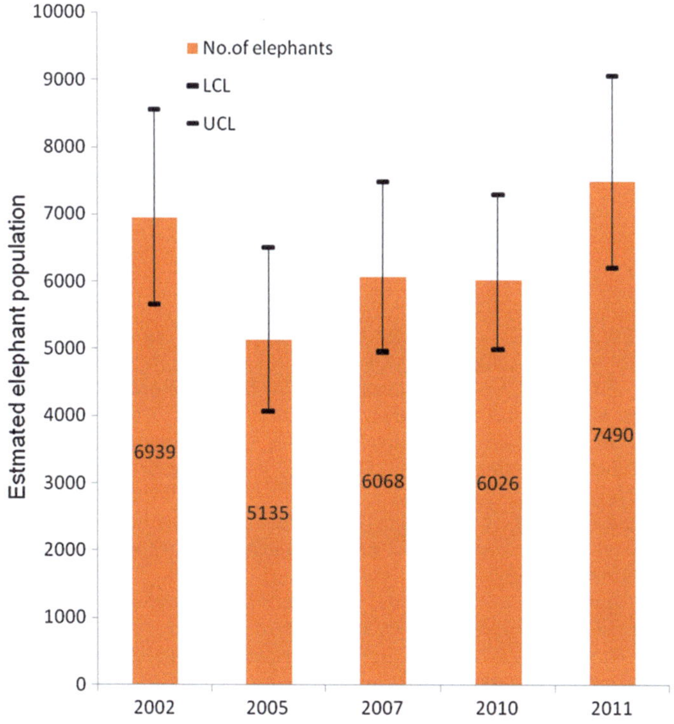

Fig. 5 Trends in estimated elephant population using line transect—dung survey

In the computation of confidence intervals for the population estimate, the standard errors of dung density alone were used. If the standard errors of the other parameters are included, the confidence intervals of the population estimates would be to an unacceptable range. This limits to reach any conclusion on population trends as to whether the elephant population is on increase or decrease (Fig. 5).

The estimates of the three methods need not agree with each other due to assumptions and limitations associated with each of the methods. The validation of the methods for elephants at landscape level is not possible as the elephants are long-ranging animals. Among the three methods used for the estimation of population of elephants, the dung survey appeared to be better method than the other two methods. However, more studies to obtain defecation rates and decay rates in varying climatic conditions will help improving the accuracy of the elephant density estimates. Proper training to field officers will also greatly improve the estimates. Despite several limitations of the methods, efforts in accurately estimating the population estimation should be continued for policy making and effective wildlife management.

Acknowledgements The authors thankfully acknowledge the excellent guidance and support provided by the Officers of the Kerala Forests and Wildlife Department. The financial assistance for the Population Estimation of Wildlife was provided by the Kerala Forests and Wildlife Department, Government of Kerala. The authors are also thankful to the reviewers for their useful suggestions which improved the readability of this chapter.

References

Barnes, R. F. W., & Barnes, K. L. (1992). Estimating decay rates of elephant dung-piles in forests. *African Journal of Ecology, 30*(4), 316–321.

Barnes, R. F. W., & Jenson, K. L. (1987). How to count elephants in forests. IUCN African Elephant and Rhino Specialist Group. *Technical Bulletin, 1*, 1–6.

Barnes, R. F. W., Asamoah-Boateng, B., Majam, J. N., & Agyei-Ohemeng, J. (1997). Rainfall and the population dynamics of elephant dung-piles in the forests of Southern Ghana. *African Journal of Ecology, 35*(1), 39–52.

Buckland, S. T., Anderson, D. R., Burnham, K. P., Laake, J. L., Borchers, D. L., & Thomas, L. (2001). *Introduction to distance sampling: Estimating abundance of biological populations*. New York: Oxford University Press.

Buckland, S. T., Anderson, D. R., Burnham, K. P., Laake, J. L., Borchers, D. L., & Thomas, L. (2004). *Advanced distance sampling*. New York: Oxford University Press Inc.

Buckland, S. T., Plumptre, A. J., Thomas, L., & Rexstad, E. A. (2010a). Line transect sampling of primates: Can animal-to-observer distance methods work? *International Journal of Primatology, 31,* 485–499.

Buckland, S. T., Plumptre, A. J., Thomas, L., & Rexstad, E. A. (2010b). Design and analysis of line transect surveys for primates. *International Journal of Primatology, 31,* 833–847.

Buckland, S. T., Rexstad, E. A., Marques, T. A., & Oedekoven, C. S. (2015). *Distance sampling: Methods and applications*. Switzerland: Springer.

Burnham, K. P., Anderson, D. R., & Laake, J. L. (1980). Estimation of density from line transect sampling of biological populations. *Wildlife Monographs, 72,* 202p.

Easa, P. S., & Jayaraman, K. (1998). *Population estimation of major mammals in the forests of Kerala—1997—A report*. Peechi: Kerala Forest Research Institute.

Easa, P. S., Sivaram, M., & Jayson, E. A. (2002). *Population estimation of major mammals in the forests of Kerala—2002* (KFRI Consultancy Report Number 09). Peechi: Kerala Forest Research Institute.

Forcard, S., Isotti, R., & Tinell, A. (2002). Line transect estimates of ungulate populations in a Mediterranean forest. *Journal of Wildlife Management, 66,* 48–58.

Jachmann, H. (1991). Evaluation of four methods for estimating elephant densities. *African Journal of Ecology, 29,* 188–195.

Jhala, Y. V., Gopal, R., & Qureshi, Q. (2008). *Status of the tigers, co-predators, and prey in India*. New Delhi: National Tiger Conservation Authority, Govt. of India; Dehradun: Wildlife Institute of India.

Jhala, Y. V., Qureshi, Q., Gopal, R., & Sinha, P. R. (2011). *Status of the tigers, co-predators, and prey in India*. New Delhi: National Tiger Conservation Authority, Govt. of India; Dehradun: Wildlife Institute of India.

KFRI. (1993). *A report on wildlife census—Kerala—1993*. Peechi: Kerala Forest Research Institute.

Lahiri-Choudhury, D. K. (1991, January 2–10). Direct count of elephants in north-east India. In U. Ramakrishnan, J. A. Santosh, & R. Sukumar (Eds.), *Censuing elephants in forests. Proceedings of an International Workshop Southern India* (Technical Report No. 2) (pp. 33–45). Bangalore: Asian Elephant Conservation Centre of IUCN/SSC Asian Elephant Specialist Group, Centre for Ecological Sciences, Indian Institute of Science.

Laing, S. E., Buckland, S., Burns, R. W., Lambie, D., & Amphien, A. (2003). Dung and nest surveys: estimating decay rates. *Journal of Applied Ecology, 40,* 1102–1111.

Marques, F. C. C., Buckland, S. T., Goffin, D., Dixon, C. E., Borchers, D. L., Mayle, B. A., & Peace, A. J. (2001). Estimating deer abundance from line transect surveys of dung: Sika deer in southern Scotland. *Journal of Applied Ecology, 38*, 349–363.

MoEF&CC. (2017). *Synchronized elephant population estimation in India—2017*. New Delhi: Project Elephant Division, Ministry of Environment, Forest and Climate Change.

MoEF&CC. (2018). http://www.moef.nic.in/division/introduction-4. Accessed September 07, 2018.

Sivaram, M., Ramachandran, K. K., Nair, P. V., & Jayson, E. A. (2006). *Population estimation of wild elephants in the elephant reserves of Kerala State—2005* (KFRI Extension Report Number 18). Peechi: Kerala Forest Research Institute.

Sivaram, M., Ramachandran, K. K., Nair, P. V., & Jayson, E. A. (2007). *Population estimation of wild elephants in the elephant reserves of Kerala State—2007* (KFRI Extension Report Number 27). Peechi: Kerala Forest Research Institute.

Sivaram, M., Ramachandran, K. K., Jayson, E. A., & Nair, P. V. (2010). *Wild elephant census of Kerala State—2010* (KFRI Extension Report Number 52). Peechi: Kerala Forest Research Institute.

Sivaram, M., Jayson, E. A., & Ramachandran, K. K. (2013). *Population estimation of major mammals in the forests of Kerala—2011* (KFRI Extension Report Number 61). Peechi: Kerala Forest Research Institute.

Sivaram, M., Ramachandran, K. K., Jayson, E. A., & Nair, P. V. (2016). Evaluation of line transect sampling technique in estimating elephant abundance in forests using dung survey. *Indian Forester, 142*(10), 959–964.

Sivaram, M., Ramachandran, K. K., Jayson, E. A., & Nair, P. V. (2018). Statistical techniques for estimating the abundance of Asiatic elephants based on dung piles. In C. Sivaperuman & K. Venkataraman (Eds.), *Indian hotspots—Vertebrate faunal diversity, conservation and management* (pp. 377–397). Singapore: Springer.

Staines, B. W., & Ratcliffe, P. R. (1987). Estimating the abundance of red deer (*Cervus elaphus* L.) and roe deer (*Capreolus capreolus* L.) and their current status in Great Britain. In *Symposia of the Zoological Society of London* (Vol. 50, pp. 131–152).

Sukumar, R. (2006). A brief review of the status, distribution and biology of wild Asian elephants. *International Zoo Yearbook, 40,* 1–8.

Thomas, L., Buckland, S. T., Rexstad, E. A., Laake, J. L., Strindberg, S., Hedley, S. L., et al. (2010). Distance software: Design and analysis of distance sampling surveys for estimating population size. *Journal of Applied Ecology, 47,* 5–14.

Varman, K. S., Ramakrishnan, U., & Sukumar, R. (1995). Direct and indirect methods of counting elephants: A comparison of results from Mudumalai Sanctuary. In J. C. Daniel & S. H. Datye (Eds.), *A week with elephants: Proceedings of the international seminar on Asiatic elephants*. New Delhi: Oxford University Press.

Watve, M. (1992). *Ecology of host-parasite interaction in a wild mammalian host community in Mudumalai, Southern India* (Ph.D. thesis). Indian Institute of Sciences, Bangalore.

White, G. C., Bartmann, R. M., Carpenter, L. M., & Garrot, R. A. (1989). Evaluation of aerial line transects for estimating deer densities. *Journal of Wildlife Management, 53,* 625–635.

Short Note: *Integrated Survey Scheme to Capture Forest Data in Bangladesh*

Mohammad Amirul Islam

Introduction

Forestry is a complex area of research in terms of assessment of available resources, management of bio-diversity and ecology, evaluation of livelihood dependency of neighboring population and estimation of the effects of climate change. In recent years, investing and investigating on forests become more important to address growing environmental threats and disaster management, emerging concept of carbon trading and assessing the contribution of forestry to gross domestic products (GDP). Recently United Nations has approved sustainable development goals (SDGs) to be implemented by 193 member states. Among the listed goals the need to collect forestry related data to measure different SDG indicators under the goal 15 has been emphasized. Where, SDG 15 aims to "Protect, restore and promote sustainable use of terrestrial ecosystems, sustainably manage forests, combat desertification, and halt and reverse land degradation and halt biodiversity loss" (GoB 2018). Furthermore, there is a growing concern to offer sound livelihood among forest dependent people living in forest fringe areas to protect wild lives and endangered species. Distribution of forest areas is sometimes of non-homogeny pattern and there are spatial variations between segments of the forest in terms of resources availability, accessibility and intensity. Hence, any survey on forestry related aspects becomes complicated, costly and difficult to maintain representativeness. Furthermore, different purposes lead to different survey designs; hence there are fewer tendencies to use a common sampling frame. Bangladesh has about 9951.34 km^2 of dense forest area in four major locations (BBS 2017a). The forest areas in Bangladesh are classified into five zones, namely, Hill, Sal, Coastal, Sundarbans and the village (Fig. 1).

In the past, there have been many attempts to conduct forest inventory either for localized forests (for Sundarbans by Revilla et al. 1998a; for coastal region by Revilla et al. 1998b; for Cox's bazar forest region by Revilla et al. 1998c; for Chittagong forest region by Revilla et al. 1998d) or for the entire country (FRA 2000; GoB 2017) or for specific plan/animal (Tiger Census by GoB 2004; Tiger Survey by GoB 2015; for

G. Chandra et al. (eds.), *Statistical Methods and Applications in Forestry and Environmental Sciences*, Forum for Interdisciplinary Mathematics,
https://doi.org/10.1007/978-981-15-1476-0

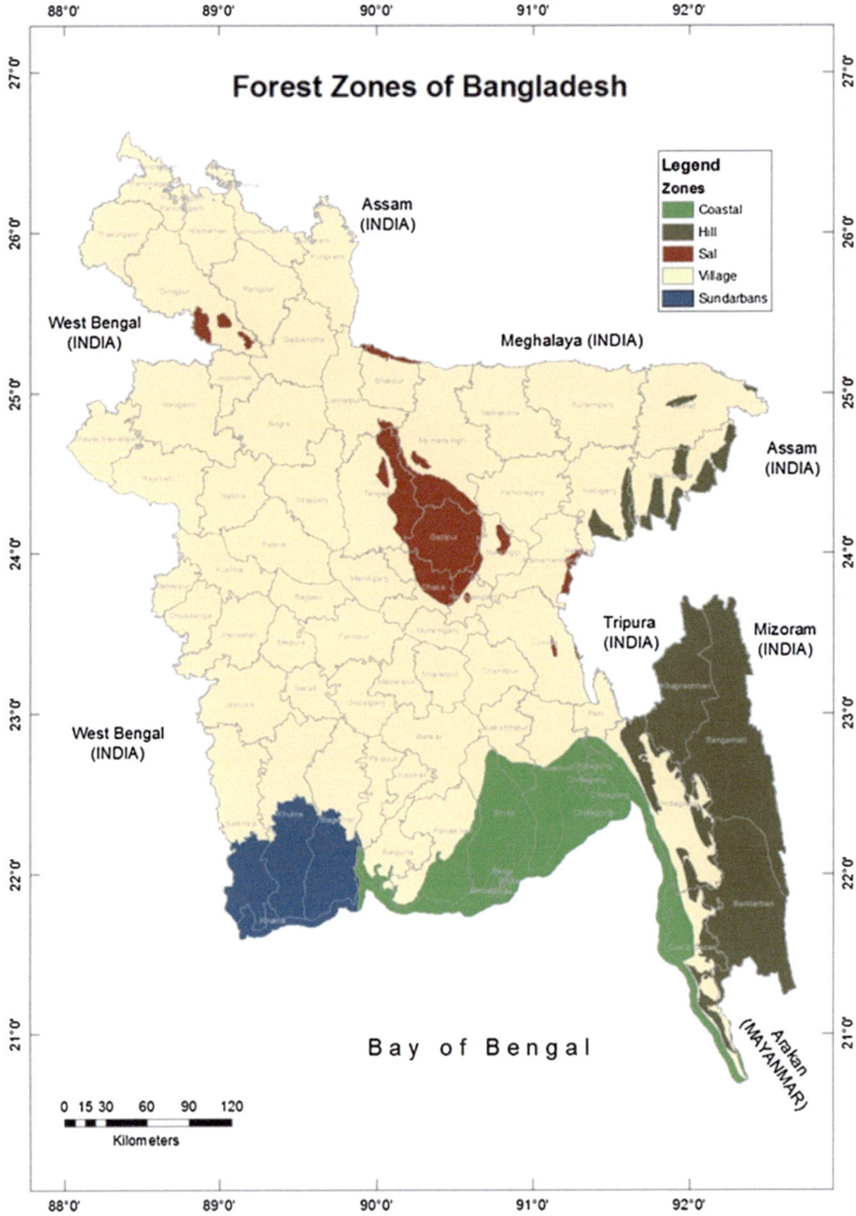

Fig. 1 Five forest zones of Bangladesh (Akhter et al. 2016)

wood and bamboo by Douglas 1981). The most elaborate and latest survey that covered the entire country and considered both resources inventory and socioeconomic component was conducted by Department of Forestry, GoB with technical support from FAO (GoB 2017). To serve the dual purposes, i.e., forest inventory and socioeconomic study, the sampling frame was based on four attributes, viz., percentage tree cover, road density, household density and water cover, for stratification. Information on the first four attributes was collected from the Population and Housing Census 2011 (BBS 2012).

Though this attempt was a gigantic one with innovative sampling scheme, it has some limitations too. These are, for example, road density sometimes may not correlate with any of the other attributes like percentage tree cover and household density. Similar is true for water cover. The arguments made to consider these two attributes may not hold in many of the cases as road planning sometimes depend on political decision and in an unplanned fashion, and water logging is a recent, unusual and irregular phenomenon. Another aspect is that the frame used is not linked with the census frame, hence will be limited in terms of trend analysis and comparability with other related studies. The main reason is that there is no permanent sampling frame proposed to repeat the survey in future. To summarize the above discussion, it may be mentioned that none of the previous surveys are efficient in terms of linking between different target parameters, keeping synergies with Population and Housing Census (to establish relationship between livelihood and forest resource management), supporting trend analysis and offering use of constant sampling frame. The following section aims to introduce an integrated survey scheme to overcome the situation. The integration refers to linking two sampling frames, namely area sampling frame and probability sampling frame to be able to collect all types of data related to forestry.

Integrated Survey Scheme

A comprehensive list of surveys/study that may be conducted based on forests includes (i) resource inventory, (ii) socioeconomic and livelihood studies of people dependent on forest products, (iii) monitoring of rare and endangered flora and fauna, (iv) assessment of loss due to natural calamity, (v) carbon footprint analysis for possible carbon trade, (vi) land degradation studies, (vii) survey on deforestation, and (viii) economic and ecological sustainability study. Besides, pilot studies on new concepts may be done in future which will require frame support to generalize the results at national level. From the above list, it is clear that different types of sampling frames required pursuing a specific survey, which will be costly and time consuming and miss the desirable qualities mentioned earlier, unless planned under an integrated approach.

The proposed integrated survey sampling scheme deals with both the area and probability sampling. Mouza, the smallest administrative unit with cadastral maps available in Bangladesh, is proposed to use as primary sampling unit (PSU) thus providing the link between the two in probability terms. This is because for forest resources related surveys, area sampling suits well while for the livelihood related surveys probability sampling is found to be better. Secondary sampling units and sampling units at higher level may be planned as needed. The proposed steps of the scheme are as follows:

1. Identify an area sampling frame (except for socioeconomic and livelihood survey) of the forest areas covering all five zones (Fig. 1). The area sampling frames may be prepared using cadastral maps and satellite imagery.
2. Create a buffer of two kilometers (or more as identified by the experts) to capture the settlements around each area sampling points (PSU) of the forests. Note that there may be some remote points where there is no locality nearby. Hence, the number of sampling points close to locality and remote ones should be pre-decided based on some characteristics so that sampling weights may be adjusted accordingly. For village forest zones this problem won't arise.
3. Identify all the mouzas inhibited by people from the selected buffer at Step 2. A probability sampling frame should be prepared using Population and Housing Census data.
4. Now for resource inventory, different sampling methods may be incorporated within PSUs. This may include, for example, ranked set sampling (McIntyre 1952) or systematic sampling (Cochran 1977) for wood resource assessment (log diameter, tree height etc.), capture-recapture methods (Barnett 2005) for animal/reptiles/fish population. Furthermore, remote sensing (Schott 2007), differential global positioning system (Monteiro et al. 2005) and radio frequency identification (Angell and Kietzmann 2006) may be considered for some purposes (for example, capturing forest affected by fire, deforestation etc.) based on dot sampling (Jinguji 2012) within the PSUs with subsequent ground trothing. With the help of normalized difference vegetation index (Crippen 1990), plant growth/yield, carbon emission etc. may be calculated.
5. For socio-economic and livelihood (including dependency on forest) study, the inhabitants in the buffer areas may be identified by a suitable sampling scheme (say, random, cluster, stratified or systematic sampling scheme where appropriate) from the probability sampling frame.
6. Results from the surveys conducted in both area and probability sampling frames may be linked in probability scale (as PSUs are common) for comparison and to devise combined index/estimates. To further elaborate, let us consider a situation where we need to know the annual use of forest resources and proportion of people living in a locality that are dependent on that resources. The former may be ascertained from area sampling scheme, the latter may be estimated using a probability sampling scheme and both may be linked in probability using their PSU enlistment.

Future surveys may be designed following the same approach and established frames (both area and probability frame) which will allow any trend analysis and comparability studies. If all the targeted surveys may be listed under a survey calendar with periodicity indicated and required resources may be mobilized, Bangladesh may able to enter into an era of forestry resource mapping and utilization keeping global warming aspects in mind. Note that National Statistics Act 2013 empowered BBS to produce and facilitate collection of all sorts of data of national interest. Different agencies may conduct surveys on their own provided that the methodology is approved by BBS. Bangladesh Strategic Plan on Agricultural and Rural Statistics (2016–2030) (BBS 2017b) advocated for integrated survey planning and calendar. Also, considering the growing demand of regular data supply on forestry related issues an integrated approach of survey sampling will be a useful guide for further development where both the Department of Forestry and BBS are currently lacking adequate number of appropriate methodologists in the field of survey sampling.

Conclusion

Forestry is an important sector contribution of which is still not properly realized in Bangladesh. Different *ad hoc* surveys and censuses provide glimpses of the sector. However, this information lacks the efficiency in comparability and trend analysis. A systematic approach of collecting information is necessary for both actual realization of the contribution of forest resources in GDP and feeding regular information for SDG monitoring. Here, a brief overview of the current data collection systems and proposed an integrated approach of surveys has been provided. The proposed approach will allow efficient resource allocation needed to collect data and guarantee standards and reliability for comparison and any trend analysis. Further thoughts may be put into the issue, especially when data requirements will be clearer, to enhance the use efficiency of the approach. This study may be a useful guide for the relevant departments in Bangladesh as well as in some other countries.

References

Akhter, M., Jalal, R., Costello, L., Rahman, L. M., & Tasnuva, U. (2016). *Zoning for tree and forest assessment in Bangladesh*. Dhaka, Bangladesh: Bangladesh Forest Department, Food and Agriculture Organization of the United Nations.

Angell, I., & Kietzmann, J. (2006). RFID and the end of cash? *Communications of the ACM, 49*(12), 90–96.

Barnett, V. (2005). Sampling in the wild. In W. A. Shewhart, S. S. Wilks, & V. Barnett (Eds.), *Environmental statistics: Methods and applications* (pp. 131–150). Chichester, UK: Wiley.

BBS (Bangladesh Bureau of Statistics). (2012). *Population and housing census 2011*. Ministry of Planning, Government of Bangladesh.

BBS (Bangladesh Bureau of Statistics). (2017a). *Yearbook of agricultural statistics—2016* (571 pp).

BBS (Bangladesh Bureau of Statistics). (2017b). *Bangladesh strategic plan on agricultural and rural statistics (2016–2030)*.

Cochran, W. G. (1977). *Sampling techniques* (3rd ed.). USA: Wiley.

Crippen, R. E. (1990). Calculating the vegetation index faster. *Remote Sensing of Environment, 34,* 71–73.

Douglas, J. J. (1981). *Consumption and supply of wood and bamboo in Bangladesh* (FAO/UNDP Project BGD/78/010). Field Document 2 (87 pp).

FRA. (2000). *Forest resources of Bangladesh, country report. Rome 2000* (Working Paper No 15) (93 pp). Rome.

GoB (Government of Bangladesh). (2004). *Tiger census 2004*. Forest Department, Government of Bangladesh.

GoB (Government of Bangladesh). (2015). *Tiger survey 2015*. Forest Department, Government of Bangladesh.

GoB (Government of Bangladesh). (2017). *The socio-economic survey of the Bangladesh forest inventory*. Forest Department, Ministry of Environment and Forest, Government of the People's Republic of Bangladesh.

GoB (Government of Bangladesh). (2018). *Sustainable development goals*. General Economic Division, Ministry of Planning, Government of Bangladesh.

Jinguji, I. (2012). *How to develop master sampling frames using dot sampling method and Google earth*. http://www.fao.org/fileadmin/templates/ess/global_strategy/PPTs/MSF_PPTs/5.MSF_Dot_sampling_method_on_Google_Earth_Jinguji.pdf.

McIntyre, G. A. (1952). A method for unbiased selective sampling using ranked sets. *Australian Journal of Agricultural Research, 3,* 385–390.

Monteiro, L. S., Moore, T., & Hill, C. (2005). What is the accuracy of DGPS? *The Journal of Navigation, 58,* 207–225.

Revilla, J. A. V., Ahmed, I. U., & Hossain, M. A. (1998a). *Final report: Forest inventory of the Sundarbans reserved forests*. Forest Resource Management Project, Forest Department, Government of Bangladesh.

Revilla, J. A. V., Ahmed, I. U., & Hossain, M. A. (1998b). *Final report: Forest inventory of the coastal afforestation divisions*. Forest Resource Management Project, Forest Department, Government of Bangladesh.

Revilla, J. A. V., Ahmed, I. U., & Hossain, M. A. (1998c). *Final report: Forest inventory of natural forest and plantations*. Cox's Bazar Forest Division, Forest Resource Management Project, Forest Department, Government of Bangladesh.

Revilla, J. A. V., Ahmed, I. U., & Hossain, M. A. (1998d). *Final report: Forest inventory of natural forest and plantations*. Chittagong Forest Division, Forest Resource Management Project, Forest Department, Government of Bangladesh.

Schott, J. R. (2007). *Remote sensing: The image chain approach* (2nd ed.). Oxford University Press.

MIX
Papier aus verantwortungsvollen Quellen
Paper from responsible sources
FSC® C105338

If you have any concerns about our products,
you can contact us on
ProductSafety@springernature.com

In case Publisher is established outside the EU,
the EU authorized representative is:
Springer Nature Customer Service Center GmbH
Europaplatz 3, 69115 Heidelberg, Germany

Printed by Libri Plureos GmbH
in Hamburg, Germany